Protozoan Parasitism

From Omics to Prevention and Control

https://doi.org/10.21775/9781910190838

Edited by

Luis Miguel de Pablos Torró

Instituto de Biotecnología y Departamento de Parasitología
Universidad de Granada
Granada
Spain

and

Jacob Lorenzo-Morales

Instituto Universitario de Enfermedades Tropicales y
Salud Pública de Canarias
Universidad de La Laguna
Tenerife
Canary Islands
Spain

Caister Academic Press

Caister Academic Press
Norfolk, UK

www.caister.com

British Library Cataloguing-in-Publication Data
A catalogue record for this book is available from the British Library

ISBN: 978-1-910190-83-8 (paperback)
ISBN: 978-1-910190-84-5 (ebook)

Cover design adapted from images provided by Luis Miguel de Pablos Torró.

Contents

Preface

In the postgenomic era, 'omics' sciences play a key role in basic and applied research. Over the last 15 years numerous protozoan parasite genomes have been sequenced, annotated and made available in public databases. In this way, genomics has laid an important foundation, on which new bricks can be added to our understanding of these diseases and opened new avenues for advancing molecular models that, in many cases, are free from the scientific community. This new information has stimulated and fuelled research in vaccine and drug discovery and new approaches for high-throughput experimentation and genetic engineering.

This book, *Protozoan Parasitism: From Omics to Prevention and Control,* provides a comprehensive account of advances and new insights into protozoan parasitic diseases with a special focus on omics sciences and their basic and clinical applications. A large portion of the book is dedicated to contributions made in the field of kinetoplastid parasites, the aetiological agents of devastating diseases such as leishmaniasis, Chagas disease or sleeping sickness. Within this part of the book (Chapters 1, 2, 3 and 5) we provide relevant information in different proteomics approaches (including glycoproteomics) together with historical perspectives in this area. The study of protozoan parasites has also benefited from new genome-editing technologies (e.g. CRISPR-Cas9, zinc finger endonucleases, cloning-free endogenous tagging) that are revolutionizing this area, described in Chapter 4. The second part of the book is dedicated to advances made in the search for high-throughput approaches for the targeting and control of amoebiasis, an infection caused by any of the amoebas belonging to the *Entamoeba* group.

All the advances described in this book are steps towards the control and prevention of these diseases. With the hope that this book will prove useful, we believe that the reader may go on to find more answers in the field of protozoan parasites.

Dr Luis Miguel de Pablos Torró
Dr Jacob Lorenzo-Morales

From Proteins to Molecular Targets: *Trypanosoma cruzi* Proteomic Insights in Drug Development

1

Giselle VF Brunoro[1], Marcelle A Caminha[1,2] and Rubem FS Menna-Barreto[2*]

[1]Laboratório de Toxinologia, Instituto Oswaldo Cruz, Fundação Oswaldo Cruz, Rio de Janeiro, Brazil.
[2]Laboratório de Biologia Celular, Instituto Oswaldo Cruz, Fundação Oswaldo Cruz, Rio de Janeiro, Brazil.

*Correspondence: rubemsadok@gmail.com

https://doi.org/10.21775/9781910190838.01

Abstract
Chagas disease is a neglected disease caused by the haemoflagellate protozoan *Trypanosoma cruzi,* which is endemic to Latin America and emerges in non-endemic areas due to immigration. The parasite presents a complex life cycle, including two hosts (invertebrate and vertebrate) and three developmental stages (epimastigotes, trypomastigotes and amastigotes). The current chemotherapy is restricted to two nitroheterocycles, nifurtimox and benznidazole, which present limited efficacy depending on the disease phase. Multidisciplinary efforts have been made to identify promising novel parasite-specific key molecules to develop nitroderivative substitutes, increase trypanocidal activity and minimize host toxicity. In this context, deep analysis of drug pharmacokinetics and mechanisms of action is crucial to improve specificity and safety. We reviewed proteomics as a valuable tool for the identification of molecular targets of *T. cruzi.* Advances in parasite proteomic maps (different developmental stages and strains) were reviewed, emphasizing the promising targets, their biological applications, and possible follow-up analysis for the development of novel anti-*T. cruzi* drugs.

Background
Chagas disease, caused by the haemoflagellate protozoan *Trypanosoma cruzi,* is a neglected sickness endemic to Latin America that emerges in non-endemic areas due to immigration. An estimated 6 million people are infected worldwide (Schmunis, 2007; WHO, 2015).

The disease is transmitted to mammals, including humans, by Triatominae insect faeces. Alternatively to vectorial transmission, oral and congenital routes, as well as blood transfusion, represent other routes of infection (Altclas *et al.*, 2008; Dias and Amato Neto, 2011; Steindel *et al.*, 2008). The clinical evolution of Chagas disease includes acute and chronic phases with typical characteristics, such as the presence of bloodstream parasitaemia in recent infections and cardiac and/or digestive impairment derived from patent inflammation induced by persistent parasitaemia (Marin-Neto *et al.*, 2009; Rassi Jr *et al.*, 2009; Rocha *et al.*, 2007; Sosa-Estani *et al.*, 2009).

The current treatment for Chagas disease is restricted to two nitroheterocycles, nifurtimox and benznidazole, that were empirically developed half a century ago (Rodriques Coura and de Castro, 2002). However, the indications of these drugs depend on the disease phase. Both compounds lead to high parasitological cure rates in acute patients, but their efficacy declines substantially in the chronic stage, reinforcing the need for alternative treatments (Jannin and Villa, 2007; Soeiro and de Castro, 2011; Urbina and Docampo, 2003). In this context, multidisciplinary efforts have been made to identify promising novel parasite-specific key molecules to minimize host toxicity (Menna-Barreto and de Castro, 2017). The *T. cruzi* life cycle includes invertebrate and vertebrate hosts and three parasite stages. Despite the clinical relevance of mammalian trypomastigotes and amastigotes, insect proliferative epimastigotes are extensively used as the experimental model due to their facility to be cultivated in axenic conditions.

Particularly for trypanosomatids, proteomics is an interesting tool to monitor protein expression because these protozoa present open reading frames in long polycistronic regions and regulate gene expression post-transcriptionally (Andrade *et al.*, 2008; Clayton and Shapira, 2007; Parodi-Talice *et al.*, 2004). In 2010, cytosolic non-translated mRNA was detected in *T. cruzi*, reinforcing the limitation of RNA-based analysis of gene expression studies in trypanosomatids (Holetz *et al.*, 2010). Therefore, large-scale proteomic technology has become increasingly attractive. The physiological role of parasite proteins and accurate genome annotations, as well as the identification of evolutive stage-specific proteins and drug resistance molecules, were assessed using high-throughput proteomics (Andrade *et al.*, 2008; Atwood *et al.*, 2005; Ferella *et al.*, 2008; Menna-Barreto *et al.*, 2010; Parodi-Talice *et al.*, 2004). We reviewed the main aspects of *T. cruzi* proteomics, focusing on potential molecular targets for alternative treatments for Chagas disease.

T. cruzi and proteomics: more than a decade of studies

In 2004, proteomic analysis of all culture-derived life forms of *T. cruzi* was performed using two-dimensional electrophoresis (2-DE) separation followed by matrix-assisted laser desorption/ionization–time of flight (MALDI-TOF) mass spectrometry peptide mass fingerprinting (Paba *et al.*, 2004). Proteome changes during *in vitro* amastigogenesis (differentiation from trypomastigotes to amastigotes) were evaluated and comparison with epimastigotes was performed. From the identified proteins (e.g. heat shock proteins, elongation factors, glycolytic enzymes, and cytoskeletal components), paraflagellar rod proteins were more abundant in trypomastigotes than in epimastigotes and amastigotes (2.4- and 15-fold, respectively), suggesting that paraflagellar rods are more prominent in the infective forms. The glycolytic enzymes pyruvate kinase and enolase 2 showed lower abundance in trypomastigotes than in amastigotes (0.7- and 0.9-fold change, respectively). However, these two proteins were more abundant (3.4- and 8.6-fold, respectively) in trypomastigotes than in epimastigotes. As previously described in *T. brucei*, the bloodstream forms are more dependent on glycolysis than the insect forms (Tielens and Van Hellemond, 1998). Recently, functional evaluation of the mitochondrial metabolism of *T. cruzi* was performed, and it was shown that the mitochondria of epimastigotes are more efficient than those of isolated bloodstream trypomastigotes (Gonçalves *et al.*, 2011). Hence, the insect forms are not as glycolytic as the bloodstream trypomastigotes.

In same year, a description of the proteome using 2-DE followed by MALDI-TOF of Dm28c epimastigotes was performed (Parodi-Talice *et al.*, 2004), and 45 proteins were identified. The functional categorization showed eleven proteins belonging to metabolic pathways of the parasite (e.g. prostaglandin F2*α* synthase, 3-phosphoglycerate kinase, cystathione β-synthase 6, and tryparedoxin peroxidase), four belonging to protein assembly and destination (e.g. heat shock protein HSP70, heat shock 70 kDa protein mitochondrial precursor, chaperonin HSP60 mitochondrial precursor, and proteasome subunit a 5), four belonging to cell cycle growth and protein synthesis (e.g. CCHC zinc finger protein, RNA-binding protein RGGm, nascent polypeptide-associated complex alpha, and retrotransposon hot spot protein), and three cytoskeleton assembly proteins (e.g. alpha-tubulin, tubulin beta chain, and beta-tubulin 1.9). The prostaglandin F2*α* synthase is an important enzyme for the antioxidant defences of the parasite and was identified in several spots in the gels, implicating a role in the detoxification process (Kubata *et al.*, 2002).

In 2005, shotgun proteomic analysis of the different *T. cruzi* life stages (Brazil strain) was performed, generating the large-scale descriptive proteome of the parasite (Atwood *et al.*, 2005). By employing liquid chromatography followed by tandem mass spectrometry (LC-MS/MS), 2784 proteins were described and nearly 30% (838 of 2784) were identified in all four life forms (epimastigotes, amastigotes, culture-derived trypomastigotes and metacyclic trypomastigotes). Considering the number of identified proteins in each stage, the metacyclic trypomastigotes had the greatest number of identified proteins with 2339 identifications, followed by culture-derived amastigotes (1871), epimastigotes (1861) and culture-derived trypomastigotes (1486). From the total dataset, 1008 proteins were identified as hypothetical proteins. Additionally, the functional classification demonstrated mainly representatives of polypeptide biosynthesis and degradation (ribosomal, proteasomal and chaperonins), and 622 identifications distributed in two major surface protein families, trans-sialidase (223 proteins) and the GP63 (399 proteins). To explain the differences in protein expression between distinct stages, the authors summarized their main findings and created several hypotheses: proteins involved in endoplasmic reticulum to Golgi trafficking were preferentially found in amastigotes, suggesting that a more active or preferred trafficking pathway is used at this stage. Because key antioxidant defence enzymes (e.g. ascorbate peroxidase, mitochondrial tryparedoxin peroxidase, trypanothione synthase, and iron superoxide dismutase) were identified with high protein coverage in the metacyclic form, a preadaptation of the parasite during the transition of epimastigotes to metacyclic trypomastigotes to the oxidative burst inside the vertebrate host, particularly in phagocytes, was suggested. The enzymes involved in the conversion of histidine to glutamate (i.e. histidine ammonia-lyase and urocanate hydratase) were abundant in the insect stage but nearly undetectable in the vertebrate stages, consistent with the idea that the parasite is well adapted to use histidine, the dominant free amino acid in the gut of the insect, as an energy source. These hypotheses remain to be confirmed and validated by orthogonal techniques.

Atwood and co-workers (2005) compared the proteome of each completely differentiated parasite form. Changes in protein levels during differentiation still needed to be determined (Atwood *et al.*, 2005). A proteomic study examining metacyclogenesis using the Dm28c strain and 2-DE coupled to MALDI-TOF mass spectrometry (MS) was performed (Parodi-Talice *et al.*, 2007). The authors identified 43 differentially expressed proteins related to metabolism, cytoskeleton flagellum organization, protein destination and biosynthesis, catabolism/proteolysis, electron transport and antioxidant defences, and hypothetical proteins (Parodi-Talice *et al.*, 2007), in accordance with previous results in epimastigotes (Parodi-Talice *et al.*, 2004). The work also suggested that the increase in glutamate dehydrogenase expression could be due to increased amino acid uptake by the parasite during nutritional stress. Additionally, the identification of protein isoforms distributed in the 2-DE gels supported the idea that post-translational modifications were an essential strategy for regulating gene expression during differentiation (Parodi-Talice *et al.*, 2007).

To explore the basic proteins of *T. cruzi*, distinct isoelectric focusing (IEF) conditions using a pH gradient from 6 to 11 (2-DE followed by MALDI-TOF) was performed to analyse epimastigote and trypomastigote (Berenice strain) samples (Magalhães *et al.*, 2008). The comparative 2-DE analysis showed higher abundance of proteins related to amino acid metabolism and dehydrogenases in epimastigotes while trans-sialidases and paraflagellar rod proteins were identified only in trypomastigotes. The use of this approach to further evaluate the alkaline proteome of *T. cruzi* during amastigogenesis was suggested by the authors. In the following year, the proteome of CL Brener strain epimastigotes, a hybrid organism derived from the genomes of *T. cruzi* I and the *T. cruzi* II, was evaluated by 2-DE followed by MALDI-TOF MSMS (Sodré *et al.*, 2009). As previously observed, tubulin and heat shock protein isoforms and prostaglandin $F_{2\alpha}$ synthase were identified in this proliferative form (Kubata *et al.*, 2002; Parodi-Talice *et al.*, 2004), and the enzyme arginine kinase of the polyamine pathway (energy supply) was identified for the first time.

T. cruzi is genetically diverse. The taxonomy of phylogenetic lineages (discrete typing units, DTUs) of this organism was recently revised according to the eco-epidemiological features, pathogenicity, and parasite biology. This classification resulted in six DTUs, *T. cruzi* I to VI (Zingales *et al.*, 2009). To compare the distinct DTUs described in *T. cruzi*, a proteome profile analysis was performed in epimastigotes from isolates 3663 (*T. cruzi* III) and 4167 (*T. cruzi* IV) and reference strains CL Brener and Dm28c using 2-DE coupled with MALDI-TOF/TOF MSMS (Kikuchi *et al.*, 2010). Of the 62 differential protein spots, 57 were related to cell metabolism, protein assembly, cellular organization, antioxidant defences, stress response, and cell-to-cell communication. The remaining five proteins were identified as hypothetical ones. The *in vitro* infectivity assays showed that the 4167 isolate was more infectious than 3663. In agreement with these results, the proteomic analysis showed overexpression of proteins associated with virulence, such as glutathione peroxidase, tryparedoxin peroxidase, S-adenosylmethionine synthetase, arginine kinase, and cruzipain. To further evaluate this hypothesis, cruzipain was analysed by flow cytometry and was found to be highly expressed in the 4167 isolate, CL Brener and Dm28c. Altogether, the correlation between the virulence and heterogeneity of the trypanosomatid cell surface composition is possible. Additional analysis of different DTUs and their implications on infection are needed to fully comprehend the crucial characteristics of parasite stocks in nature. The same rationale was employed to compare three strains of DTU TcI [Colombiana (COL), SylvioX10/4 (SYL) and the culture clone (TCC)] with distinct biological behaviour in terms of parasitaemia and tissue effects (Zago *et al.*, 2016). Trypomastigotes of the COL and SYL strains showed enhanced cytoskeleton remodelling proteins that are implicated in flagellar motility and antioxidant pathway proteins. The higher abundance of cytosolic and mitochondrial tryparedoxin peroxidases, tryparedoxin, and iron superoxide dismutase was confirmed by Western blot. Furthermore, these two strains were resistant to treatment with stable oxidants (hydrogen peroxide and peroxynitrite) and evaded ROS-mediated macrophage cytotoxicity, demonstrating that independently of phylogenetic lineage, the parasite uses its oxidative resistance machinery to ensure survival.

The *T. cruzi* life cycle is also diverse, and in regard to environmental conditions for survival, growth and transmission, the physiological characteristics of the vertebrate and invertebrate hosts are heterogeneous. The success of infection is dependent on the capacity of the parasite to adapt; for instance, the temperature in the insect midgut is 28°C while the mammalian bloodstream is 37°C, demonstrating that an adaptation must occur for parasite survival. To study this type of stress, Ninoa strain epimastigotes were exposed to heat shock stress at 37 and 42°C and compared with control incubation at 28°C. The proteome of each condition was evaluated by 2-DE and MALDI-TOF MS (Pérez-Morales *et al.*, 2012). Twenty-four proteins with differential abundance relative to the control were identified. Changes in proteins of cell metabolism (i.e. higher abundance of transitional endoplasmic reticulum ATPase, glutamate dehydrogenase, and ATP synthase), cell defence (i.e. higher abundance of heat shock protein 70 and thiol transferase), protein fate (including ubiquitin hydrolase), and several hypothetical proteins were detected. Additionally, the temperature increase resulted in the overexpression of surface molecules related to infectivity and virulence, suggesting an association between the stress response and infection. The novelty of this work was the description of proteins other than chaperonins involved in the *T. cruzi* heat shock stress response.

In 2012, shotgun proteomics was employed to analyse the metacyclogenesis process in *T. cruzi* (de Godoy *et al.*, 2012). The authors studied epimastigotes, three intermediary stages (epimastigotes subjected to nutritional stress for 30 minutes, 2 hours or 12 hours), and metacyclic trypomastigotes using LC online with mass spectrometer (MS/MS). This work quantified roughly 3000 proteins involved in metacyclogenesis, resulting in one of the most comprehensive quantitative *T. cruzi* proteomics studies to date. One of the striking findings of this work was the higher abundance of 18 isoforms of trans-sialidase in trypomastigotes. These data are in accordance with previous studies demonstrating the role of cyclic-AMP in the induction of metacyclogenesis (Gonzales-Perdomo *et al.*, 1988). Another noteworthy finding was the overexpression of cytoskeletal proteins (e.g. tubulin, actin, dyneins, and paraflagellar rod components) upon differentiation to trypomastigotes. Important morphological changes, such as the displacement of the kinetoplast, in the flagellar pocket and the flagellar attachment zone must occur during metacyclogenesis, corroborating to the increase in cytoskeletal protein levels. Opposite to expectations, four small nuclear ribonucleoproteins (involved in the splicing process) and five histones were more abundant in the non-proliferative form, indicating an increase in gene regulation in metacyclic trypomastigotes. However, this hypothesis must be tested experimentally. Additionally, translation-related proteins including ribosomal proteins, tRNA synthetase, initiation, and elongation factors, were downregulated in the metacyclic forms, as previously described by Marchini *et al.* (2011). Although these results expand the knowledge regarding the regulation of such pathways, the reason for this down-regulation remains unclear.

The proteome of bloodstream trypomastigotes was studied in 2015 in a work by our group (Brunoro *et al.*, 2015). The proteome of this parasite stage (Y strain) responsible for infection in mammalian hosts comprised 5901 proteins. Compare to previous proteomic studies including all parasite forms, 2202 proteins were exclusively identified in the bloodstream form. These proteins are related to surface proteins (e.g. mucins, GPI-anchor transamidase, phospholipases), non-classical secretion pathways, cytoskeletal dynamics (ARP 2/3), cell cycle and transcription (elongation factors), proteolysis, redox metabolism, biosynthetic pathways, bioenergetics, protein folding (chaperones), cell signalling (diacylglycerol kinase, phosphatidylinositol kinase), vesicular traffic (snare and rab proteins), DNA repair, and cell death. The authors discussed the possible contribution of this proteomic evaluation of the bloodstream form to comprehension of disease pathogenesis in the vertebrate hosts.

The first proteomic study aiming to identify *T. cruzi* protein localization was performed on the insect forms of the CL Brener strain using LC-MS/MS (Ferella *et al.*, 2008). After the enrichment of organelles in fractions, 38 novel proteins that were not previously detected in the whole epimastigote proteome were identified. For further characterization of the cellular compartment location of five hypothetical proteins, immunofluorescence assays were performed. Two proteins (one predicted to be a mitochondrial protein and the other with sequence similarity to dehydrogenase) were localized in the endoplasmic reticulum, one was a putative metal transporter in acidocalcisomes, one presented a porin domain in the mitochondrion, and one was a potential transporter in cytosolic vesicles. The authors identified nearly all the enzymes of the tricarboxylic acid cycle and the iron–sulfur subunit of succinate dehydrogenase, which were not previously detected in the analysis of whole epimastigotes (Atwood *et al.*, 2005). This work demonstrated the importance of subcellular fraction analysis because it provides subcellular localization and the identification of many proteins, including those that might not have been previously identified using other approaches. Subsequently, the reservosome sub-proteome was analysed for the first time (Sant'Anna *et al.*, 2009). This organelle is an epimastigote-exclusive acidic organelle that represents the endpoint of the endocytic pathway and stores lipids and proteins. They are not considered typical lysosomes due to the absence of lysosomal molecular markers (e.g. acid phosphatases) and their particular pH (6.0, in contrast to lysosomal pH 5.0). After fractionation monitored by transmission electron microscopy, reservosome fractions from Dm28c epimastigotes were analysed by 2D-LC-MSMS. A total of 709 reservosome proteins were identified. Of these, parasite proteins, such as cruzipain, serine carboxypeptidase, ABC transporters, and protein tyrosine phosphatase, were previously described in reservosomes, and a novel isoform of P-type H^+-ATPase was detected for the first time. The identification of additional hydrolases, such as cysteine peptidases, α-mannosidases, acid phosphatases, calpains, lipases, and serine carboxypeptidase, suggests that reservosomes are the main site of lysosomal hydrolases, although other typical lysosomal markers of other organisms have not been found.

Continuing the subproteome evaluation, the detergent-solubilized membrane fraction was analysed by two-dimensional liquid chromatography (2D-LC) coupled with tandem mass spectrometry (MSMS) in *T. cruzi* epimastigotes and metacyclic trypomastigotes (G strain) (Cordero *et al.*, 2009). Employing the neutral detergent Triton X-114, 280 and 98 proteins were identified in the metacyclic form and as epimastigotes, respectively. Approximately 60% of the identified membrane proteins presented predicted post-translational modifications (PTM), e.g. GPI-anchor, myristoylation or prenylation. Additionally, the proteins identified in the proliferative epimastigotes were related to metabolic pathways, in accordance with the energy requirements for protozoa mitosis. The infective metacyclic trypomastigotes expressed a large repertoire of surface glycoproteins (e.g. GP82 and GP90), which have been associated with host-cell adhesion and invasion. In 2013, the cell surface subproteome of epimastigotes was characterized using two new approaches: protein surface trypsinization of intact living (Shave) and biotin labelling of surface proteins followed by affinity chromatography purification (Queiroz *et al.*, 2013). Both approaches identified 2095 proteins, and the ontology categorization revealed 643 and 550 predicted transmembrane domains in the biotinylated and Shave samples, respectively. Although the identified protein lists were poorly overlapped, known features of this insect stage, such as glycolysis/gluconeogenesis and arginine and proline metabolism pathways, were detected in both. The authors continued the investigation by applying the same methodology to study the plasma membrane subproteome of human-host parasite forms (Queiroz *et al.*, 2014a), showing a few stage-specific enzymes, such as Ser/Thr protein kinases, MAPK (only detected in axenic amastigotes), and superoxide dismutase (only detected in trypomastigotes). Overall, the identification of several enzymes involved in carbohydrate and lipid metabolism suggested that amastigotes interfere in host-cell metabolism for intracellular self-maintenance and present key molecules of amastigogenesis. On the other hand, trypomastigotes expressed enzymes associated with oxidative stress defence, important virulence factors.

Over its life cycle, *T. cruzi* alternates between proliferative and non-proliferative stages, which have an influence on the control of transcription. The chromatin proteome of trypomastigotes and epimastigotes revealed the presence of not only known nuclear proteins (RNA-binding proteins, ribosomal proteins) and histones but also cytoskeleton and carbohydrate pathway proteins (e.g. fructose-bisphosphate aldolase, glyceraldehyde 3-phosphate dehydrogenase, hexokinase, and pyruvate dehydrogenase E1) (de Jesus *et al.*, 2017). The most striking differences between the compared parasite forms were higher protein diversity and exclusive identification of proteins associated with DNA dynamics (e.g. DNA topoisomerases) in epimastigotes and the enrichment of histones (in particular, the histone H2B variant) in trypomastigotes. Further investigation of histone PTMs in these two parasite forms confirmed known modifications and showed 44 new modifications, including the acetylation of the lysines of histone H2A and methylation of lysine 23 of histone H3, which was enriched in non-proliferative parasites (de Jesus *et al.*, 2016). In proliferative epimastigotes, the predominant expressed PTMs were methylation of lysine 76 of histone H3 and phosphorylation of serine 23 of H2B. The variable extent of histone modification in distinct stages reflects specific modulations of gene transcription and DNA replication according to the proliferative status. Some of these modifications, such as multiple modifications at the H2A C-terminal and modification at histone H1, H3T29, and H2BS63, appear trypanosome specific. More recently, a work describing 119 modification sites of histones with 13 distinct modification types, such as citrullination, ubiquitination, hydroxylation, phosphorylation and lysine acylation (glutarylation, crotonylation, 2-hydroxyisobutyrylation, malonylation, and succinylation), was published (Picchi *et al.*, 2017). This publication extended the knowledge on the PTM of *T. cruzi* histones, but their biological significance requires further investigation. Subproteome studies of the nuclear fraction provided new insight into the role of the chromatin-related proteins and histone modifications involved in the control of gene expression and protozoa cycle regulation in different stages of this pathogen.

Improvements in the *T. cruzi* cell disruption protocol based on hypo-osmotic and sonication lysis and centrifugation for shotgun LC-MSMS analysis were also proposed (Silva Galdino *et al.*, 2014). The protein extract (TcS12) presented 1153 total identifications, mainly from the cytoplasm, kinetoplast–mitochondrion complex, nucleus, cell membrane/surface, and extracellular compartment. By employing mild chemical and surfactant-free extraction, the authors used this method to further study cell-free translation system, i.e. the integrity of the cellular structure for protein synthesis was maintained. In 2015, the nuclear proteome of *T. cruzi* epimastigotes (CL Brener strain) was analysed by 2D-LC-MSMS after isolation of the cellular subfraction by two steps of density gradient centrifugation and confirmation by phase-contrast and fluorescence microscopy (dos Santos Júnior *et al.*, 2015). Of the 864 identified proteins, 275 were not detected in the reported global proteome analyses of Atwood and colleagues (2005). As expected, proteins involved in nuclear and cell division, ribosome biosynthesis, chromatin organization, and RNA/DNA binding were found. Proteins typically found in cell compartments other than the nucleus, such as enzymes of carbohydrate and lipid metabolism and membrane proteins involved in host–parasite interactions (mucins and TS), were also identified. The analysis of this enriched subfraction revealed new and atypical nuclear proteins, demonstrating the importance of subproteome characterization.

As discussed previously, another adverse environmental condition that *T. cruzi* is exposed to during its life cycle is the osmotic variation between mammalian blood and the vector midgut. The association between the acidocalcisomes and the contractile vacuole complex for the regulation of the parasite cell volume under hypo-osmotic stress was previously published (Rohloff and Docampo, 2008). Nevertheless, the structural and physiological implications of parasite homeostasis have been poorly studied. To identify the protein composition of the contractile vacuole complexes and its physiological roles, Ulrich and co-workers (2011) analysed the subproteome of the contractile vacuole complex of CL Brener strain epimastigotes using one-dimensional gel electrophoresis followed by LC-MS/MS (Ulrich *et al.*, 2011). This compartment fraction revealed 220 proteins. Some peptides mapped to the 74 hypothetical proteins and at least 39 members of the dispersed gene family 1, demonstrating that several components of this gene family are concurrently expressed in epimastigotes. Additionally, amastins, calpains, transport-related and intracellular proteins (e.g. small G proteins), transmembrane glycoproteins, and vacuolar-H^+-pyrophosphatases (i.e. acidocalcisome marker) were detected. The identified proteins that were previously localized in the contractile vacuole of other organisms and those that could have a potential role in this compartment were selected for validation by ultrastructural immunolocalization. The data showed that six proteins (Rab11, Rab32, AP180, ATPase subunit B, VAMP1, and phosphate transporters) were predominantly found in the vacuole bladder, while TcSNARE and calmodulin were distributed in the spongiome. The results demonstrated that subproteome analysis was essential for the identification of 109 novel proteins in this parasite stage and for the determination of protein localization and function. However, the total purity of each subcellular fraction is an important limitation of this approach. Therefore, further morphological and biochemical assays must be performed to confirm the data.

The first proteomic study focused on the analysis of *T. cruzi* glycoproteins was performed by Atwood and colleagues in 2006. They applied hyphenated LC separation-MSMS analysis in culture-derived trypomastigotes (Brazil strain) (Atwood *et al.*, 2006). To improve the method resolution, the parasites were initially subjected to subcellular fractionation.

Each fraction was subjected to lectin affinity capture, and the isolated glycoproteins were isotopically labelled. A total of 29 glycoproteins were identified, of which 11 were specifically detected in this pathogen at the protein level for the first time. These experimental data demonstrated that mucin-associated surface proteins and dispersed gene family members displayed N-linked glycan post-translational modifications (PTM). In 2017, the *T. cruzi* glycoproteome was further investigated in epimastigotes and trypomastigotes using glycopeptide enrichment coupled to LC-MSMS (Alves *et al.*, 2017). A total of 1309 N-glycosylation sites were identified in 690 proteins, of which 170 glycoproteins were exclusive to epimastigotes and 334 were exclusive to trypomastigotes (MASP and mucin family members). The results also identified 10 and 97 unique O-linked peptides in epimastigotes and trypomastigotes, respectively.

Phosphorylation is another essential PTM that is crucial for the regulation of cell signal transduction networks and biological processes. The involvement of kinases in parasite biology could represent an interesting possibility for drug intervention. More than a decade ago, the study of phosphoproteins in *T. cruzi* epimastigotes (Y strain and Dm28 clone) by phosphoprotein/phosphopeptide fractionation followed by LC-MSMS analysis was explored (Marchini *et al.*, 2011; Nakayasu *et al.*, 2009). The previous analysis detected 237 phosphopeptides from 119 distinct proteins, which were categorized into biological process ontologies such as parasite motility, metabolism, ion transport, and differentiation (Nakayasu *et al.*, 2009). In 2011, Marchini and colleagues revealed 753 phosphoproteins with 2572 phosphorylation sites and reported consensus phosphorylation sequence motifs. Recently, the analysis of differential phosphorylation over *in vitro* amastigogenesis showed 165 mono-phosphorylated and 18 multi-phosphorylated peptides (Queiroz *et al.*, 2014b). A total of 46 membrane proteins were modulated, and the most striking phosphorylation differences were observed after 2 h incubation in acidic medium (several kinases and phosphatases that are part of the signalling cascade).

Additional insights into glycosylation and phosphorylation of *T. cruzi* epimastigote (Y strain) proteins were obtained by 2-DE gel staining with Pro-Q-Emerald and Pro-Q-Diamond fluorescent dyes, respectively (Beghini *et al.*, 2012). A total of 15 glycoproteins were identified, including tubulin, heat shock proteins, peroxiredoxin, tryparedoxin peroxidase, prostaglandin $F_{2\alpha}$ synthase, ATPase subunits, sterol 24-c-methyltransferase, and tyrosine aminotransferase. Furthermore, the identified phosphoprotein subset comprised 14-3-3 protein, initiation factor, enolase, glutamine synthetase, prolyl oligopeptidase, aminoacylase, pyrroline-5-carboxylate synthetase, actin, tubulin, paraflagellar rod protein, chaperonins, distinct hypothetical proteins, peroxiredoxin, proteasome subunit, and spermidine synthase, totalling 22 protein identifications.

Another known protein modification in *T. cruzi* is N-myristoylation. Recent data showed that approximately 0.44% of this parasite proteome is N-myristoylated (Roberts and Fairlamb, 2016). By analysing the N-myristoylated-enriched subfraction by bead-based click-chemistry coupled with LC-MSMS, 56 proteins were identified, 32 of which were confirmed in two additional orthogonal experiments (stable isotopic labelling with amino acids in cell culture [SILAC]). Although the majority of these identified proteins are uncharacterized, *T. cruzi* protein myristoylation is involved in membrane trafficking and cellular transport (ADP-ribosylation factors), cell signalling and regulation of metabolic processes (AMP-activated protein kinase and protein phosphatase 2C), cytoskeletal remodelling and protein turnover (calpains and the proteasome).

Bayer-Santos and colleagues (2013) explored the parasite secretome using electron microscopy, immunochemistry and proteomics. After 2D-LC-MSMS, 102 proteins were uniquely identified in epimastigotes while 22 were only identified in metacyclic trypomastigotes. Both datasets identified 243 common proteins. Altogether, 367 distinct proteins were found and classified into gene ontologies. The functional categories presenting the higher number of representatives were nucleic acid binding, suggesting a role in intercellular communication and gene expression regulation; protein and amino acid metabolism; stress responses by oxidation and reduction, which are likely to be involved in intracellular survival and cellular protection against free radicals; and transcription and protein synthesis proteins that might play auxiliary roles in pathogenesis. Additionally, proteins involved in host–parasite interactions (e.g. calreticulin, prolyl oligopeptidase, GP82 and GP90 glycoproteins, and trans-sialidases) were also identified. In addition to identification, the authors performed label-free quantification, showing an enrichment of glycoprotein 82 kDa and calcium-binding protein in the vesicle fraction compared to the vesicle-free fraction. Additionally, secreted exovesicles of trypomastigotes and amastigotes were investigated via immunoprecipitation with immune sera of patients with Chagas disease followed by LC-MSMS (Bautista-López *et al.*, 2016). Of the 766 identified proteins, approximately 10% were differentially expressed. The majority of the identified *T. cruzi* proteins belonged to the family of trans-sialidases. Furthermore, proteins that were secreted only in trypomastigotes/amastigotes included amastigote surface proteins 2 and 3, proteasome subunits, transporter proteins, polyubiquitin protein, and complement regulatory proteins. These reports expended the proteome knowledge of the *T. cruzi* secretome and provided evidence that the pathogen uses two mechanisms of vesicle secretion.

The characterization of the protein translational responses of *T. cruzi* to ionizing radiation was performed using 2D-DIGE gel separation followed by MSMS (Vieira *et al.*, 2014). This organism is resistant to these adverse environmental conditions, and by comparing the ionized and the non-ionized epimastigotes (CL Brener strain), the authors revealed 215 identifications, representing a total of 53 proteins. The number of identified spots of each protein was representative of the presence of different isoforms generated by processing and/or PTM. This work raised new questions about how *T. cruzi* modulates protein expression to survive the ionization stress conditions.

In an investigation of a mixture of *T. cruzi* genotypes and their role in the transmission and severity of the disease, the interaction of two genotypes, one fast- growing and highly virulent (P209 cl1 stock) and the other slow-growing and mildly virulent (Mn cl2 stock), was performed (Machin *et al.*, 2014). Epimastigotes from the mixed (after 0, 7, and 14 days) and isolated phenotypes were analysed by 2D-DIGE followed by MSMS identification. Taking P209 cl1 as reference, 103 protein spots were differentially expressed, of which 40 were identified as proteolytic, metabolic, biosynthetic, structural, or stress proteins. Heat shock proteins and tubulins comprised approximately 50% of the identifications. Seventy-six of the 103 differential spots showed changes on day 0, demonstrating that the interaction started quickly. The interactome revealed overexpression and underexpression of proteins, suggesting cross-stimulation and inhibition between these genotypes. Another interactome study focused on the redox interactome of *T. cruzi* was performed (Arias *et al.*, 2015). Using affinity chromatography followed by SDS-PAGE and mass spectrometry, 16 putative partners of tryparedoxin II were identified in epimastigotes. The antioxidant system (cytoplasmic and mitochodrial peroxiredoxin), energy metabolism (F0F1-type ATP synthase, amidinotransferase), cytoskeleton (α-, β, ε-tubulin, I/6 autoantigen, and radial spoke protein), and protein translation processes (elongation factor 2) were identified. Most of these proteins are cytosolic and have homologues involved with redox signalling in other organisms.

The investigation of the mechanisms of mitochondrial antioxidant defence through the elucidation of partners of mitochondrial tryparedoxin peroxidase was performed by co-precipitation followed by 2DE-LC-MSMS (Peloso *et al.*, 2016). Although, proteins involved in stress response, oxidoreduction, and thiol transfer were identified under physiological conditions, the main difference in the protein profile under stress was the detection of structural proteins. Additionally, proteins involved in amino acid metabolism, microtubule-associated processes, stress response (e.g. glutathione peroxidase, and spermidine synthase), and protein polymerization were also found in the latter conditions. The involvement of tryparedoxin peroxidase in protein and small molecule metabolism, protein folding, and redox-signalling cascades throughout the cell opens new perspectives on the functions exerted by this enzyme. Fig. 1.1 summarizes the main findings in *T. cruzi* proteomics.

Trypanocidal compounds: mechanism of action and resistance

Many research groups have made efforts towards the development of new trypanocidal compounds with high activity, low side effects, proven chronic infection efficacy, and low costs (Soeiro and de Castro, 2011). To improve the selectivity of these compounds, the mechanism of action should involve enzymes and/or biochemical pathways that occur exclusively in *T. cruzi*. Proteomics is a valuable tool for the identification of interesting molecular targets in protozoa (Menna-Barreto and Perales, 2014; Menna-Barreto *et al.*, 2014).

Despite the promising applications, only three proteomic analyses have been published concerning drug mechanisms of action. In the first study, Andrade and co-workers (2008) analysed proteins involved in Y strain epimastigote resistance to the drug benznidazole by 2-DE and MALDI-TOF/TOF. For parasitic infections, the development of drug resistance represents a major clinical barrier to efficient chemotherapy. In Chagas disease, the increase in benznidazole resistance has been associated with large genotypic diversity, enhancing the therapeutic failure. Thirty-six distinct proteins were modulated by the induction of benznidazole resistance, such as calpain and other peptidases (peptidase M20, metallo-carboxipeptidase and mitochondrial peptidases), cyclophilin A, glutamate dehydrogenase, tyrosine amino transferase, nucleoside diphosphate kinase, iron superoxide dismutase, peroxiredoxin and a hypothetical protein. In susceptible epimastigotes, an increase in prostaglandin F2α synthase expression was observed, indicating an anion radical detoxification pathway as part of the mode of action of benznidazole (Murta *et al.*, 2006). Recently, using a chemical proteomic approach, *T. cruzi* proteins that bind directly to benznidazole were assessed. The specific non-covalent proteins, such as aldo-ketoreductase family members, were found to interact with the nitroderivative immobilized in a solid matrix (Trochine *et al.*, 2014). This enzyme is associated with nitroreductase type- I and old yellow enzyme, which were previously related to benznidazole reduction. Furthermore, subsequent studies in chemical proteomics could reveal proteins involved in drug reductive activation and/or resistance mechanisms.

In 2010, our group published the second proteomic work on trypanocidal drugs describing the mechanism of action of three naphthoimidazoles derived from beta-lapachone, the

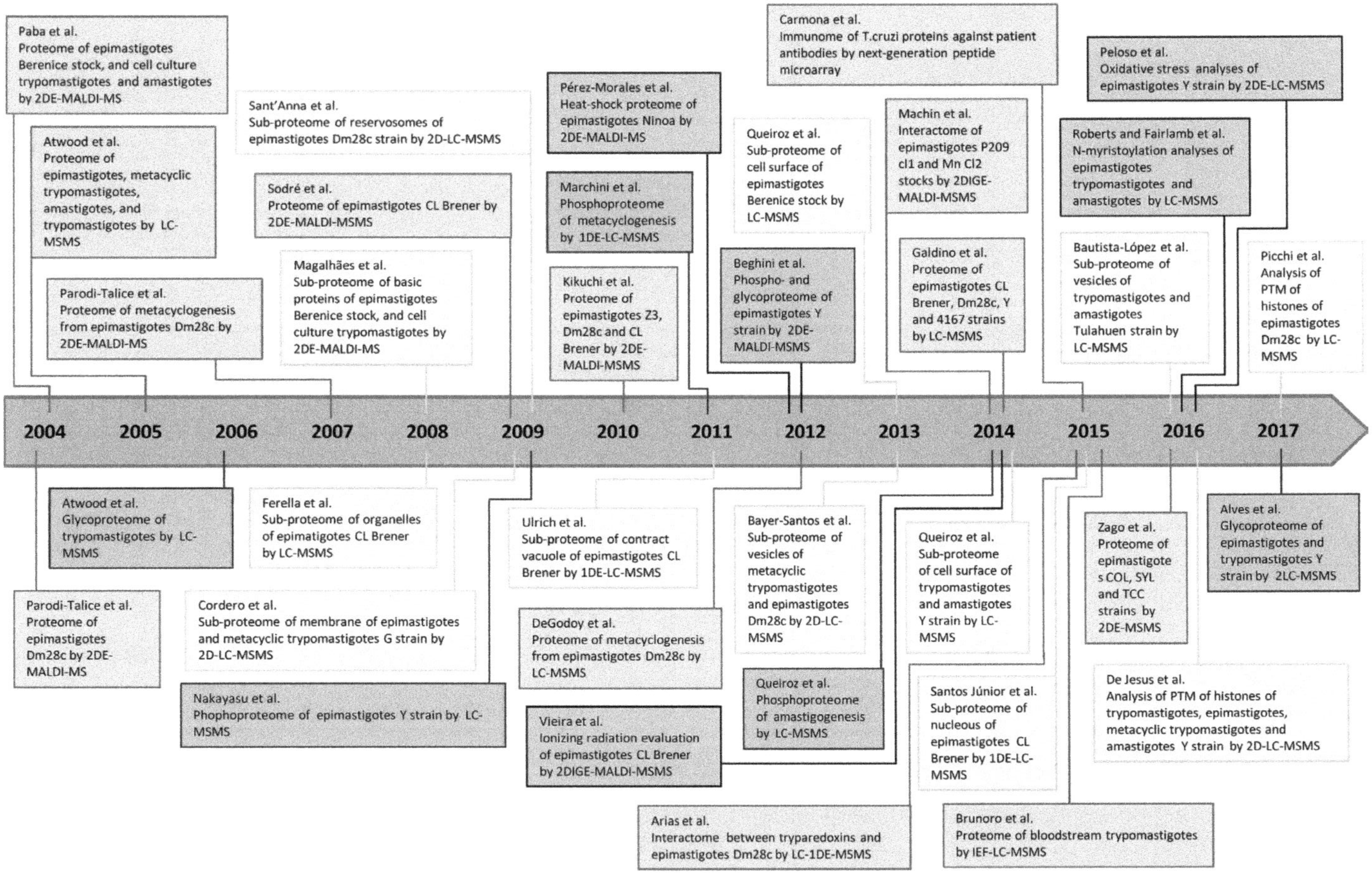

Figure 1.1 Timeline of proteomic studies in *T. cruzi*.

most promising trypanocidal compounds among the many naphthoquinones and their derivatives screened between 1997 and 2013 (Silva Júnior *et al.*, 2014). Using 2-DE and MALDI-TOF/TOF approach, Y strain epimastigotes treated with naphthoimidazoles showed mitochondrial proteins as the most frequently differentially expressed proteins, confirming the previous ultrastructural data (Menna-Barreto *et al.*, 2005, 2007, 2010). Among the proteins involved in the epimastigote response to treatment with these compounds, tyrosine aminotransferase, elongation factors, glutamamyl carboxypeptidase, cystathionine beta-synthase, sterol 24-c-methyltransferase, tubulin, proteasomal subunits and chaperonins were also detected (Table 1.1) (Menna-Barreto *et al.*, 2010). The overexpression of trypanothione synthetase was observed upon treatment with the three naphthoimidazoles; however, these compounds do not possess redox properties that could lead to the generation of reactive oxygen species. The explanation of the increase of the levels of this antioxidant enzyme under these experimental conditions is still under investigation in our laboratory.

Recently, in continuation of our proteomic analysis of naphthoimidazoles, we investigated the effect of these trypanocidal compounds on *T. cruzi* trypomastigotes purified from the mammalian bloodstream, the most clinically relevant parasite form. In this third work, sample quantity was a crucial limitation. Because the protein yield of non-proliferative trypomastigotes isolated from mice blood was not high, a different approach was employed to minimize the number of infected animals per experiment. A 2-DE approach with high sensitivity, reproducibility and increased linear dynamic range for protein spot comparison was necessary (Alban *et al.*, 2003), and 2D-DIGE was selected. We could reduce the protein amount 10-fold per gel compared to previous epimastigote research (Brunoro *et al.*, 2016; Menna-Barreto *et al.*, 2010).

The treatment of trypomastigotes with naphthoimidazoles led to modulation of the abundance of 61 protein spots, with 34 spots being identified. Among the differentially abundant

Table 1.1 Molecular targets of naphthoimidazoles in *T. cruzi* epimastigotes and bloodstream trypomastigotes

			Spot modulation[a]					
			Epi[b]			BT[c]		
Molecular function	Gene ID	Protein	N1	N2	N3	N1	N2	N3
Protein folding/ stress response	gi|123667	Heat shock protein 85	↓↑[d]	↓	↓	↓	↓	–
	gi|205278870	Mitochondrial heat shock	↓	–	–	–	–	–
	gi|3023478	Heat shock protein 60	–	↓	↓	↑	–	–
	gi|70885659	Chaperonin	–	–	–	↑	–	–
	gi|205278868	Heat shock protein 70	–	–	–	↓↑[d]	↓	–
	gi|70876223	Heat shock 70 kDa protein	–	–	–	↓	↓	–
	gi|119394469	70 kDa heat shock protein	–	–	–	–	↓	–
	gi|407842744	Glucose-regulated protein 78	–	–	–	↑	–	–
Cell motility and structure	gi|71397525	Alpha-tubulin	–	–	↓	↓	–	–
	gi|18568139	Beta-tubulin	↓	↓	↓	–	↓	–
	gi|162179	Major paraflagellar rod protein	–	–	–	↑	–	–
	gi|2209137	Paraflagellar rod component	–	–	–	↑	–	–

Table 1.1 Continued

Molecular function	Gene ID	Protein	Spot modulation[a]					
			Epi[b]			BT[c]		
			N1	N2	N3	N1	N2	N3
Energetic metabolism	gi|71665461	Enolase	–	–	↑	↑	–	–
	gi|71420903	Pyruvate dehydrogenase E1 beta subunit	–	–	↓	–	–	–
	gi|71660419	Cytochrome C oxidase subunit IV	–	↓	–	–	–	–
	gi|407849036	Succinyl-CoA ligase [GDP-forming] beta-chain	–	–	–	↓	–	–
Amino acid metabolism	gi|71402529	Glutamamyl carboxypeptidase	↓	–	–	–	–	–
	gi|13241624	Cystathionine beta-synthase 6	–	↓	–	–	–	–
	gi|1168606	Tyrosine aminotransferase	–	–	↓	–	–	–
	gi|348658746	Asparagine synthetase A, partial	–	–	–	↑	–	–
Protein metabolism	gi|1929445	Elongation factor 1-alpha	↓	–	–	–	–	–
	gi|205278864	Elongation factor 2	–	↓	–	–	–	↑
	gi|71405953	Proteasome alpha 7 subunit	↓	–	↓	–	–	–
Oxidative stress response/ metabolism	gi|16588445	Trypanothione synthase	↑	↑	↑	–	–	–
	gi|4388655	Peroxiredoxin	–	↓	–	–	–	–
	gi|71405176	Peptide methionine sulfoxide reductase	–	–	–	↓	–	–
Others	gi|71661424	Activated protein kinase C receptor	–	–	↓	–	↓	–
	gi|71407834	IgE-dependent histamine-releasing factor	–	–	↓	–	–	–
	gi|71399140	Sterol 24-c-methyltransferase	–	↓	↓	–	–	–
	gi|70874663	Guanine deaminase	–	–	–	↑	–	–
	gi|407410583	RAB GDP dissociation inhibitor alpha	–	–	–	↑	–	–
	gi|70870795	Vacuolar ATP synthase subunit B	–	–	–	↑	–	–
	gi|407844351	Arginine kinase	–	–	–	–	↓	–
	gi|70883145	Hypothetical protein, conserved	–	–	–	↑	–	–
	gi|71409986	Hypothetical protein	–	↓	–	–	–	–
	gi|71651427	Hypothetical protein	↓	↓	–	–	–	–
	gi|557861434	Hypothetical protein TCDM_14147	–	–	–	↓	–	–
	gi|70871170	Hypothetical protein, conserved	↑	↑	–	–	–	–

[a]The arrows ↑ and ↓ represent, respectively, higher and lower intensity of the protein spot in the treated (N1, N2, or N3) sample than in the control. The dashes (–) represent the lack of protein spot detection in the respective sample.
[b]Proliferative epimastigotes analysed by 2D-PAGE and MALDI-TOF/TOF (Menna-Barreto *et al.*, 2010).
[c]Bloodstream trypomastigotes analysed by 2D-DIGE and MALDI-TOF/TOF (Brunoro *et al.*, 2016).
[d]Protein identified in different spots (Menna-Barreto *et al.*, 2010; Brunoro *et al.*, 2016).

proteins were tubulin isoforms, enolase, guanine deaminase, arginine kinase, asparagine synthetase, activated protein kinase C receptor, elongation factor 2, heat shock proteins, hypothetical proteins, paraflagellar rod components, RAB GDP dissociation inhibitor, succinyl-CoA ligase, ATP synthase subunit B and methionine sulfoxide reductase (Table 1.1) (Brunoro *et al.*, 2016). As previously observed in the insect form, naphthoimidazoles induced differential expression of a large number of mitochondrial proteins in trypomastigotes, corroborating the morphological and biochemical data showing this organelle as the primary target of these compounds (Menna-Barreto *et al.*, 2005, 2007). Both epimastigotes and bloodstream trypomastigotes treated with the three compounds presented important down-regulation of the tubulin levels, but no ultrastructural effects were previously detected in the cytoskeletal structures, such as flagella and subpellicular cage. Our hypothesis is that only the tyrosinated tubulin isoform (directly related to labile microtubules) is affected by treatment, which was reinforced by ELISA assays with anti-tyrosinated and anti-acetylated tubulin antibodies (Menna-Barreto *et al.*, 2010). The reduction of labile microtubules could lead to impairment of vesicular trafficking and/or parasite mitosis in the case of epimastigotes (Brunoro *et al.*, 2016; Menna-Barreto *et al.*, 2005, 2007, 2010).

On the other hand, some proteins were differentially abundant in the bloodstream form but not in epimastigotes treated with naphthoimidazoles due to the significant variation between both parasite forms and their protein contents (Atwood *et al.*, 2005). Some enzymes from different pathways, including asparagine synthetase, peptide methionine sulfoxide reductase, guanine deaminase and arginine kinase, and cytoskeleton-associated proteins were differentially detected only in trypomastigotes (Brunoro *et al.*, 2016). These data reinforced the complexity of the mechanisms of action of naphthoimidazoles, involving modulation of protein synthesis, transduction signalling, energetic metabolism and parasite cellular trafficking, enabling future trypanocidal strategies. A comparative analysis among the differential proteins identified after the treatment with benznidazole or naphthoimidazoles in epimastigotes and/or trypomastigotes could be observed in Fig. 1.2.

Molecular drug targets and promising antigens for vaccine development

As previously mentioned, the limited efficacy of benznidazole, especially in chronic Chagas disease, has stimulated the search for alternative drugs (Nwaka and Hudson, 2006). The ideal compound would interfere with crucial and exclusive *T. cruzi* pathways present in clinically relevant parasite forms from different subpopulations, improve trypanocidal activity and selectivity, and reduce side effects to the host (Soeiro *et al.*, 2010). Pre-clinical *in vitro* studies on the mechanisms of action of the selected drugs based on the accumulated knowledge about *T. cruzi* biology and biochemistry represent an Achilles' heel for the development of novel trypanocidal compounds. Proteomic studies have been conducted and have identified promising molecular targets in the parasite, as we will discuss below (Table 1.2).

Moreover, the development of Chagas diagnostic tests and/or vaccines depends on the discovery of novel antigens located in the parasite surface (Cordero *et al.*, 2009). The *T. cruzi* glycoproteomic approach led to the identification of 11 trypomastigote-specific glycoproteins (mucin-associated surface proteins, dispersed gene family proteins) also characterizing N- and O-linked glycan residues on the parasite surface (Atwood *et al.*, 2006). Recently,

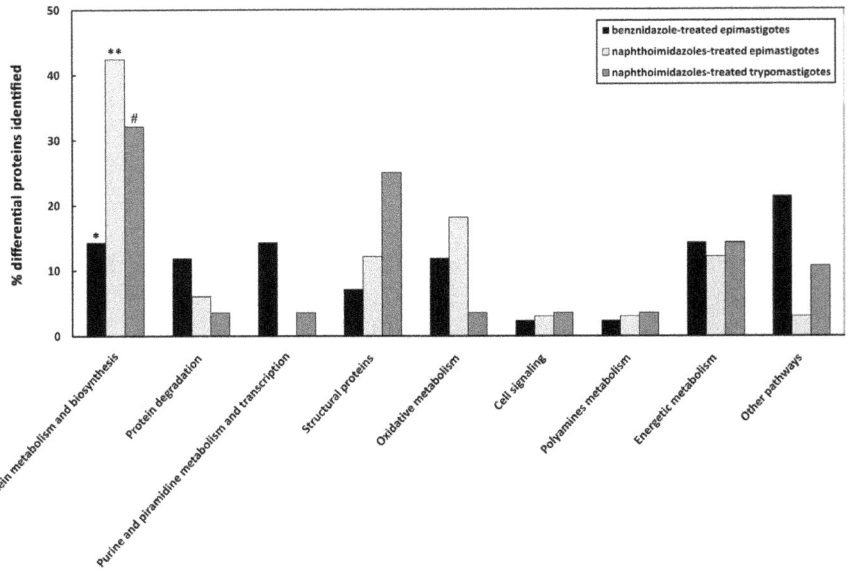

Figure 1.2 Percentage of differential proteins identified in proteomic studies of anti-*T. cruzi* drugs. *2D-PAGE analysis of epimastigotes resistant or susceptible to benznidazole (Andrade *et al.*, 2008); **2D-PAGE evaluation of the treatment of epimastigotes with naphthoimidazoles (Menna-Barreto *et al.*, 2010); #DIGE analysis of naphthoimidazoles-treated bloodstream trypomastigotes (Brunoro *et al.*, 2016).

Y strain trypomastigotes were thoroughly assessed by LC-MS/MS to identify potential antigens for vaccines (Nakayasu *et al.*, 2012). Forty-five highly immunogenic epitopes were identified, and glycosylphosphatidylinositol-anchored surface proteins involved in pathogenesis represented 14% of the non-redundant sequences analysed. The compiled data from these two studies produced a promising collection of vaccine antigens to be validated in near future.

More recently, the 'immunome', i.e. the use of next-generation peptide microarrays to characterize serological antibody specificities and to map antigens, was analysed in *T. cruzi* (Carmona *et al.*, 2015). This novel large-scale antibody-based approach was used to determine which antigens (15 amino acid monomer peptides) from the 457 protein sequences of CL Brener strain are recognized by antibodies from the serum of Chagas disease patients. The work revealed 2031 disease-specific peptides (linear epitopes) and 97 novel parasite antigens, almost doubling the number of known antigens and corresponding to 15% of parasite proteins targeted by the humoral immune response in these patients. Characterization of antibody responses to *T. cruzi* proteins showed new peptidic markers for Chagas disease, contributing to the discovery of potential serological biomarkers for diagnosis, follow-up and/or disease treatment.

The *T. cruzi* plasma membrane also participates in the first contact with the host cell, and the protozoa surface molecules are crucial for the recognition and entry of the parasite. The blockage of a molecule essential for protozoa internalization is a potential strategy for drug intervention. Platelet-activating factor (PAF) is involved in *T. cruzi* differentiation, indirectly

Table 1.2 Potential molecular drug targets for the treatment of Chagas disease found in proteomic studies

Target	Biological process	Protein function	Reference
Platelet-activating factor receptor	Parasite differentiation	G protein-coupled receptor signal transduction	Kawano *et al.* (2011)
Cruzipain	Host cell invasion and immune modulation	Cyteine protease	Sant'Anna *et al.* (2009), Kikuchi *et al.* (2010), Queiroz *et al.* (2014b), Brunoro *et al.* (2016)
Calpains	Parasite growth, viability, infectivity and differentiation	Cyteine protease	Atwood *et al.* (2005), Andrade *et al.* (2008), Sant'Anna *et al.* (2009), Nakayasu *et al.* (2009), Kikuchi *et al.* (2010), Beghini *et al.* (2012), Queizon *et al.* (2013), Brunoro *et al.* (2016)
Glycogen synthase kinase 3	Parasite growth	Kinase activity	Nakayasu *et al.* (2009), Queiroz *et al.* (2014b)
Arginine kinase	Parasite survival, growth, viability and infectivity	Kinase activity	Andrade *et al.* (2008), Sodré *et al.* (2009), Kikuchi *et al.* (2010), Beghini *et al.* (2012), Queiroz *et al.* (2014b), Brunoro *et al.* (2016)
Tyrosine aminotransferase	Parasite viability	Amino acids and proteins biosynthesis	Andrade *et al.* (2008), Sodré *et al.* (2009), Sant'Anna *et al.* (2009), Nakayasu *et al.* (2009), Kikuchi *et al.* (2010), Menna-Barreto *et al.* (2010), Beghini *et al.* (2012), Queiroz *et al.* (2013), Brunoro *et al.* (2016)
Prostaglandin $F^{2\alpha}$ synthase	Parasite survival	Lipid biosynthesis and drug detoxification	Parodi-Talice *et al.*, (2004), Andrade *et al.*, (2008), Sodré *et al.*, (2009), Menna-Barreto *et al.*, 2010), Queiroz *et al.*, (2013), Brunoro *et al.*, (2016)
Trypanothione synthetase, peroxiredoxins, tryparedoxins and superoxide dismutase	Parasite survival	Oxidative metabolism	Paba *et al.* (2004), Parodi-Talice *et al.* (2004), Andrade *et al.* (2008), Sodré *et al.* (2009), Kikuchi *et al.* (2010), Menna-Barreto *et al.* (2010), Beghini *et al.* (2012), Queiroz *et al.* (2013, 2014b), Arias *et al.* (2015), Brunoro *et al.* (2016)
Sterol 24-C-methyltransferase	Parasite viability	Ergosterol biosynthesis	Sant'Anna *et al.* (2009), Menna-Barreto *et al.* (2010), Beghini *et al.* (2012), Brunoro *et al.* (2016)

Table 1.2 Continued

Target	Biological process	Protein function	Reference
Glutamate dehydrogenase, phosphoenolpyruvate carboxykinase, pyruvate phosphate dikinase, enolase, hexokinase and fructose biphosphate aldolase	Parasite viability	Energetic metabolism	Paba *et al.* (2004), Andrade *et al.* (2008), Sodré *et al.* (2009), Sant'Anna *et al.* (2009), Nakayasu *et al.* (2009), Kikuchi *et al.* (2010), Menna-Barreto *et al.* (2010), Beghini *et al.* (2012), Queiroz *et al.* (2013, 2014b), Brunoro *et al.* (2016)
I/6 autoantigen	Parasite viability	Cell structure	Nakayasu *et al.* (2009), Sant'Anna *et al.* (2009), Beghini *et al.* (2012), Brunoro *et al.* (2016)
Nucleoside phosphorylase	Parasite viability	Nucleotide salvage	Beghini *et al.* (2012), Queiroz *et al.* (2013), Brunoro *et al.* (2016)
Chaperones	Parasite survival and viability	Protein folding and stress response	Parodi-Talice *et al.* (2004), Andrade *et al.* (2008), Sodré *et al.* (2009), Sant'Anna *et al.* (2009), Kikuchi *et al.* (2010), Menna-Barreto *et al.* (2010), Beghini *et al.* (2012), Queiroz *et al.* (2013), Brunoro *et al.* (2016)

impacting infection, as demonstrated by treatment with PAF antagonist. Despite the bioinformatics approach employed to identify the parasite PAF receptor, further experiments are necessary to analyse the twenty-nine hypothetical proteins that share G protein-coupled receptor features with this receptor (Kawano *et al.*, 2011). Cruzipain is a cysteine protease that is highly distributed on the *T. cruzi* surface and is also associated with parasite entry into the host cell, justifying the search for specific inhibitors (Kikuchi *et al.*, 2010). Additional cysteine peptidases that are potential targets due to sequence differences from humans are calpains, cytoplasmic calcium-regulated proteases that are involved in signal transduction and cytoskeletal assembly (Goll *et al.*, 2003). Metagenomic evidence pointed to many calpain genes in trypanosomatids, expressed in different *T. cruzi* forms, with biological function related to infectivity and differentiation (Andrade *et al.*, 2008; Atwood *et al.*, 2005; Ennes-Vidal *et al.*, 2010, 2011; Ersfeld *et al.*, 2005; Kikuchi *et al.*, 2010; Sant'Anna *et al.*, 2009; Sodré *et al.*, 2009).

Differences between the host and parasite signalling pathways could also represent a good opportunity for drug inhibition. In this context, phosphorylation presents a fundamental step to trigger distinct parasite processes. *T. cruzi* phosphoproteome indicates kinases, specifically glycogen synthase kinase 3, as excellent targets (Nakayasu *et al.*, 2009; Queiroz *et al.*, 2014b). Kikuchi and colleagues (2010) also pointed to arginine kinase, which is responsible for the conversion of L-arginine to phosphoarginine, a crucial product for antioxidant defences, bioenergetics and infectivity, as a drug target, due to its high abundance in two virulent strains isolated from triatomine insects from the Brazilian Amazon region. These data were reinforced by the absence of arginine kinase in mammalian hosts (Kikuchi *et al.*, 2010; Sodré *et al.*, 2009).

Another target suggested by proteomics research was glutamate dehydrogenase, an enzyme involved in L-glutamate deamination to alpha-ketoglutarate, which is a fundamental carbon source for proliferative epimastigote forms (Sodré *et al.*, 2009). In protein metabolism and amino acid biosynthesis pathways, tyrosine aminotransferase, an enzyme responsible for methionine, tyrosine, tryptophan and phenylalanine turnover, was modulated by treatment of epimastigotes with benznidazole or naphthoimidazoles (Andrade *et al.*, 2008; Menna-Barreto *et al.*, 2010), suggesting impairment of the amino acids balance as a promising unexplored drug target.

Since 2012, our group has been investigating possible drug target candidates with low or no sequence similarity to minimize potential side effects (Beghini *et al.*, 2012; Brunoro *et al.*, 2015; Menna-Barreto and Perales, 2014; Menna-Barreto *et al.*, 2014). In the following paragraphs, we discuss some promising molecular targets based on our identifications in Y strain epimastigotes and bloodstream trypomastigotes.

Global proteomic evaluation of epimastigotes treated with naphthoimidazoles or benznidazole indicated oxidative stress as an important part of the trypanocidal mechanism. Different antioxidant enzymes were modulated by treatment, reinforcing the hypothesis relating ROS misbalance to parasite lysis (Andrade *et al.*, 2008; Menna-Barreto *et al.*, 2010). It was previously demonstrated that protozoa resistance to reactive oxygen species increased virulence and infectivity, pointing to the redox system as a crucial step for successful infection (Piacenza *et al.*, 2009). The reduction observed in the abundance of prostaglandin F2α synthase after treatment (Andrade *et al.*, 2008; Menna-Barreto *et al.*, 2010), together with its presence in distinct strains (Murta *et al.*, 2006; Parodi-Talice *et al.*, 2004; Sodré *et al.*, 2009), makes this enzyme an interesting candidate for drug intervention. In other trypanosomatids, prostaglandin F2α synthase participates in NADPH-catalysed prostaglandin production, and its role has been associated with drug detoxification (Kubata *et al.*, 2000). Among other antioxidant enzymes, trypanothione synthetase, peroxiredoxins, tryparedoxins and superoxide dismutases are interesting in trypanosomatids (Andrade *et al.*, 2008; Arias *et al.*, 2015; Irigoín *et al.*, 2008; Menna-Barreto *et al.*, 2010), and their inhibition could guide to reactive oxygen species overproduction and parasite death.

On the other hand, ergosterol biosynthesis steps have been extensively analysed in pathogenic trypanosomatids due to their absence in mammals. Squalene synthase and 24-C-methyltransferase, enzymes responsible for the intermediate steps of ergosterol production, were less abundant in epimastigotes treated with naphthoimidazoles, indicating the potential of this pathway in the development of anti-*T. cruzi* therapeutics (Barros-Alvarez *et al.*, 2014; Menna-Barreto *et al.*, 2010). The fluidity of the *T. cruzi* plasma membrane depends on the ergosterol levels, and the reduction of this sterol culminates in parasite lysis (Beghini *et al.*, 2012; Menna-Barreto *et al.*, 2010; Urbina, 2010).

Owing to the significant differences between *T. cruzi* and *Homo sapiens* substrate specificity and amino acid sequences, gluconeogenesis enzyme phosphoenolpyruvate carboxykinase, which converts oxaloacetate into phosphoenolpyruvate, is a promising target for the development of new trypanocidal drugs (Beghini *et al.*, 2012; Cazzulo, 1994; Trapani *et al.*, 2001).

Described at the protein level for the first time in 2012 by Beghini and colleagues, the calcium-binding microtubule-associated protein named I/6 autoantigen has a predicted biological function associated with cross-linking to microtubule filaments, but it deserves further investigation. Interference with this structural protein could cause down-regulation in cytoskeleton-dependent processes, such as cellular trafficking and parasite division, leading to *T. cruzi* death.

Other interesting pathways for future drug intervention include nucleotide salvage pathways, specifically the enzyme from purine metabolism nucleoside phosphorylase. Nucleoside recovery after nucleotide degradation depends on the purine and pyrimidine salvage pathways, and this enzyme was detected in trypanosomatids using different methods (Beghini *et al.*, 2012; Miller *et al.*, 1987; Silva *et al.*, 2011). Therefore, nucleoside phosphorylase inhibition could be a promising strategy.

In addition to proteomic and biochemical characterization of molecular targets, ultrastructural analysis of treated parasites supports the description of potential targets in early pre-clinical steps. Promising target organelles and/or structures are based on their presence in clinically relevant forms of the parasite (trypomastigotes and amastigotes), as well as remarkable differences (or even absence) in host mammalian cells. *T. cruzi* non-canonical organelles include the mitochondrion, cytoskeletal-related structures, acidocalcisomes/contractile vacuole complex and glycosomes (Menna-Barreto and de Castro, 2017). Subsequent subfractional proteomic analysis of these parasite structures has greatly increased the number of identified specific organelle proteins, enhancing the possibility of the identification of specific molecular drug targets (Ferella *et al.*, 2008; Ulrich *et al.*, 2011). However, more multidisciplinary studies associating phenotypic and proteomic analysis must be performed to better describe the novel potential target candidates for drug intervention.

In 2015, the first description of the bloodstream trypomastigote proteomic map (Y strain) identified interesting proteins that could be analysed as drug targets and/or antigens (Brunoro *et al.*, 2015). Among the surface proteins, several transporters (amino acids, nucleobases, cations), mucins (in special trans-sialidases), and GPI-anchor transamidase phospholipases (A1 and C), were identified and need to be carefully selected for investigations of parasite-host cell interactions and for the search for potential antigens. Chaperones, such as heat shock proteins and calreticulin, are abundant in a large majority of *T. cruzi* forms and are particularly overexpressed under stress conditions (Andrade *et al.*, 2008; Atwood *et al.*, 2005; Brunoro *et al.*, 2015; de Godoy *et al.*, 2012; Menna-Barreto *et al.*, 2010; Pérez-Morales *et al.*, 2012). The inhibition of the chaperones' activity in trypomastigotes constantly subjected to extreme temperature variation and to the host immune response could be a promising unexplored strategy to kill the parasite. At the top of the list of abundant proteins in the bloodstream form are four enzymes from glycolysis/gluconeogenesis pathways (pyruvate phosphate dikinase, enolase, hexokinase and fructose-bisphosphate aldolase), reinforcing the dependence of bloodstream trypomastigotes on glycolysis (Clayton and Michels, 1996; Gonçalves *et al.*, 2011). The citric acid cycle enzyme aconitase is among the most abundant proteins in the bloodstream form; however, it is poorly studied in this protozoa. Blockage of the main ATP source of the infective parasite form is a potential trypanocidal approach. Different splicing-related elongation factors were identified in trypomastigotes, indicating particularities in parasite post-transcriptional regulation. Better comprehension of this molecular control could create new possibilities for the development of specific inhibitors.

Past, present and future of Chagas disease chemotherapy

Although parasitological cure can be achieved by benznidazole in a large majority of congenital cases and adult acute or early chronic infections (Urbina, 2010), the limitation and variable efficacy in late chronic patients, as well as differences in drug resistance among *T. cruzi* strains (Filardi and Brener, 1987), justify the search for novel and alternative compounds. A precise and clear comprehension of the mechanism of the trypanocidal action of the current clinical drug remains necessary (Salomao *et al.*, 2016). Two trypanocidal mechanisms have been proposed for benznidazole, both involving the participation of nitroreductases. The involvement of type II nitroreductase-producing reactive species is the first, but no hydrogen peroxide or superoxide ions generation were detected. Thus, benznidazole reduction occurred through the covalent modification of macromolecules and/or through an anaerobic route using nitroreduction derivatives (Díaz de Toranzo *et al.*, 1988; Moreno *et al.*, 1982; Polak and Richle, 1978). The second mechanism involved benznidazole activation by oxygen-insensitive type-I nitroreductase through amine production via nitroso and hydroxylamine intermediates (Hall *et al.*, 2012; Wilkinson and Kelly, 2009; Wilkinson *et al.*, 2008). Type-I nitroreductase was presented in this chapter as old yellow enzyme, and its biological function is associated with the metabolization of nitroheterocycles in the parasite (Kubata *et al.*, 2002; Murta *et al.*, 2006). However, a decrease in low-molecular-weight thiols in treated parasites, together with the lack of reactive species production observed after treatment with antiproliferative doses, sustains the discussion about the involvement of oxidative stress in the anti-*T. cruzi* mechanism of benznidazole (Boiani *et al.*, 2010). Another possibility for the trypanocidal mechanism of benznidazole is the formation of DNA-binding adducts of glyoxal and/or other cytotoxic metabolites (Hall *et al.*, 2012). Furthermore, the high toxicity to mammalian hosts from the non-specificity of the mode of action represents an important problem (Urbina, 2015).

Among the real alternatives proposed for current Chagas disease treatment, the sterol 14α-demethylase (CYP51) is the most interesting target based on *in vitro* and *in vivo* data. Despite the clinical trial failure of the E1224 inhibitor and posaconazole, an international consortium has been developing new azolic and non-azolic inhibitors (Torrico, 2013; Molina *et al.*, 2014). The high-content screening investigation showed that CYP51 inhibitors are more active and more highly strain-sensitive than benznidazole, substantially decreasing intracellular infection *in vitro* but not eliminating the parasite (Moraes *et al.*, 2014). Additionally, high efficacy of the combination of benznidazole and CYP51 inhibitors is expected (Gaspar *et al.*, 2015; Salomao *et al.*, 2016), representing an alternative for the treatment of acute and chronic patients. Doses and treatment period optimization, associated with cure improvement, substantiate the use of new combinatorial therapies (Bahia *et al.*, 2014). A clinical trial of the combined treatment of benznidazole and E1224 is planned, and in next few years, new directions of Chagas disease chemotherapy could be suggested (Chatelain and Ioset, 2011).

The commonly employed phenotypic approaches have suggested that *T. cruzi* organelles, structures and/or biochemical pathways are affected by a great variety of drugs (Menna-Barreto and de Castro, 2017; Menna-Barreto *et al.*, 2009; de Souza and Rodrigues, 2009; Vannier-Santos and De Castro, 2009). The interference with mitochondrial metabolism is among the most recurrent parasite alterations caused by drug treatment, followed by impairment of lipid biosynthesis and interaction with DNA (De Souza *et al.*, 2006; Menna-Barreto and de Castro, 2017; Menna-Barreto *et al.*, 2005, 2007, Santa-Rita *et al.*, 2000, 2006; Soeiro *et al.*, 2010; de Souza and Rodrigues, 2009; Vannier-Santos and De Castro, 2009; Weatherly *et al.*, 2009). However, the molecular mechanisms of action of trypanocidal compounds, including benznidazole, have been poorly investigated. In this framework, large-scale proteomics is a powerful and neglected tool. Few proteomic studies have attempted to identify the mechanisms of action involved in the trypanocidal activity of compounds. Of the three related publications, only one was performed on the infective parasite form (trypomastigotes isolated from mice bloodstream) (Brunoro *et al.*, 2016). Once the molecular differences among *T. cruzi* forms are known (Atwood *et al.*, 2005; Brunoro *et al.*, 2015), due to the challenges that Chagas disease chemotherapy represents, additional proteomic studies about trypanocidal drugs must be performed to provide parasite cell biology information, leading to the introduction of novel strategies for drug design.

The first assessment of the bloodstream trypomastigote proteomic profile in 2015 represents an important landmark for the discovery of promising molecular targets in the clinically relevant parasite form. Brunoro and colleagues (2015) identified a huge number of proteins (17,394 peptides and 3716 *T. cruzi* non-redundant proteins), providing a large databank for subsequent analysis. Proteins from a wide variety of biochemical pathways and processes were described, some for the first time in the infective form. A comparative analysis with other proteomic maps of trypomastigotes (culture-derived and metacyclic forms) identified more than 2000 proteins that were exclusively expressed in the bloodstream form, including structural proteins, surface molecules and proteins involved in various biochemical pathways, such as cell signalling, protein folding, bioenergetics, biosynthesis and DNA repair. Additional studies must be performed to develop specific inhibitors of this parasite form, which is responsible for infection dissemination in patients.

It is not uncommon that inhibitors designed for specific molecular targets do not bind or even exhibit the predicted biological effect (Chatelain and Ioset, 2011; Gilbert *et al.*, 2011; Sykes and Avery, 2013). The best strategy is the development of secure trypanocidal drugs in the established pipeline that were selected based on *in vivo* activity (Romanha *et al.*, 2010) using rational and phenotypic screening (Alonso-Padilla *et al.*, 2015). Finally, standardization of parasite strains, animal models and treatment protocols must be performed to improve the search for chemotherapeutic substitutes to benznidazole due to the poor translation of results from animal models to humans (Chatelain and Konar, 2015).

Huge multidisciplinary studies including medicinal chemistry, biochemical, bioinformatic, proteomic, molecular and cell biology groups using large-scale automated screenings must be conducted (Andriani *et al.*, 2011; Annang *et al.*, 2015; Bettiol *et al.*, 2009; Peña *et al.*, 2015). Moreover, the pharmacokinetics must be determined, and extensive analysis of the mechanisms of action has to be performed in non-human primates to minimize drugs unspecificity and to increase safety. The validation of the molecular targets identified by proteomics and/or bioinformatics is mandatory in follow-up studies (Andrade *et al.*, 2008; Menna-Barreto *et al.*, 2010).

References

Alban, A., David, S.O., Bjorkesten, L., Andersson, C., Sloge, E., Lewis, S., and Currie, I. (2003). A novel experimental design for comparative two-dimensional gel analysis: two-dimensional difference gel electrophoresis incorporating a pooled internal standard. Proteomics 3, 36–44. https://doi.org/10.1002/pmic.200390006.

Alonso-Padilla, J., Cotillo, I., Presa, J.L., Cantizani, J., Peña, I., Bardera, A.I., Martín, J.J., and Rodriguez, A. (2015). Automated high-content assay for compounds selectively toxic to *Trypanosoma cruzi* in a myoblastic cell line. PLOS Negl. Trop. Dis. *9,* e0003493. https://doi.org/10.1371/journal.pntd.0003493.

Altclas, J.D., Barcan, L., Nagel, C., Lattes, R., and Riarte, A. (2008). Organ transplantation and Chagas disease. JAMA *299,* 1134. https://doi.org/10.1001/jama.299.10.1134-a.

Alves, M.J., Kawahara, R., Viner, R., Colli, W., Mattos, E.C., Thaysen-Andersen, M., Larsen, M.R., and Palmisano, G. (2017). Comprehensive glycoprofiling of the epimastigote and trypomastigote stages of *Trypanosoma cruzi*. J. Proteomics *151,* 182–192. https://doi.org/10.1016/j.jprot.2016.05.034.

Andrade, H.M., Murta, S.M., Chapeaurouge, A., Perales, J., Nirdé, P., and Romanha, A.J. (2008). Proteomic analysis of *Trypanosoma cruzi* resistance to Benznidazole. J. Proteome Res. *7,* 2357–2367. https://doi.org/10.1021/pr700659m.

Andriani, G., Chessler, A.D., Courtemanche, G., Burleigh, B.A., and Rodriguez, A. (2011). Activity in vivo of anti-*Trypanosoma cruzi* compounds selected from a high throughput screening. PLOS Negl. Trop. Dis. *5,* e1298. https://doi.org/10.1371/journal.pntd.0001298.

Annang, F., Pérez-Moreno, G., García-Hernández, R., Cordon-Obras, C., Martín, J., Tormo, J.R., Rodríguez, L., de Pedro, N., Gómez-Pérez, V., Valente, M., *et al.* (2015). High-throughput screening platform for natural product-based drug discovery against 3 neglected tropical diseases: human African trypanosomiasis, leishmaniasis, and Chagas disease. J. Biomol. Screen. *20,* 82–91. https://doi.org/10.1177/1087057114555846.

Arias, D.G., Piñeyro, M.D., Iglesias, A.A., Guerrero, S.A., and Robello, C. (2015). Molecular characterization and interactome analysis of *Trypanosoma cruzi* tryparedoxin II. J. Proteomics *120,* 95–104. https://doi.org/10.1016/j.jprot.2015.03.001.

Atwood, J.A., Minning, T., Ludolf, F., Nuccio, A., Weatherly, D.B., Alvarez-Manilla, G., Tarleton, R., and Orlando, R. (2006). Glycoproteomics of *Trypanosoma cruzi* trypomastigotes using subcellular fractionation, lectin affinity, and stable isotope labeling. J. Proteome Res. *5,* 3376–3384. https://doi.org/10.1021/pr060364b.

Atwood, J.A., Weatherly, D.B., Minning, T.A., Bundy, B., Cavola, C., Opperdoes, F.R., Orlando, R., and Tarleton, R.L. (2005). The *Trypanosoma cruzi* proteome. Science *309,* 473–476. https://doi.org/10.1126/science.1110289.

Bahia, M.T., Diniz, L. de F., and Mosqueira, V.C. (2014). Therapeutical approaches under investigation for treatment of Chagas disease. Expert Opin. Investig. Drugs *23,* 1225–1237. https://doi.org/10.1517/13543784.2014.922952.

Barros-Alvarez, X., Gualdrón-López, M., Acosta, H., Cáceres, A.J., Graminha, M.A., Michels, P.A., Concepción, J.L., and Quiñones, W. (2014). Glycosomal targets for anti-trypanosomatid drug discovery. Curr. Med. Chem. *21,* 1679–1706. https://doi.org/ 10.2174/09298673113209990139.

Bautista-López, N.L., Ndao, M., Camargo, F.V., Nara, T., Annoura, T., Hardie, D.B., Borchers, C.H., and Jardim, A. (2016). Characterization and diagnostic application of *Trypanosoma cruzi* trypomastigote excreted-secreted antigens shed in extracellular vesicles released from infected mammalian cells. J. Clin. Microbiol. https://doi.org/10.1128/JCM.01649-16.

Bayer-Santos, E., Aguilar-Bonavides, C., Rodrigues, S.P., Cordero, E.M., Marques, A.F., Varela-Ramirez, A., Choi, H., Yoshida, N., da Silveira, J.F., and Almeida, I.C. (2013). Proteomic analysis of *Trypanosoma cruzi* secretome: characterization of two populations of extracellular vesicles and soluble proteins. J. Proteome Res. *12,* 883–897. https://doi.org/10.1021/pr300947g.

Beghini, D.G., Ferreira, A.T. da S., Almeida, V.C. de, Caminha, M.A., Silva-Jr, F.P., Perales, J., and Menna-Barreto, R.F.S. (2012). New insights in *Trypanosoma cruzi* proteomic map: further post-translational modifications and potential drug targets in Y strain epimastigotes. J. Integr. OMICS 2. https://doi.org/10.5584/jiomics.v2i2.107.

Bettiol, E., Samanovic, M., Murkin, A.S., Raper, J., Buckner, F., and Rodriguez, A. (2009). Identification of three classes of heteroaromatic compounds with activity against intracellular *Trypanosoma cruzi* by chemical library screening. PLOS Negl. Trop. Dis. *3,* e384. https://doi.org/10.1371/journal.pntd.0000384.

Boiani, M., Piacenza, L., Hernández, P., Boiani, L., Cerecetto, H., González, M., and Denicola, A. (2010). Mode of action of nifurtimox and N-oxide-containing heterocycles against *Trypanosoma cruzi*: is oxidative stress involved? Biochem. Pharmacol. *79,* 1736–1745. https://doi.org/10.1016/j.bcp.2010.02.009.

Brunoro, G.V., Caminha, M.A., Ferreira, A.T., Leprevost, F. da V., Carvalho, P.C., Perales, J., Valente, R.H., and Menna-Barreto, R.F. (2015). Reevaluating the *Trypanosoma cruzi* proteomic map: the shotgun description of bloodstream trypomastigotes. J. Proteomics *115,* 58–65. https://doi.org/10.1016/j.jprot.2014.12.003.

Brunoro, G.V., Faça, V.M., Caminha, M.A., Ferreira, A.T., Trugilho, M., de Moura, K.C., Perales, J., Valente, R.H., and Menna-Barreto, R.F. (2016). Differential Gel Electrophoresis (DIGE) Evaluation of Naphthoimidazoles Mode of Action: A Study in *Trypanosoma cruzi* Bloodstream Trypomastigotes. PLOS Negl. Trop. Dis. *10*, e0004951. https://doi.org/10.1371/journal.pntd.0004951.

Carmona, S.J., Nielsen, M., Schafer-Nielsen, C., Mucci, J., Altcheh, J., Balouz, V., Tekiel, V., Frasch, A.C., Campetella, O., Buscaglia, C.A., *et al.* (2015). Towards High-throughput Immunomics for Infectious Diseases: Use of Next-generation Peptide Microarrays for Rapid Discovery and Mapping of Antigenic Determinants. Mol. Cell. Proteomics *14*, 1871–1884. https://doi.org/10.1074/mcp.M114.045906.

Cazzulo, J.J. (1994). Intermediate metabolism in *Trypanosoma cruzi*. Bioenerg, J. Biomembr. *26*, 157–165. https://doi.org/10.1016/0169-4758(94)90010-8.

Chatelain, E., and Ioset, J.R. (2011). Drug discovery and development for neglected diseases: the DNDi model. Drug Des. Devel. Ther. *5*, 175–181. https://doi.org/10.2147/DDDT.S16381.

Chatelain, E., and Konar, N. (2015). Translational challenges of animal models in Chagas disease drug development: a review. Drug Des. Devel. Ther. *9*, 4807–4823. https://doi.org/10.2147/DDDT.S90208.

Clayton, C., and Shapira, M. (2007). Post-transcriptional regulation of gene expression in trypanosomes and leishmanias. Mol. Biochem. Parasitol. *156*, 93–101. https://doi.org/10.1016/j.molbiopara.2007.07.007.

Clayton, C.E., and Michels, P. (1996). Metabolic compartmentation in African trypanosomes. Parasitol. Today *12*, 465–471. https://doi.org/10.1016/S0169-4758(96)10073-9.

Cordero, E.M., Nakayasu, E.S., Gentil, L.G., Yoshida, N., Almeida, I.C., and da Silveira, J.F. (2009). Proteomic analysis of detergent-solubilized membrane proteins from insect-developmental forms of *Trypanosoma cruzi*. J. Proteome Res. *8*, 3642–3652. https://doi.org/10.1021/pr800887u.

Rodriques Coura, J., and de Castro, S.L. (2002). A critical review on Chagas disease chemotherapy. Mem. Inst. Oswaldo Cruz *97*, 3–24. https://doi.org/10.1590/S0074-02762002000100001.

De Souza, E.M., Menna-Barreto, R., Araújo-Jorge, T.C., Kumar, A., Hu, Q., Boykin, D.W., and Soeiro, M.N.C. (2006). Antiparasitic activity of aromatic diamidines is related to apoptosis-like death in *Trypanosoma cruzi*. Parasitology *133*, 75–79. https://doi.org/10.1017/S0031182006000084.

Dias, J.C., and Amato Neto, V. (2011). [Prevention concerning the different alternative routes for transmission of *Trypanosoma cruzi* in Brazil.] Rev. Soc. Bras. Med. Trop. *44* (Suppl. 2), 68–72. https://doi.org/ 10.1590/S0037-86822011000800011.

Díaz de Toranzo, E.G., Castro, J.A., Franke de Cazzulo, B.M., and Cazzulo, J.J. (1988). Interaction of benznidazole reactive metabolites with nuclear and kinetoplastic DNA, proteins and lipids from *Trypanosoma cruzi*. Experientia *44*, 880–881. https://doi.org/10.1007/BF01941187.

Ennes-Vidal, V., Menna-Barreto, R.F., Santos, A.L., Branquinha, M.H., and d'Avila-Levy, C.M. (2010). Effects of the calpain inhibitor MDL28170 on the clinically relevant forms of *Trypanosoma cruzi in vitro*. J. Antimicrob. Chemother. *65*, 1395–1398. https://doi.org/10.1093/jac/dkq154.

Ennes-Vidal, V., Menna-Barreto, R.F., Santos, A.L., Branquinha, M.H., and d'Avila-Levy, C.M. (2011). MDL28170, a calpain inhibitor, affects *Trypanosoma cruzi* metacyclogenesis, ultrastructure and attachment to Rhodnius prolixus midgut. PLOS ONE *6*, e18371. https://doi.org/10.1371/journal.pone.0018371.

Ersfeld, K., Barraclough, H., and Gull, K. (2005). Evolutionary relationships and protein domain architecture in an expanded calpain superfamily in kinetoplastid parasites. J. Mol. Evol. *61*, 742–757. https://doi.org/10.1007/s00239-004-0272-8.

Ferella, M., Nilsson, D., Darban, H., Rodrigues, C., Bontempi, E.J., Docampo, R., and Andersson, B. (2008). Proteomics in *Trypanosoma cruzi* – localization of novel proteins to various organelles. Proteomics *8*, 2735–2749. https://doi.org/10.1002/pmic.200700940.

Filardi, L.S., and Brener, Z. (1987). Susceptibility and natural resistance of *Trypanosoma cruzi* strains to drugs used clinically in Chagas disease. Trans. R. Soc. Trop. Med. Hyg. *81*, 755–759. https://doi.org/10.1016/0035-9203(87)90020-4.

Gaspar, L., Moraes, C.B., Freitas-Junior, L.H., Ferrari, S., Costantino, L., Costi, M.P., Coron, R.P., Smith, T.K., Siqueira-Neto, J.L., McKerrow, J.H., *et al.* (2015). Current and Future Chemotherapy for Chagas Disease. Curr. Med. Chem. *22*, 4293–4312. https://doi.org/10.2174/0929867322666151015120804.

Gilbert, I.H., Leroy, D., and Frearson, J.A. (2011). Finding new hits in neglected disease projects: target or phenotypic based screening? Curr. Top. Med. Chem. *11*, 1284–1291. https://.doi.org/10.2174/156802611795429176.

de Godoy, L.M., Marchini, F.K., Pavoni, D.P., Rampazzo, R. de C., Probst, C.M., Goldenberg, S., and Krieger, M.A. (2012). Quantitative proteomics of *Trypanosoma cruzi* during metacyclogenesis. Proteomics *12*, 2694–2703. https://doi.org/10.1002/pmic.201200078.

Goll, D.E., Thompson, V.F., Li, H., Wei, W., and Cong, J. (2003). The Calpain System. Physiol. Rev. *83*, 731–801. https://doi.org/10.1152/physrev.00029.2002.

Gonçalves, R.L., Barreto, R.F., Polycarpo, C.R., Gadelha, F.R., Castro, S.L., and Oliveira, M.F. (2011). A comparative assessment of mitochondrial function in epimastigotes and bloodstream trypomastigotes of *Trypanosoma cruzi*. J. Bioenerg. Biomembr. *43*, 651–661. https://doi.org/10.1007/s10863-011-9398-8.

Gonzales-Perdomo, M., Romero, P., and Goldenberg, S. (1988). Cyclic AMP and adenylate cyclase activators stimulate *Trypanosoma cruzi* differentiation. Exp. Parasitol. *66*, 205–212. https://doi.org/10.1016/0014-4894(88)90092-6.

Hall, B.S., Meredith, E.L., and Wilkinson, S.R. (2012). Targeting the substrate preference of a type I nitroreductase to develop antitrypanosomal quinone-based prodrugs. Antimicrob. Agents Chemother. *56*, 5821–5830. https://doi.org/10.1128/AAC.01227-12.

Holetz, F.B., Alves, L.R., Probst, C.M., Dallagiovanna, B., Marchini, F.K., Manque, P., Buck, G., Krieger, M.A., Correa, A., and Goldenberg, S. (2010). Protein and mRNA content of TcDHH1-containing mRNPs in *Trypanosoma cruzi*. FEBS J. *277*, 3415–3426. https://doi.org/10.1111/j.1742-4658.2010.07747.x.

Irigoín, F., Cibils, L., Comini, M.A., Wilkinson, S.R., Flohé, L., and Radi, R. (2008). Insights into the redox biology of *Trypanosoma cruzi*: Trypanothione metabolism and oxidant detoxification. Free Radic. Biol. Med. *45*, 733–742. https://doi.org/10.1016/j.freeradbiomed.2008.05.028.

Jannin, J., and Villa, L. (2007). An overview of Chagas disease treatment. Mem. Inst. Oswaldo Cruz 102 (Suppl. 1), 95–97. https://doi.org/10.1590/S0074-02762007005000106.

de Jesus, T.C., Calderano, S.G., Vitorino, F.N., Llanos, R.P., Lopes, M.C., de Araújo, C.B., Thiemann, O.H., Reis, M.D., Elias, M.C., and Chagas da Cunha, J.P. (2017). Quantitative Proteomic Analysis of Replicative and Nonreplicative Forms Reveals Important Insights into Chromatin Biology of *Trypanosoma cruzi*. Mol. Cell. Proteomics *16*, 23–38. https://doi.org/10.1074/mcp.M116.061200.

de Jesus, T.C., Nunes, V.S., Lopes, M. de C., Martil, D.E., Iwai, L.K., Moretti, N.S., Machado, F.C., de Lima-Stein, M.L., Thiemann, O.H., Elias, M.C., *et al.* (2016). Chromatin Proteomics Reveals Variable Histone Modifications during the Life Cycle of *Trypanosoma cruzi*. J. Proteome Res. *15*, 2039–2051. https://doi.org/10.1021/acs.jproteome.6b00208.

Kawano, D.F., Silva, V.B., Jorge, D.M., Silva, C.H., and Carvalho, I. (2011). Search for a platelet-activating factor receptor in the *Trypanosoma cruzi* proteome: a potential target for Chagas disease chemotherapy. Mem. Inst. Oswaldo Cruz *106*, 957–967. https://doi.org/10.1590/S0074-02762011000800010.

Kikuchi, S.A., Sodré, C.L., Kalume, D.E., Elias, C.G., Santos, A.L., de Nazaré Soeiro, M., Meuser, M., Chapeaurouge, A., Perales, J., and Fernandes, O. (2010). Proteomic analysis of two *Trypanosoma cruzi* zymodeme 3 strains. Exp. Parasitol. *126*, 540–551. https://doi.org/10.1016/j.exppara.2010.06.005.

Kubata, B.K., Duszenko, M., Kabututu, Z., Rawer, M., Szallies, A., Fujimori, K., Inui, T., Nozaki, T., Yamashita, K., Horii, T., *et al.* (2000). Identification of a novel prostaglandin f(2alpha) synthase in *Trypanosoma brucei*. J. Exp. Med. *192*, 1327–1338. https://doi.org/10.1084/jem.192.9.1327.

Kubata, B.K., Kabututu, Z., Nozaki, T., Munday, C.J., Fukuzumi, S., Ohkubo, K., Lazarus, M., Maruyama, T., Martin, S.K., Duszenko, M., *et al.* (2002). A key role for old yellow enzyme in the metabolism of drugs by *Trypanosoma cruzi*. J. Exp. Med. *196*, 1241–1251. https://doi.org/10.1084/jem.20020885.

Machin, A., Telleria, J., Brizard, J.P., Demettre, E., Séveno, M., Ayala, F.J., and Tibayrenc, M. (2014). *Trypanosoma cruzi*: gene expression surveyed by proteomic analysis reveals interaction between different genotypes in mixed in vitro cultures. PLOS ONE *9*, e95442. https://doi.org/10.1371/journal.pone.0095442.

Magalhães, A.D., Charneau, S., Paba, J., Guércio, R.A., Teixeira, A.R., Santana, J.M., Sousa, M.V., and Ricart, C.A. (2008). *Trypanosoma cruzi* alkaline 2-DE: Optimization and application to comparative proteome analysis of flagellate life stages. Proteome Sci. *6*, 24. https://doi.org/10.1186/1477-5956-6-24.

Marchini, F.K., de Godoy, L.M., Rampazzo, R.C., Pavoni, D.P., Probst, C.M., Gnad, F., Mann, M., and Krieger, M.A. (2011). Profiling the *Trypanosoma cruzi* phosphoproteome. PLOS ONE *6*, e25381. https://doi.org/10.1371/journal.pone.0025381.

Marin-Neto, J.A., Rassi, A., Avezum, A., Mattos, A.C., Rassi, A., Morillo, C.A., Sosa-Estani, S., Yusuf, S., and BENEFIT Investigators. (2009). The BENEFIT trial: testing the hypothesis that trypanocidal therapy is beneficial for patients with chronic Chagas heart disease. Mem. Inst. Oswaldo Cruz *104* (Suppl. 1), 319–324. https://doi.org/10.1590/S0074-02762009000900042.

Menna-Barreto, R.F., and de Castro, S.L. (2017). Clear Shot at Primary Aim: Susceptibility of *Trypanosoma cruzi* Organelles, Structures and Molecular Targets to Drug Treatment. Curr. Top. Med. Chem *17*, 1212–1234. https://doi.org/10.2174/1568026616666161025161858.

Menna-Barreto, R.F., and Perales, J. (2014). The expected outcome of the *Trypanosoma cruzi* proteomic map: a review of its potential biological applications for drug target discovery. Subcell. Biochem. *74*, 305–322. https://doi.org/10.1007/978-94-007-7305-9_13.

Menna-Barreto, R.F., Henriques-Pons, A., Pinto, A.V., Morgado-Diaz, J.A., Soares, M.J., and De Castro, S.L. (2005). Effect of a beta-lapachone-derived naphthoimidazole on *Trypanosoma cruzi*: identification of target organelles. J. Antimicrob. Chemother. *56*, 1034–1041. https://doi.org/10.1093/jac/dki403.

Menna-Barreto, R.F., Corrêa, J.R., Pinto, A.V., Soares, M.J., and de Castro, S.L. (2007). Mitochondrial disruption and DNA fragmentation in *Trypanosoma cruzi* induced by naphthoimidazoles synthesized from beta-lapachone. Parasitol. Res. *101*, 895–905. https://doi.org/10.1007/s00436-007-0556-1.

Menna-Barreto, R.F., Corrêa, J.R., Cascabulho, C.M., Fernandes, M.C., Pinto, A.V., Soares, M.J., and De Castro, S.L. (2009). Naphthoimidazoles promote different death phenotypes in *Trypanosoma cruzi*. Parasitology *136*, 499–510. https://doi.org/10.1017/S0031182009005745.

Menna-Barreto, R.F., Beghini, D.G., Ferreira, A.T., Pinto, A.V., De Castro, S.L., and Perales, J. (2010). A proteomic analysis of the mechanism of action of naphthoimidazoles *in Trypanosoma cruzi* epimastigotes *in vitro*. J. Proteomics 73, 2306–2315. https://doi.org/10.1016/j.jprot.2010.07.002.

Menna-Barreto, R.F., Belloze, K.T., Perales, J., and Silva-Jr, F.P. (2014). Proteomic and bioinformatic analysis of *Trypanosoma cruzi* chemotherapy and potential drug targets: new pieces for an old puzzle. Curr. Drug Targets *15*, 255–271. https://doi.org/10.2174/13894501113146660218.

Miller, R.L., Sabourin, C.L., and Krenitsky, T.A. (1987). *Trypanosoma cruzi* adenine nucleoside phosphorylase. Purification and substrate specificity. Biochem. Pharmacol. *36*, 553–560. https://doi.org/10.1016/0006-2952(87)90366-2.

Molina, I., Gómez i Prat, J., Salvador, F., Treviño, B., Sulleiro, E., Serre, N., Pou, D., Roure, S., Cabezos, J., Valerio, L., *et al.* (2014). Randomized trial of posaconazole and benznidazole for chronic Chagas' disease. N. Engl. J. Med. *370*, 1899–1908. https://doi.org/10.1056/NEJMoa1313122.

Moreno, S.N., Docampo, R., Mason, R.P., Leon, W., and Stoppani, A.O. (1982). Different behaviors of benznidazole as free radical generator with mammalian and *Trypanosoma cruzi* microsomal preparations. Arch. Biochem. Biophys. *218*, 585–591. https://.doi.org/10.1016/0003-9861(82)90383-6.

Murta, S.M., Krieger, M.A., Montenegro, L.R., Campos, F.F., Probst, C.M., Avila, A.R., Muto, N.H., de Oliveira, R.C., Nunes, L.R., Nirdé, P., *et al.* (2006). Deletion of copies of the gene encoding old yellow enzyme (TcOYE), a NAD(P)H flavin oxidoreductase, associates with in vitro-induced benznidazole resistance in *Trypanosoma cruzi*. Mol. Biochem. Parasitol. *146*, 151–162. https://doi.org/10.1016/j.molbiopara.2005.12.001.

Nakayasu, E.S., Gaynor, M.R., Sobreira, T.J., Ross, J.A., and Almeida, I.C. (2009). Phosphoproteomic analysis of the human pathogen *Trypanosoma cruzi* at the epimastigote stage. Proteomics *9*, 3489–3506. https://doi.org/10.1002/pmic.200800874.

Nakayasu, E.S., Sobreira, T.J., Torres, R., Ganiko, L., Oliveira, P.S., Marques, A.F., and Almeida, I.C. (2012). Improved proteomic approach for the discovery of potential vaccine targets in *Trypanosoma cruzi*. J. Proteome Res. *11*, 237–246. https://doi.org/10.1021/pr200806s.

Nwaka, S., and Hudson, A. (2006). Innovative lead discovery strategies for tropical diseases. Nat. Rev. Drug Discov. *5*, 941–955. https://doi.org/10.1038/nrd2144.

Paba, J., Santana, J.M., Teixeira, A.R., Fontes, W., Sousa, M.V., and Ricart, C.A. (2004). Proteomic analysis of the human pathogen *Trypanosoma cruzi*. Proteomics *4*, 1052–1059. https://doi.org/10.1002/pmic.200300637.

Parodi-Talice, A., Durán, R., Arrambide, N., Prieto, V., Piñeyro, M.D., Pritsch, O., Cayota, A., Cerveñansky, C., and Robello, C. (2004). Proteome analysis of the causative agent of Chagas disease: *Trypanosoma cruzi*. Int. J. Parasitol. *34*, 881–886. https://doi.org/10.1016/j.ijpara.2004.05.002.

Parodi-Talice, A., Monteiro-Goes, V., Arrambide, N., Avila, A.R., Duran, R., Correa, A., Dallagiovanna, B., Cayota, A., Krieger, M., Goldenberg, S., *et al.* (2007). Proteomic analysis of metacyclic trypomastigotes undergoing *Trypanosoma cruzi* metacyclogenesis. J. Mass Spectrom. *42*, 1422–1432. https://doi.org/10.1002/jms.1267.

Peloso, E.F., Dias, L., Queiroz, R.M., Leme, A.F., Pereira, C.N., Carnielli, C.M., Werneck, C.C., Sousa, M.V., Ricart, C.A., and Gadelha, F.R. (2016). *Trypanosoma cruzi* mitochondrial tryparedoxin peroxidase is located throughout the cell and its pull down provides one step towards the understanding of its mechanism of action. Biochim. Biophys. Acta *1864*, 1–10. https://doi.org/10.1016/j.bbapap.2015.10.005.

Peña, I., Pilar Manzano, M., Cantizani, J., Kessler, A., Alonso-Padilla, J., Bardera, A.I., Alvarez, E., Colmenarejo, G., Cotillo, I., Roquero, I., *et al.* (2015). New compound sets identified from high throughput phenotypic screening against three kinetoplastid parasites: an open resource. Sci. Rep. *5*, 8771. https://doi.org/10.1038/srep08771.

Pérez-Morales, D., Lanz-Mendoza, H., Hurtado, G., Martínez-Espinosa, R., and Espinoza, B. (2012). Proteomic analysis of *Trypanosoma cruzi* epimastigotes subjected to heat shock. J. Biomed. Biotechnol. *2012*, 902803. https://doi.org/10.1155/2012/902803.

Piacenza, L., Alvarez, M.N., Peluffo, G., and Radi, R. (2009). Fighting the oxidative assault: the *Trypanosoma cruzi* journey to infection. Curr. Opin. Microbiol. *12*, 415–421. https://doi.org/10.1016/j.mib.2009.06.011.

Picchi, G.F., Zulkievicz, V., Krieger, M.A., Zanchin, N.T., Goldenberg, S., and de Godoy, L.M. (2017). Post-translational Modifications of *Trypanosoma cruzi* Canonical and Variant Histones. J. Proteome Res. *16*, 1167–1179. https://doi.org/10.1021/acs.jproteome.6b00655.

Polak, A., and Richle, R. (1978). Mode of action of the 2-nitroimidazole derivative benznidazole. Ann. Trop. Med. Parasitol. *72*, 45–54. https://.doi.org/10.1080/00034983.1978.11719278.

Queiroz, R.M., Charneau, S., Motta, F.N., Santana, J.M., Roepstorff, P., and Ricart, C.A. (2013). Comprehensive proteomic analysis of *Trypanosoma cruzi* epimastigote cell surface proteins by two complementary methods. J. Proteome Res. *12*, 3255–3263. https://doi.org/10.1021/pr400110h.

Queiroz, R.M., Charneau, S., Bastos, I.M., Santana, J.M., Sousa, M.V., Roepstorff, P., and Ricart, C.A. (2014a). Cell surface proteome analysis of human-hosted *Trypanosoma cruzi* life stages. J. Proteome Res. *13*, 3530–3541. https://doi.org/10.1021/pr401120y.

Queiroz, R.M., Charneau, S., Mandacaru, S.C., Schwämmle, V., Lima, B.D., Roepstorff, P., and Ricart, C.A. (2014b). Quantitative proteomic and phosphoproteomic analysis of *Trypanosoma cruzi* amastigogenesis. Mol. Cell. Proteomics *13*, 3457–3472. https://doi.org/10.1074/mcp.M114.040329.

Rassi Jr, A., Rassi, A., and Marin-Neto, J.A. (2009). Chagas heart disease: pathophysiologic mechanisms, prognostic factors and risk stratification. Mem. Inst. Oswaldo Cruz *104* (Suppl. 1), 152–158. https://doi.org/10.1590/S0074-02762009000900021.

Roberts, A.J., and Fairlamb, A.H. (2016). The N-myristoylome of *Trypanosoma cruzi*. Sci. Rep. *6*, 31078. https://doi.org/10.1038/srep31078.

Rocha, M.O., Teixeira, M.M., and Ribeiro, A.L. (2007). An update on the management of Chagas cardiomyopathy. Expert Rev. Anti Infect. Ther. *5*, 727–743. https://doi.org/10.1586/14787210.5.4.727.

Rohloff, P., and Docampo, R. (2008). A contractile vacuole complex is involved in osmoregulation in *Trypanosoma cruzi*. Exp. Parasitol. *118*, 17–24. https://doi.org/ 10.1016/j.exppara.2007.04.013.

Romanha, A.J., Castro, S.L., Soeiro, M. de N., Lannes-Vieira, J., Ribeiro, I., Talvani, A., Bourdin, B., Blum, B., Olivieri, B., Zani, C., *et al.* (2010). *In vitro* and *in vivo* experimental models for drug screening and development for Chagas disease. Mem. Inst. Oswaldo Cruz *105*, 233–238. https://doi.org/10.1590/S0074-02762010000200022.

Salomao, K., Menna-Barreto, R.F.S., and de Castro, S.L. (2016). Stairway to Heaven or Hell? Perspectives and Limitations of Chagas Disease Chemotherapy. Curr. Top. Med. Chem. *16*, 2266–2289. https://doi.org/10.2174/1568026616666160413125049.

Sant'Anna, C., Nakayasu, E.S., Pereira, M.G., Lourenço, D., de Souza, W., Almeida, I.C., and Cunha-E-Silva, N.L. (2009). Subcellular proteomics of *Trypanosoma cruzi* reservosomes. Proteomics *9*, 1782–1794. https://doi.org/10.1002/pmic.200800730.

Santa-Rita, R.M., Santos Barbosa, H., Meirelles, M.N., and de Castro, S.L. (2000). Effect of the alkyl-lysophospholipids on the proliferation and differentiation of *Trypanosoma cruzi*. Acta Trop. *75*, 219–228. https://doi.org/10.1016/S0001-706X(00)00052-8.

Santa-Rita, R.M., Barbosa, H.S., and de Castro, S.L. (2006). Ultrastructural analysis of edelfosine-treated trypomastigotes and amastigotes of *Trypanosoma cruzi*. Parasitol. Res. *100*, 187–190. https://doi.org/10.1007/s00436-006-0250-8.

dos Santos Júnior, A. de C., Kalume, D.E., Camargo, R., Gómez-Mendoza, D.P., Correa, J.R., Charneau, S., de Sousa, M.V., de Lima, B.D., and Ricart, C.A. (2015). Unveiling the *Trypanosoma cruzi* Nuclear Proteome. PLOS ONE *10*, e0138667. https://doi.org/10.1371/journal.pone.0138667.

Schmunis, G.A. (2007). Epidemiology of Chagas disease in non-endemic countries: the role of international migration. Mem. Inst. Oswaldo Cruz *102* (Suppl. 1), 75–85. https://doi.org/10.1590/S0074-02762007005000093.

Silva, R.G., Vetticatt, M.J., Merino, E.F., Cassera, M.B., and Schramm, V.L. (2011). Transition-state analysis of *Trypanosoma cruzi* uridine phosphorylase-catalyzed arsenolysis of uridine. J. Am. Chem. Soc. *133*, 9923–9931. https://doi.org/10.1021/ja2031294.

Silva Galdino, T., Menna-Barreto, R.F., Britto, C., Samudio, F., Brandão, A., and Kalume, D.E. (2014). Cell disruption using a different methodology for proteomics analysis of *Trypanosoma cruzi* strains. Anal. Biochem. *448*, 1–8. https://doi.org/10.1016/j.ab.2013.11.010.

Silva Júnior, E.N. da, Jardim, G.A.M., Menna-Barreto, R.F.S., and Castro, S.L. de (2014). Anti-*Trypanosoma cruzi* Compounds: Our Contribution for the Evaluation and Insights on the Mode of Action of Naphthoquinones and Derivatives. J. Braz. Chem. Soc. https://doi.org/10.5935/0103-5053.20140180.

Sodré, C.L., Chapeaurouge, A.D., Kalume, D.E., de Mendonça Lima, L., Perales, J., and Fernandes, O. (2009). Proteomic map of *Trypanosoma cruzi* CL Brener: the reference strain of the genome project. Arch. Microbiol. *191*, 177–184. https://doi.org/10.1007/s00203-008-0439-6.

Soeiro, M. de N., and de Castro, S.L. (2011). Screening of Potential anti-*Trypanosoma cruzi* Candidates: In Vitro and In Vivo Studies. Open Med. Chem. J. *5*, 21–30. https://doi.org/10.2174/1874104501105010021.

Soeiro, M. de N.C., Daliry, A., Silva, C.F., Batista, D.G.J., Souza, E.M., Salomão, K., Menna-Barreto, R.F.S., and De Castro, S.L. (2010). Electron microscopy approaches for the investigation of the cellular targets of trypanocidal agents in *Trypanosoma cruzi*. In Microscopy: Science, Technology, Applications and Education, (Spain: A. Mendez-Villas, J. Diaz), pp. 191–203.

Sosa-Estani, S., Viotti, R., and Segura, E.L. (2009). Therapy, diagnosis and prognosis of chronic Chagas disease: insight gained in Argentina. Mem. Inst. Oswaldo Cruz *104* (Suppl. 1), 167–180. https://doi.org/10.1590/S0074-02762009000900023.

de Souza, W., and Rodrigues, J.C. (2009). Sterol Biosynthesis Pathway as Target for Anti-trypanosomatid Drugs. Interdiscip. Perspect. Infect. Dis. *2009*, 642502. https://doi.org/10.1155/2009/642502.

Steindel, M., Kramer Pacheco, L., Scholl, D., Soares, M., de Moraes, M.H., Eger, I., Kosmann, C., Sincero, T.C., Stoco, P.H., Murta, S.M., *et al.* (2008). Characterization of *Trypanosoma cruzi* isolated from humans, vectors, and animal reservoirs following an outbreak of acute human Chagas disease in Santa Catarina State, Brazil. Diagn. Microbiol. Infect. Dis. *60*, 25–32. https://doi.org/10.1016/j.diagmicrobio.2007.07.016.

Sykes, M.L., and Avery, V.M. (2013). Approaches to protozoan drug discovery: phenotypic screening. J. Med. Chem. *56*, 7727–7740. https://doi.org/10.1021/jm4004279.

Torrico, F. (2013). E1224–Results of proof of concept clinical trial in patients with chronic indeterminate Chagas disease. In: Proceedings of the 62nd Annual Meeting of the American Society of Tropical Medicine and Hygiene; November 13–17, 2013; Washington, DC. Proof-of-Concept Study of E1224 to Treat Adult Patients With Chagas Disease. Drugs for Neglected Diseases Initiative. Available: http://clinicaltrials.gov/. Identifier: NCT01489228.

Trapani, S., Linss, J., Goldenberg, S., Fischer, H., Craievich, A.F., and Oliva, G. (2001). Crystal structure of the dimeric phosphoenolpyruvate carboxykinase (PEPCK) from *Trypanosoma cruzi* at 2 A resolution. J. Mol. Biol. *313*, 1059–1072. https://doi.org/10.1006/jmbi.2001.5093.

Trochine, A., Alvarez, G., Corre, S., Faral-Tello, P., Durán, R., Batthyany, C.I., Cerecetto, H., González, M., and Robello, C. (2014). *Trypanosoma cruzi* chemical proteomics using immobilized benznidazole. Exp. Parasitol. *140*, 33–38. https://doi.org/10.1016/j.exppara.2014.03.013.

Ulrich, P.N., Jimenez, V., Park, M., Martins, V.P., Atwood, J., Moles, K., Collins, D., Rohloff, P., Tarleton, R., Moreno, S.N., *et al.* (2011). Identification of contractile vacuole proteins in *Trypanosoma cruzi*. PLOS ONE *6*, e18013. https://doi.org/10.1371/journal.pone.0018013.

Urbina, J.A. (2010). Specific chemotherapy of Chagas disease: relevance, current limitations and new approaches. Acta Trop. *115*, 55–68. https://doi.org/10.1016/j.actatropica.2009.10.023.

Urbina, J.A. (2015). Recent clinical trials for the etiological treatment of chronic chagas disease: advances, challenges and perspectives. J. Eukaryot. Microbiol. *62*, 149–156. https://doi.org/10.1111/jeu.12184.

Urbina, J.A., and Docampo, R. (2003). Specific chemotherapy of Chagas disease: controversies and advances. Trends Parasitol. *19*, 495–501. https://doi.org/10.1016/j.pt.2003.09.001.

Vannier-Santos, M.A., and De Castro, S.L. (2009). Electron microscopy in antiparasitic chemotherapy: a (close) view to a kill. Curr. Drug Targets *10*, 246–260. https://doi.org/10.2174/138945009787581168.

Vieira, H.G., Grynberg, P., Bitar, M., Pires, S. da F., Hilário, H.O., Macedo, A.M., Machado, C.R., de Andrade, H.M., and Franco, G.R. (2014). Proteomic analysis of *Trypanosoma cruzi* response to ionizing radiation stress. PLOS ONE *9*, e97526. https://doi.org/10.1371/journal.pone.0097526.

Weatherly, D.B., Boehlke, C., and Tarleton, R.L. (2009). Chromosome level assembly of the hybrid *Trypanosoma cruzi* genome. BMC Genomics *10*, 255. https://doi.org/10.1186/1471-2164-10-255.

WHO (2015). Chagas disease in Latin America: an epidemiological update based on 2010 estimates. Releve Epidemiol. Hebd. *90*, 33–43.

Wilkinson, S.R., and Kelly, J.M. (2009). Trypanocidal drugs: mechanisms, resistance and new targets. Expert Rev. Mol. Med. *11*, e31. https://doi.org/10.1017/S1462399409001252.

Wilkinson, S.R., Taylor, M.C., Horn, D., Kelly, J.M., and Cheeseman, I. (2008). A mechanism for cross-resistance to nifurtimox and benznidazole in trypanosomes. Proc. Natl. Acad. Sci. U.S.A. *105*, 5022–5027. https://doi.org/10.1073/pnas.0711014105.

Zago, M.P., Hosakote, Y.M., Koo, S.J., Dhiman, M., Piñeyro, M.D., Parodi-Talice, A., Basombrio, M.A., Robello, C., and Garg, N.J. (2016). TcI Isolates of *Trypanosoma cruzi* Exploit the Antioxidant Network for Enhanced Intracellular Survival in Macrophages and Virulence in Mice. Infect. Immun. *84*, 1842–1856. https://doi.org/10.1128/IAI.00193-16.

Zingales, B., Andrade, S.G., Briones, M.R., Campbell, D.A., Chiari, E., Fernandes, O., Guhl, F., Lages-Silva, E., Macedo, A.M., Machado, C.R., *et al.* (2009). A new consensus for *Trypanosoma cruzi* intraspecific nomenclature: second revision meeting recommends TcI to TcVI. Mem. Inst. Oswaldo Cruz *104*, 1051–1054. https://doi.org/10.1590/S0074-02762009000700021

High-throughput Screenings: New Breeze in Anti-*Trypanosoma cruzi* Drug Discovery

2

Julio Alonso-Padilla[1]* and Ana Rodríguez[2]

[1]Barcelona Institute for Global Health (ISGlobal), Hospital Clinic–Universitat de Barcelona, Barcelona, Spain.
[2]Department of Microbiology, New York University, New York, NY, United States of America.

Corresponding author: julio.a.padilla@isglobal.org

https://doi.org/10.21775/9781910190838.02

Abstract

Development of therapeutic solutions for Chagas disease, which is caused by the protozoan *Trypanosoma cruzi*, has been historically neglected. Despite being the parasitic disease of highest impact in America, there is no vaccine in sight and the only two available drugs for treatment, benznidazole and nifurtimox, display severe toxicity. These drugs have good efficacy against acute *T. cruzi* infections and are relatively well tolerated by young patients. Nonetheless, the acute stage of the disease is mostly asymptomatic and largely remains undiagnosed and untreated. Cardiac and/or digestive life-threatening symptoms arise in the chronic stage, which is when diagnosis is generally achieved. At this stage, the efficacy of current drugs is diminished and their side effects prompt to ≈ 20% of treatment discontinuation. Thus there is an urgent need for more efficacious and less toxic therapies for the estimated 7 million chronically infected patients. To meet this challenge, massive screenings have been recently performed over large chemical collections to identify new chemical structures fitted with anti-*T. cruzi* specific activity. The parasite's biological complexity and the lack of detailed metabolic and biochemical information have favoured a phenotypic over a target-based drug discovery strategy. The first fruits of these high-throughput screening campaigns are now arriving at the pre-clinical steps of the drug developmental pathway.

Introduction

The global strategy for the control of neglected tropical diseases (NTDs) relies on chemotherapy (Hotez *et al.*, 2016). In the majority of cases, the available treatments for them have severe side effects, questioned efficacy, emergent resistance and yet unknown mechanisms of action (MoA). However, in spite of a desperate need of improved therapies, no major investments have been historically made on their research and development (R&D) as there is little incentive to invest in products to a market that cannot pay (Pedrique *et al.*, 2013). But big pharmaceutical companies are the best (and almost only) gifted with high-quality large chemical collections and the required equipment and expertise to interrogate them for the discovery, development, and production of new drugs.

At present there is a momentum of political and financial will to tackle NTDs impact, which has involved the establishment of global alliances such as the London Declaration in 2012 and the World Health Assembly resolution [WHA 66.12] (Hotez *et al.*, 2016). Besides, non-profit public–private partnerships, such as Drugs for Neglected Diseases Initiative (DNDi) and the Global Health Innovative Technology Fund (GHIT), as well as academia-industry collaborative frameworks (e.g. GSK's Tres Cantos Open Lab Foundation) have been established to promote and fund NTDs drug discovery (Katsuno *et al.*, 2015; Olliaro *et al.*, 2015). These efforts have yielded an increase on R&D, fuelled by investments by philanthropic institutions and cooperation governmental agencies (Olliaro *et al.*, 2015; Pedrique *et al.*, 2013).

However, drugs and vaccines available against NTDs are yet insufficient and the overall picture has not improved in decades. In a 2013 article that analysed the pipeline of drugs and vaccines for neglected diseases – a term that englobes NTDs, plus malaria, tuberculosis, diarrhoeal diseases, and other diseases of poverty – in the 2000–2011 period, concluded that barely 1% (4 out of 336) of the new chemical entities (NCEs) registered were indicated for them (3 for malaria and 1 for diarrhoeal diseases; Fig. 2.1) (Pedrique *et al.*, 2013). A

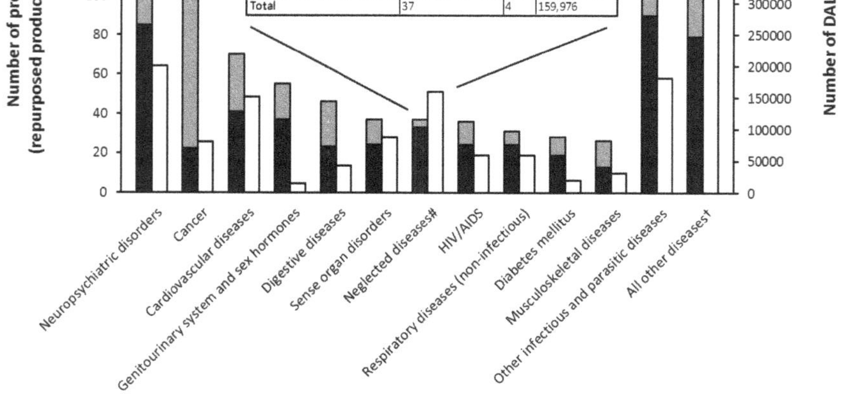

Figure 2.1 New therapeutic products for neglected diseases (be it repurposed or newly discovered chemical entities [NCEs]) registered during the 2000–2011 interval. *, 'All new products' refers to both repurposed and NCEs, registered. #, 'Neglected diseases' stands for those depicted in the inner table; †, 'All other diseases' stands for: maternal and perinatal disorders, nutritional deficiencies, congenital abnormalities, skin diseases, endocrine disorders, oral diseases, and injuries. DALYs, disease-adjusted life years. Numbers for the figure and the inner table were adapted from Pedrique *et al.* (2013) Tables 1 and 2.

previous study, dated 11 years before, had arrived to the same numbers when analysing the approved chemical entities in the period of 1975 to 1999 (16/1396) (Trouiller *et al.*, 2002). By now, despite recent increases in investment, the picture has not changed much. The number of drugs to fight neglected infections (either NCEs or repurposed) is still overwhelmed by the diseases burden (expressed as disability-adjusted life years, DALYs) (see Fig. 2.1). Nonetheless, it is expected that recent efforts in mass high-throughput screening (HTS) of large compound libraries against neglected pathogens will eventually yield an increase in drug registrations (Pedrique *et al.*, 2013).

Chagas disease (American trypanosomiasis)

Chagas disease is caused by infection with the protozoan parasite *Trypanosoma cruzi* (*T. cruzi*; order Kinetoplastida; family Trypanosomatidae). First described by the Brazilian physician Carlos Chagas in 1909, it is within the list of 17 NTDs of the WHO. This same organization estimates there are ≈7 million people infected with *T. cruzi* worldwide (www.who.int/mediacentre/factsheets/fs340/en/). Most of them are in Latin America, where triatomine vectors (order Hemiptera; family Reduviidae) that transmit the infection are endemic (Gascon and Pinazo, 2015). In recent years acute outbreaks have been related to consumption of vector-mediated parasite contamination of food or drink (Robertson *et al.*, 2016). Besides, vector-independent transmission routes, like organ transplant, blood transfusion, and from mother to child have also been described (Gascon and Pinazo, 2015). These are of relevance also in non-endemic regions (e.g. Australia, Japan, Canada and Europe), where the disease has been globalized in the last decades due to migration flows from endemic regions (Gascon and Pinazo, 2015).

Two stages can be distinguished in Chagas disease clinical progression. Firstly, a short acute stage (8–12 weeks) during which parasitaemia can be directly detected (Bern *et al.*, 2011; Gascon and Pinazo, 2015). Nonetheless, at this point the infection largely goes undiagnosed because the symptomatology is mostly non-existent or courses as a mild flu. Surpassed the acute phase, the disease enters in an indeterminate chronic stage that may span 10–30 years. Around 70% of patients will not manifest any clinical symptoms, but in the remaining 30% heart and/or gastrointestinal tract disruptions will cause severe disability and potentially lead to death (Bern *et al.*, 2011; Gascon and Pinazo, 2015).

Chagas disease is the parasitic infection of highest impact in America. With no prospects of an effective vaccine in sight, chemotherapy and vector control remain as the only tools to fight the disease. The two available chemotherapies, benznidazole (BNZ) and nifurtimox (NFX) (Fig. 2.2), have been in use since 1966 and 1970 respectively, despite presenting

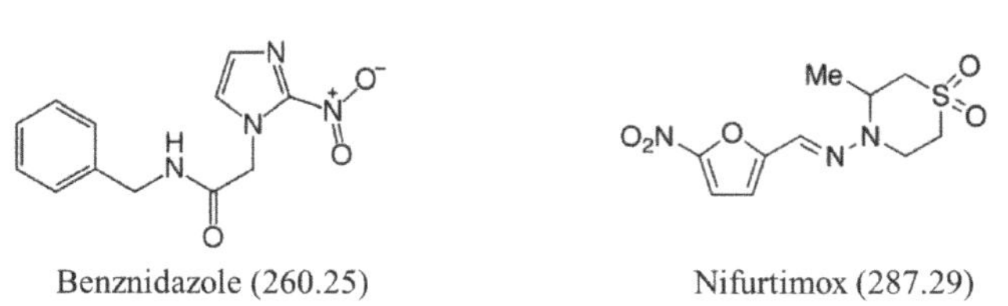

Figure 2.2 Commercially available drugs to treat Chagas disease: benznidazole and nifurtimox. Both are nitroheterocyclic compounds with a nitro group linked to an imidazole or furan ring, respectively. Their corresponding molecular weights are indicated in brackets (g/mol).

severe associated toxicity (Wilkinson and Kelly, 2009). Nowadays BNZ is produced in Argentina by ELEA under the name Abarax, and in Brazil by LAFEPE; whereas NFX is commercialized by Bayer as Lampit. A BNZ regime spans up to 60 days during which 5 mg/kg of drug are taken daily. The dosage is higher in the case of NFX with 8 mg/kg/day (Gascon and Pinazo, 2015). Both are orally administered and show good efficacy against the acute stage of the disease (Gascon and Pinazo, 2015; Wilkinson and Kelly, 2009). But, as aforementioned, this acute stage is mostly asymptomatic and undiagnosed. Diagnosis is achieved when the onset of symptoms is evident, and treatment dispensed then. Unluckily, BNZ and NFX efficacy diminish the longer a person has been infected and their side effects often prompt to treatment discontinuation (Pinazo *et al.*, 2013; Wilkinson and Kelly, 2009). Thus, new, safer, and more efficacious chemotherapies are urgently needed for the chronic phase of the disease.

A hint of *T. cruzi* biology

There are four phenotypically distinct parasite life stages. In its natural life cycle, epimastigotes replicate in the insect vector gut and become metacyclic trypomastigotes upon migration to its hindgut. These are the first-in mammalian infective form. Despite vectors are haematophagous, metacyclic trypomastigotes are excreted in the insect's faeces upon a blood-meal that usually occurs at night (Bern *et al.*, 2011). Then they gain access to the host tissues either through the mucosa or through skin micro-abrasions self-inflicted by the host when scratching the vector bite site (Bern *et al.*, 2011). Once inside, metacyclic trypomastigotes will invade host cells and transform into amastigotes. These are the round-shaped, intracellular replicative *T. cruzi* life-form in mammalian hosts. By binary fission amastigotes number will grow until crowding the host cell cytoplasm. Then, their transformation into flagellated trypomastigotes will mean the burst of the cell, and these motile forms will take their business elsewhere. If escaping from host immunity vigilance, trypomastigotes may re-infect other host cells and carry on with the intra-mammalian 'minor' cycle; or they may be picked up from the bloodstream by a feeding triatomine vector thus closing the 'major' biological cycle of the parasite (Bern *et al.*, 2011) (and check it graphically at www.cdc.gov/parasites/chagas/biology.html).

From a drug discovery standpoint, mammalian infective amastigotes and trypomastigotes are the key stages to target (Chatelain, 2015). Fortunately, standard techniques for *in vitro* growth of all four *T. cruzi* life forms and induction of their transitions are mastered in many laboratories. This knowledge has allowed the performance of stage-specific studies (Atwood *et al.*, 2005; Minning *et al.*, 2009). It has also granted mass production of trypomastigotes and amastigotes for drug screening campaigns.

Similarly to what has occurred with many other organisms, next generation sequencing and genome-related -omics have also touched *T. cruzi* research. Genomes of various strains are now available, and transcriptomic, proteomic and metabolomics data have been generated [visit the TriTrypDB resource at http://tritrypdb.org/tritrypdb/; Aslett *et al.*, (2010)]. These efforts definitely represent a breakthrough in the research on this parasite; and while analysis and curation of the wealth of data available is still challenging, bioinformatics is jumping in the arena of neglected diseases investigation.

What can all these data mean for anti-*T. cruzi* drug discovery? BNZ and NFX have been (are) used for decades unknowingly of their targets or MoA. Nonetheless, along the process of finding safer and more efficient therapeutic solutions, identification of targets and

MoA will clearly have an impact on Chagas disease drug discovery. Omics-related subjects like chemoinformatics and chemogenomics are now contributing to that goal (Ekins *et al.*, 2015; Khare *et al.*, 2015b; Magarinos *et al.*, 2012). In summary, mass data acquisition and its appropriate management is expected to speed up the achievement of safer and more efficient drugs. Taking into account the huge amount of information generated and managed, high-throughput screening (HTS) campaigns of large chemical libraries may well be regarded as a drug discovery-omics derivate. But what is a HTS campaign?

High-throughput screening

High-throughput screening sits at the early steps of the drug discovery pipeline, yielding *in vitro* identified compounds with the desired activity, also called 'hits', that represent starting points for further drug design. It is said to be made as a 'campaign' as it involves several actors and a well-organized plan of intertwined sequential events in order to be performed. Since the amount of compounds to screen can be enormous, HTS makes use of robotics, data processing and control software, liquid handling devices, and sensitive detectors that allow the performance of millions of tests from diverse libraries of chemical compounds, antibodies or siRNA (Fig. 2.3). Certainly automation plays an important part in the process, particularly in ultra-HTS campaigns (> 100,000 compounds tested per day), to handle liquids, transport assay-plates and read lots of them repetitively and even simultaneously. Assay readout can be made by fluorescence or luminescence detection or by microscopy image acquisition and analysis. High content microscopy screening relies on the latter and it is gaining pace within HTS assays because it allows the quantitative detection of several cellular features in a single experiment (Zanella *et al.*, 2010).

The assay support format is the microplate, generally of 384 wells or higher (1536 or 3456 wells) where each well receives the biological specimen that is going to be tested. As

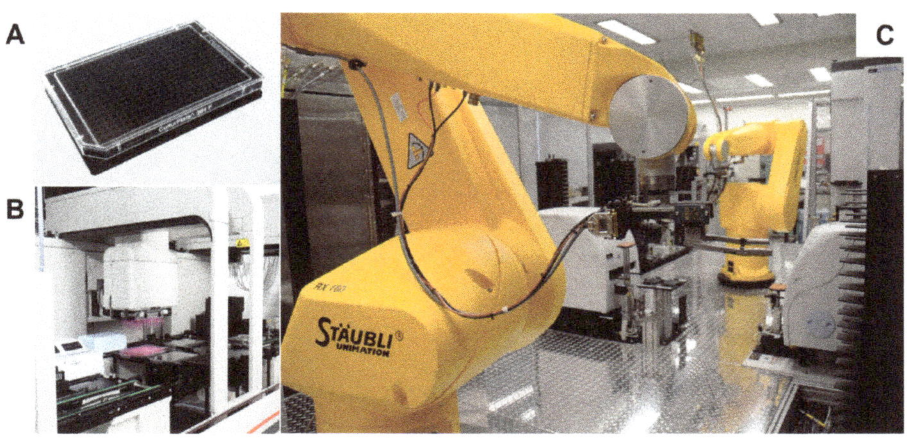

Figure 2.3 HTS elements: (A) the experimental support is the microplate, generally of 384 wells or higher (1536 or 3456 wells); (B) fully or semi-automated liquid-handlers are required to manage with accuracy very little volumes over and over dozens of plates per run; (C) robots are important in the process of management and analysis of many assay-plates repetitively and simultaneously.

regards the anti-*T. cruzi* HTS campaigns, this specimen was mammalian host cells infected with trypomastigotes so that inhibition of their intracellular replication (as amastigotes) could be measured. Each assay plate includes a series of negative and positive control wells. In the case of inhibition type assays, like those performed against *T. cruzi*, negative control is the minimum signal data and positive control the maximum one (untreated specimens).

Quality control (QC) is a key feature considering the millions of test determinations made in a particular campaign. The statistical parameter that measures reliability of an assay and ensures whether it is ready to be performed under high-throughput conditions is the Z' (Zhang *et al.*, 1999). To calculate it the following formula must be resolved: $Z' = 1 - [(3\sigma_{c+} + 3\sigma_{c-})/|\mu_{c+} - \mu_{c-}|]$, where σ_{c+} = standard deviation (SD) of positive control, σ_{c-} = SD of negative control, μ_{c+} = mean of positive control and μ_{c-} = mean of negative control. It takes into account the mean signals of control wells, as well as their standard deviations in order to determine the assay dynamic range window (Zhang *et al.*, 1999). Besides the aforementioned Z' parameter, features such as signal-to-background and signal-to-noise ratio, assay window and its variability ratio are also evaluated and regularly checked during the campaign to control the assay performance throughout it. Importantly too, assay-plates are controlled and corrected for patterns derived from liquid dispensing errors or induced by evaporation by means of computational statistical analysis with specifically dedicated algorithms (Iversen *et al.*, 2012).

Phenotypic versus target-based screening

Two types of screening approaches can be distinguished: target-based and phenotypic screening. A target-based drug discovery approach stands on the basis that the library is screened for hits against a specific gene-encoded biological target. Therefore, this kind of HTS depends on the previous identification of that relevant target (Lee *et al.*, 2012). Despite acknowledgement of the target holds great advantage in terms of subsequent improvement of the hits by means of medicinal chemistry and structure–activity relationship (SAR) studies, it cannot be the choice in anti-*T. cruzi* screening. Same as with other parasitic infections, research on Chagas disease has been (is) ill-funded. As a result there are only a few identified targets available (Herrera *et al.*, 2016; Holloway *et al.*, 2009), despite the parasite is biologically complex and its protein coding sequences (CDS) exceed 10K (Atwood *et al.*, 2005; El-Sayed *et al.*, 2005). In any case, target-based screening does not provide information on whether the tested compounds are membrane permeable and would be able to reach their target in the parasite. This represents a major drawback, as inability to cross biological membranes to reach their target will result in the loss of any hits further down in the discovery path (Sykes and Avery, 2013).

In contrast, whole cell-based phenotypic screening is target-blind; thus, it will probably identify hits with unknown targets. Nonetheless, target identification for effective compounds is now frequently successful and broadly reported in the literature (Lee *et al.*, 2012). At the same time, a phenotypic approach will be a better predictor of *in vivo* activity, since the compounds must traverse whatever cell membranes encountered to be active. This redounds in an overall faster and more probable *in vitro* to *in vivo* translation of the hits, which ultimately will save developmental costs. This is particularly important in the low investment field of neglected diseases drug discovery. Therefore, phenotypic screening is the preferred strategy for HTS campaigns against malaria, tuberculosis, and NTDs (Ballell *et al.*, 2013; Gamo *et al.*, 2010; Peña *et al.*, 2015). Overall, its success over target-based approaches is reflected by the higher percentage of NCEs approved by the FDA derived from phenotypic campaigns in recent years (Lee *et al.*, 2012).

Anti-*T. cruzi* HTS campaigns

Screening of large chemical libraries has been feasible thanks to the availability of transgenic parasites. These have revolutionized Chagas disease drug discovery. Engineered to express reporter enzymatic activities, their implementation has simplified the *in vitro* and *in vivo* work allowing the development of biological assays with high sensitivity and reproducibility. Transgenic parasites have transformed the previously time-consuming, labour-intensive, observer biased visual microscopic counting, into faster, unbiased (colorimetric, fluorometric, or luminescence-based) measurements (Alonso-Padilla and Rodríguez, 2014; Rodriguez and Tarleton, 2012).

The most popular transgenic *T. cruzi* for *in vitro* screening has been the robust Tulahuen strain expressing an *Escherichia coli* β-galactosidase gene (Buckner *et al.*, 1996). Generously shared by the lab of origin, the Tulahuen-β-galactosidase strain has been core in the performance of all HTS against *T. cruzi*, from studies screening small libraries (Annang *et al.*, 2015; Bettiol *et al.*, 2009) to the ultra-HTS campaigns done at GSK (Tres Cantos, Spain) (Peña *et al.*, 2015), and at the Genomics Institute of the Novartis Research Foundation (San Diego, USA) (Khare *et al.*, 2016). Other three recombinant β-galactosidase strains were developed at the same lab from the University of Washington in Seattle (USA) (Buckner *et al.*, 2001). Other transgenes have been also cloned into *T. cruzi*, like that encoding the tandem tomato protein to make fluorescent parasites (Canavaci *et al.*, 2010), or luciferase genes for luminescence-based detection of parasites (Andriani *et al.*, 2011; Canavaci *et al.*, 2010; Hyland *et al.*, 2008; Lewis *et al.*, 2014, 2016). These have been mostly used for *in vivo* screening by bio-imaging, an innovation that has introduced a paradigm shift in the drug discovery progression cascade (Rodriguez and Tarleton, 2012).

Amongst the HTS performed against *T. cruzi*, despite target-based approaches have been described (Holloway *et al.*, 2009), most projects have relied on a phenotypic screening strategy (Alonso-Padilla and Rodríguez, 2014; Annang *et al.*, 2015; Khare *et al.*, 2016; Neitz *et al.*, 2015; Peña *et al.*, 2015). Since phenotypic screening is aimed at detection of compounds with inhibitory activity against the whole parasite mammalian life-stages, any MoA would remain unknown. Notwithstanding that the inhibition might be mediated by disruption of the parasite's membranes, by compounds interfering with its metabolism, or due to the inhibition of a protein implicated in its replication, the output of a phenotypic screening will be to flag inhibitory chemical entities (structures) as potential drug starting points. This will be the 'undercooked' harvest of the primary assay. However, there are various hurdles that complicate the interpretation of results, such as that compounds may be self-fluorescent and confuse the assay readout if it is based on fluorescence intensity detection (FLINT); they may also be cytotoxic to the host cell, thus precluding growth of the parasite inside; they may just inhibit the reporter activity carried by the transgenic parasites (their growth surrogate) yet leave their replication untouched; or it might be that the compound contains impurities that interfere with its own biological activity. That is why secondary assays are fundamental for the selection of hits to be progressed. These may include host cell cytotoxicity assays, as well as orthogonal assays that do not rely on the reporter enzymatic activity to demonstrate that parasite growth inhibition is occurring (i.e. image-based assays). We will walk through these concepts with GSK's *T. cruzi* HTS program as example.

Primary assays

T. cruzi life stages in the mammalian host must be the target of any Chagas disease drug discovery efforts. Primarily the amastigote replicative stage that thrives within host cells – they are obliged to replicate intracellularly. Thus, biological assays to test inhibitors against them must include *T. cruzi*-infected mammalian cells as hosts. The first assay for screening of drugs that used a transgenic *T. cruzi* was developed by Buckner *et al.* (1996)and involved the Tulahuen-β-galactosidase strain and NIH-3T3 mouse fibroblasts as host cells. Some years later Bettiol *et al.* modified that protocol to reduce the assay time span from 7 to 4 days of incubation with compounds, while preserved the 96-well plate format and the colorimetric readout based on chlorophenol red-β-D-galactopyranoside (CPRG) reagent (Bettiol *et al.*, 2009). These assays served as the basis to develop the first HTS-adapted anti-*T. cruzi* biological assay, which was used to screen a NIH Molecular Library of more than 300K compounds (Carmody *et al.*, 2010a; Germain *et al.*, 2011). It employed 384-well microplates and the assay readout was luminescence-based, mediated by the reagent Gal-Screen (Carmody *et al.*, 2010a). The primary assay used in the GSK HTS campaign to screen at single dose (5 μM) a 1.8 million compounds chemical library also departed from Bettiol and co-worker's. This time, modifications involved that compounds were pre-dispensed in the 1536-well microtitre plates, biological reagents (NIH-3T3 and trypomastigotes; MOI = 1) were used directly after thawing, and the detection system chosen was FLINT by addition of the reagent resorufin-β-D-galactopyranoside (Peña *et al.*, 2015). Assay time remained at 4 days and countermeasures were taken to cope with potential evaporation issues since the assay volume was miniaturized to 6 μl per well.

Indeed, the GSK anti-*T. cruzi* HTS primary assay targeted intracellular amastigotes growing in NIH-3T3. But, since compounds were already present in the plate when biological reagents were delivered, active compounds could potentially also interfere with free swimming trypomastigotes and/or the host–parasite interactions required for parasite cell entry. In the other ultra-HTS against *T. cruzi*, performed at the Genomics Institute of the Novartis Research Foundation, primary assay protocol to screen 3 million compounds (at 6.7 μM) involved dispensing host cells first (NIH-3T3), then infecting them with *T. cruzi* Tulahuen-β-galactosidase trypomastigotes (MOI = 1), and finally pouring in the compounds [see Supplementary info by (Khare *et al.*, 2016)]. Assay was read with a luminescent reagent (BetaGlo; Promega). Remarkably, HTS assays that used 384-well or higher microplate format relied on fluorescent or luminescent readouts as they allow a more sensitive detection under miniaturization conditions.

In the GSK HTS screen, the primary assay output were 130,678 hits (7.7% hit rate), established upon statistical cut off (response + 3SD) at those compounds inhibiting parasite growth > 27% (Fig. 2.4). But due to the very high number of compounds for further progression, this yield was reduced to 45,100 hits by increasing the statistical cut off up to > 45% inhibition and already selecting compounds with predicted certain physical properties. These included a calculated Property Forecast Index (cPFI) < 8 and that the number of aromatic rings (#A-rings) ≤ 4. Both cPFI and #A-rings inform about the charge, hydrophobicity and solubility of a given compound, which will influence its further absorption and distribution *in vivo*. Under these premises it is possible to choose the optimal compounds

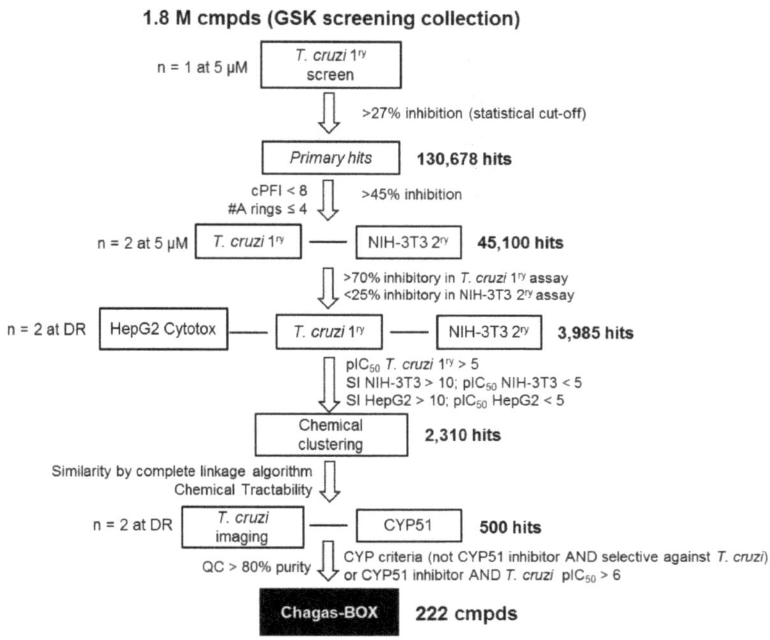

Figure 2.4 Data of GSK's HTS campaign to identify inhibitors of *T. cruzi* growth. The cascade led to 222 confirmed non-cytotoxic anti-*T. cruzi* compounds called the Chagas Box. (Figure adapted from Peña *et al*., 2015).

for progression, which contributes to save costs and time in their subsequent design and development process (Young *et al.*, 2011).

Secondary assays

As mentioned earlier, inhibitors identified by a phenotypic primary screening that involves host cells and intracellular growing parasites may or may not be parasite specific. Compounds disrupting host cells will preclude growth of amastigotes due to a host cytotoxicity effect. Therefore, assays to determine whether inhibitors drained from anti-*T. cruzi* primary screen really are *T. cruzi*-specific must be used too. Thus, NIH-3T3 cytotoxicity assays were included in all HTS campaigns to screen primary hits out against host cells' viability (Bettiol *et al.*, 2009; Carmody *et al.*, 2010a; Khare *et al.*, 2015b; Peña *et al.*, 2015).

In the GSK HTS, NIH-3T3 cytotoxicity assay mirrored the primary one in terms of assay format, number of cells per well, assay volume per well, and incubation time. It obviously lacked the parasites, and the readout was based on CellTiter-Glo Luminescent Cell Viability Assay (Promega) luminescent reagent instead. This assay helped to reduce the available 45,100 hits down to 3985 hits, considering thresholds of < 25% inhibitory effect on NIH-3T3 cells and > 70% inhibition at the confirmatory primary assay (Fig. 2.4). Other cytotoxicity assay commonly performed during HTS screening is based on human hepatocytes (HepG2), which might anticipate *in vitro*, potentially occurring liver toxicity (Moeller *et al.*, 2012). A HepG2 cytotoxicity assay was included in the GSK HTS when primary screen confirmed compounds were taken to dose–response (DR) testing stage (Fig. 2.4).

Whatever the mammalian cytotoxicity assay made, when compounds are tested at DR, the activity readouts obtained from the series of concentration points allow the calculation of their CC_{50}, which is a type of half-maximal inhibitory concentration (IC_{50}). In parallel, compounds are also assayed at DR on the parasite assay, which allows calculation of their IC_{50}. These parameters will respectively determine the potency of a given compound against host cells and against the parasites. They can also be indicated with a logarithmic scale [i.e. $pIC_{50} = -\log(IC_{50})$] to be shown with molar concentration (mol/l or M) and facilitate direct comparison between many compounds. The ratio between the CC_{50} and IC_{50} is known as selectivity index (SI) and determines whether a hit is specifically toxic against the parasite cells saving the host cells from that effect. SI cut-off may vary between campaigns. It was set at > 5 at (Khare *et al.*, 2016), and > 10 at the GSK HTS (Fig. 2.4). In the latter, from 3985 hits assayed at DR, upon considering SI and hits chemical clustering, 2310 were selected for further consideration (Fig. 2.4).

Image-based assays

In recent years, optics technological advancements in the form of automated high-content microscopy as well as user-friendly image analysis software have allowed the implementation of image-based medium to high-throughput screening assays (Zanella *et al.*, 2010). For example, several of them are now available at various sites to screen drugs against *T. cruzi* (see Table 2.1). Similarly to HTS based on transgenic parasites, image-based automatic quantification is not observer-biased, provides reproducible and accurate readings and saves valuable lab time. Other advantages provided by anti-*T. cruzi* high content assays are (i) their image-based readout allows determination of parasite and host cell toxicities in a single assay, and (ii) they do not rely on transgenic strains and virtually any cell infective strain (including relevant clinical isolates) could be used (Alonso-Padilla and Rodríguez, 2014). Besides a host cytotoxicity readout (number of host cells), *T. cruzi* image-based assays render two parasite-specific readouts (% of infected host cells and number of amastigotes per infected cell).

In the different assays developed so far, several combinations of parasite strains and host cell lines have been used (Table 2.1). Attaching again to the GSK HTS campaign as example, an image-based assay was included within the progression cascade to further confirm by an orthogonal methodology the hits that had passed through DR testing stage (Fig. 2.4). Beyond host cell's cytotoxicity, compounds inhibitory effect might have been due to interference with the readout methodology used in the primary screening.

What other assays have been developed?

A few more assays have been implemented to aid prioritize the best possible HTS hits against *T. cruzi*. For instance, one was developed to triage compounds targeting the parasite's sterol 14α-demethylase (CYP51) (Riley *et al.*, 2015). The rationale to do so stands on the failure of azole drugs that target CYP51, such as posaconazole and ravuconazole, as monotherapy in *in vivo* models of *T. cruzi* infection (Francisco *et al.*, 2015; Khare *et al.*, 2015a) and in clinical trials (Riley *et al.*, 2015). Thus, a biochemical assay based on recombinant CYP51 expressed in bacteria was set up to guide exploration of active chemical series that do not exert their inhibition through that *T. cruzi* enzyme. This assay was included in the GSK HTS campaign and together with the orthogonal image-based assay contributed to finally shape the disclosed Chagas-BOX of 222 hits (Peña *et al.*, 2015).

Table 2.1 List of image-based anti-*T. cruzi* assays ordered accordingly to publication date

Reference	Host cells (no. cells/well)[a]	Strain (MOI)	Protocol (order into plate)	Time (hours)[b]	Stain (host cell + *T. cruzi* DNA)	Microscope	Image analysis software	Z'[c]
Engel *et al*. (2010)	Huh7 (1,000)	CA-I/72 (0.5)	Seed cells – infect – @ 16 hpi Cmpds in	72	DAPI	IN Cell Analyser 1000 (GE Healthcare)	IN Cell Workstation 3.5 (GE Healthcare)	-
Engel *et al*. (2010)	BESM (2,000)	CA-I/72 (5)	Seed cells – infect – @ 16 hpi Cmpds in	72	DAPI	IN Cell Analyser 1000 (GE Healthcare)	IN Cell Workstation 3.5 (GE Healthcare)	0.45
Moraes *et al*. (2014)	U2OS (700–1,000)	Multiple strains[d]	Seed cells (24 hours) – infect (24 hours) – Cmpds	72[e]–96	Draq5	Operetta (Perkin-Elmer)	IPK proprietary software and algorithm	-
Alonso-Padilla *et al*. (2015)	H9c2 (2,500)[f]	Tulahuen-βGal (1)	Cmpds – add infected-cells	72	Draq5	Opera (Perkin-Elmer)	Acapella (Perkin-Elmer)	0.6
Neitz *et al*. (2015)	C2C12 (500)	CA-I/72 (5)	Cmpds – add cells + trypomastigotes	72	DAPI	IN Cell Analyser 2000 (GE Healthcare)	IN Cell Developer Toolbox 1.9 (GE Healthcare)	> 0.3
Khare *et al*. (2015b)	3T3 (1,000)	CL Brener (10 or 20)[g]	Seed cells (O/N) – Infect (6 hours) – @ 24 hpi Cmpds in	48	SYBR Green	Opera (Perkin-Elmer)	CellProfiler 2.1.0	-
Sykes and Avery (2015)	3T3 (1,000)	Tulahuen (5)	Seed cells (24 hours) – infect (24 hours) – Cmpds in	48	Hoescht 33342 + CellMask Green	Operetta/Opera (Perkin-Elmer)	Assay Language Interface/Harmony	> 0.5
De Rycker *et al*. (2016)[h]	Vero (4,000)[d]	Silvio X10/7 A1 (5)	Cmpds – add infected-cells	72	Hoescht 33342	Operetta (Perkin-Elmer)	Columbus system (Perkin-Elmer)	0.8[i]

[a]All but first reference (which employed a 96-well) employed a 384-well microplate assay format.
[b]Refers to time exposure of the infected cells to the compounds.
[c]Amastigotes-per-cell readout Z' value must be > 0.5 to ensure excellent assay reproducibility under high-throughput conditions (Zhang *et al*., 1999).
[d]Strains used were (their corresponding MOIs in brackets): Dm28c and Y (4); Era cl2, 92–80 cl2 and Brener (20); Tulahuen (15); Arma13 cl1 (40).
[e]Only the assay with *T. cruzi*-Dm28c were exposed for 72 h; the other strains were incubated for 96 h with compounds.
[f]Number of cells refers to those already infected for 18h that were dispensed per well for assay incubation with compounds.
[g]A MOI = 10 was used with the wild type *T. cruzi*-CL whereas a MOI = 20 was used with the GNF7686-resistant mutant strain.
[h]Moraes *et al*. and de Rycker et al. assays served as the basis to respectively develop time-kill, and static-cidal and rate-of-kill hits prioritization assays.
[i]Robust z-factor calculated according to the equation given in De Rycker *et al.* (2016).

When compared *in vitro* to nitroheterocyclic drugs like benznidazole and nifurtimox, azoles have shown to exert their *T. cruzi* growth inhibition in a slower fashion and have been deemed as static compounds (De Rycker *et al.*, 2016). This observation was shown with a static-cidal assay developed from an image-based high content assay (De Rycker *et al.*, 2016). Departing from the same image-based assay as basis, a rate-of-kill assay was also described. Both, static-cidal and rate-of-kill have been designed to flag those hits that will kill the parasite. In fact, sterile *in vitro* anti-parasitic activity in macrophages has been proposed to be key criteria for further compound progression (Cal *et al.*, 2016). Nonetheless, the microscopy-based assay proposed by Cal *et al.* relies on Giemsa parasite staining and visual counting of parasites, which precludes its standardization to high-throughput conditions. Notably, authors suggest that values obtained by microscopic determination are more meaningful that those obtained by reported-based detection (i.e. colorimetric readout of *T. cruzi* linked β-galactosidase activity), because any surviving amastigote can be detected by microscopic counting, even if it is 'temporarily' knocked out by the drugs action. Furthermore, by prolonging the assay to 11 days long and performing wash-outs, independently of the drug concentration used, the inability of CYP51 inhibitors (several azoles) to fully clear amastigotes presence was demonstrated (Cal *et al.*, 2016). This has also been observed to occur when testing by high content a panel of strains with representatives of genotypes I to VI against those drugs (Moraes *et al.*, 2014). The variability of azoles' efficacy against the distinct strains showed that strain diversity must be definitely taken into account in any *T. cruzi* drug discovery pathway (Zingales *et al.*, 2014). This parasite has a wide genetic diversity, and occurrence of co-infections and co-existence of various genotypes in the same patient over time appears to be a common fact (Burgos *et al.*, 2010; Martinez-Perez *et al.*, 2016). Thus, it will be fundamental that any progressed hits are active against a wide range of parasite strains.

Finally, anti-trypomastigote stage specific assays have been as well designed. This parasite stage does not replicate by itself as it needs to invade cells and transform into amastigotes to do so. Nonetheless, together with amastigotes make the mammalian infective parasitic stages and it is considered as well a drug target of interest. Thereby, trypomastigote lysis assays have usually accompanied *T. cruzi* growth inhibition ones (Bettiol *et al.*, 2009; Buckner *et al.*, 1996; Khare *et al.*, 2015a). Destruction of trypomastigotes by compounds has been analysed by assessing a loss of their transgenic activity (Bettiol *et al.*, 2009; Buckner *et al.*, 1996) and by means of reagents like CellTiter-Glo that provide a biological readout of the health status of cells measuring their ATP-contain (Khare *et al.*, 2015a).

The importance of sharing – where can the hits be found?

As has been shown in a recent international collaborative work with the Medicines for Malaria Venture Malaria Box, public availability of compounds chemical structures and the sharing of results between the drug discovery community are fundamental towards a faster and safer progression in the drug developmental path against neglected diseases (Van Voorhis *et al.*, 2016). It is also highlighted in that article the importance of a tight interaction between academic labs, where the parasites' biology knowledge resides, and the industry that holds the resources and expertise to develop new drugs. Thus, the great relevance of establishing academia–industry collaborative frame works to bridge the gap. In the case of *T. cruzi* (and other kinetoplastids like *Leishmania* species and *Trypanosoma brucei*) this has been facilitated in recent years mainly thanks to monetary inputs by organizations such as GHIT and the Wellcome Trust.

All information from the Broad Institute's *T. cruzi* HTS upon NIH Molecular Libraries is publicly accessible; from the description of the campaign assays (Carmody *et al.*, 2010a), to subsequent compound-focused sub-studies (Carmody *et al.*, 2010b,c; Carmody *et al.*, 2010d). Furthermore, appropriately linked to the former, all retrieved chemical structures can be checked in Pubchem. Similarly, in accordance with its early drug discovery open innovation policy, already followed with the *Plasmodium falciparum* and *Mycobacterium tuberculosis* HTS campaigns (Ballell *et al.*, 2013; Gamo *et al.*, 2010), the 222 hits from the *T. cruzi* GSK HTS have been released in what is called the Chagas-BOX (Peña *et al.*, 2015). Although not yet retrievable via PubChem repository, physical samples are available to collaborators upon request. Unfortunately, other efforts, like the one led by Novartis, have not been publicly disclosed.

Next drug discovery steps

Phenotypic screening rather than target based approaches appear to be better suited to widen the pipeline for neglected diseases, including Chagas disease. However, further down in the drug discovery cascade, target deconvolution and MoA determination are largely advisable in order to maximize the chances of marketing a product. Their identification grants a better controlled medicinal chemistry space and swifter prediction of potential drug adverse effects. Chemical optimization of hits can be precluded by ignoring whether their potential activity gain or loss is due to a biochemical interaction or due to cellular permeability issues. Thus, identification of the target represents a huge advantage towards the performance of SAR studies.

Such target deconvolution can be sought experimentally, for example by chemical genetics as described by Khare and co-workers (Khare *et al.*, 2015b). Departing from a pan-kinetoplastid inhibitor (from the Novartis HTS campaigns), they grew *T. cruzi in vitro* exposed to increasing amounts of it in order to evolve resistant mutants that were then whole-genome sequenced. Once the resistance-conferring mutations were identified, the putative target was experimentally demonstrated in a biochemical assay (Khare *et al.*, 2015b).

Nonetheless, it might not be possible to rescue any mutants if the target is essential to the parasite, or it might not be possible to identify the target if there is a high number of unannotated CDS in the *T. cruzi* genome for this target. Interestingly, bioinformatics could be of help by assigning functions to hypothetical proteins by comparative genomics, and linking ligands to potential targets by chemoinformatics-driven approaches (Gregori-Puigjané *et al.*, 2012). These *in silico* steps will guide wet-lab experimental binding confirmation and MoA determination, something that has already been initiated based on a recombinant yeast system (Ottilie *et al.*, 2017).

Due to its importance, major emphasis should be placed now in *T. cruzi* functional genomics, proteomics, and metabolomics to identify the most promising targets of HTS campaigns' harvest. Acknowledgement of the biological targets will greatly facilitate the medicinal chemistry work saving time and resources and allowing a quicker development of the most efficient compounds. Then, once a potential target is identified, another way informatics can aid the drug discovery process is by virtual screening (VS). This is an *in silico* screening methodology based on computational techniques that identifies the chemical structures that will most likely bind to a particular drug target (Perez-Pineiro *et al.*, 2009). The approach has already been applied to screen chemical libraries for inhibitors of *T. cruzi* enzymes and tagging chemical series of interest within them (Maccari *et al.*, 2011; Wiggers *et al.*, 2013).

Summary

Drug development is a long and quite often frustrating path and HTS stands at its beginning. Even earlier on sits the biological and clinical understanding of the pathogen and the disease it causes. A deep knowledge of both subjects and their interconnections will contribute to develop the most adequate assays towards a safe and sound hits progression down the path. Hitherto, historically ill-funded research on neglected diseases, NTDs in particular, has very negatively impacted discovery and development of drugs against them (Fig. 2.1). Fortunately, the tide is changing and the recent monetary increment is allowing much-needed academia–industry synergies. A major example of this collaborative framework has been the performance of phenotypic screening campaigns against *T. cruzi* (Neitz *et al.*, 2015; Peña *et al.*, 2015). These will find hits and leads, but drug candidates will ultimately result from the combined outcome of chemistry, metabolism, pharmacokinetics and *in vivo* preclinical studies that will maximize the chances to successfully reach the clinic.

Although results from clinical trials of posacozanole and E1224 have been disappointing, approximately 90% of drugs fail in clinical trials despite successfully completing *in vitro* and *in vivo* tests (Hay *et al.*, 2014). Taking the optimistic message, clinical trials are now under way against Chagas disease [Alonso-Padilla and Rodríguez (2014), and see www.dndi.org/2016/clinical-trials/clinical-trials-chagas/]. Until now, drugs used in clinical trials against Chagas disease have been developed for other diseases and tested for repurposing. In the future, mostly thanks to the performance of HTS campaigns, drug candidates for Chagas will likely be new chemical entities developed for this purpose. In order to minimize the attrition rate and make the most of the available funding, the focus must now be set on the election and performance of the most relevant *in vivo* assays to translate *in vitro* results to *in vivo*. Besides, knowledge, as far detailed as possible, on the retrieved hits potential targets and MoA will be fundamental for successfully developing them into drugs available for patients.

References

Alonso-Padilla, J., Cotillo, I., Presa, J.L., Cantizani, J., Peña, I., Bardera, A.I., Martín, J.J., and Rodriguez, A. (2015). Automated high-content assay for compounds selectively toxic to *Trypanosoma cruzi* in a myoblastic cell line. PLOS Negl. Trop. Dis. *9*, e0003493. https://doi.org/10.1371/journal.pntd.0003493.

Alonso-Padilla, J., and Rodríguez, A. (2014). High throughput screening for anti-*Trypanosoma cruzi* drug discovery. PLOS Negl. Trop. Dis. *8*, e3259. https://doi.org/10.1371/journal.pntd.0003259.

Andriani, G., Chessler, A.D., Courtemanche, G., Burleigh, B.A., and Rodriguez, A. (2011). Activity in vivo of anti-*Trypanosoma cruzi* compounds selected from a high throughput screening. PLOS Negl. Trop. Dis. *5*, e1298. https://doi.org/10.1371/journal.pntd.0001298.

Annang, F., Pérez-Moreno, G., García-Hernández, R., Cordon-Obras, C., Martín, J., Tormo, J.R., Rodríguez, L., de Pedro, N., Gómez-Pérez, V., Valente, M., *et al.* (2015). High-throughput screening platform for natural product-based drug discovery against 3 neglected tropical diseases: human

African trypanosomiasis, leishmaniasis, and Chagas disease. J Biomol Screen *20*, 82–91. https://doi.org/10.1177/1087057114555846.

Aslett, M., Aurrecoechea, C., Berriman, M., Brestelli, J., Brunk, B.P., Carrington, M., Depledge, D.P., Fischer, S., Gajria, B., Gao, X., *et al.* (2010). TriTrypDB: a functional genomic resource for the Trypanosomatidae. Nucleic Acids Res. *38*, D457–462.

Atwood, J.A., Weatherly, D.B., Minning, T.A., Bundy, B., Cavola, C., Opperdoes, F.R., Orlando, R., and Tarleton, R.L. (2005). The *Trypanosoma cruzi* proteome. Science *309*, 473–476.

Ballell, L., Bates, R.H., Young, R.J., Alvarez-Gomez, D., Alvarez-Ruiz, E., Barroso, V., Blanco, D., Crespo, B., Escribano, J., González, R., *et al.* (2013). Fueling open-source drug discovery: 177 small-molecule leads against tuberculosis. ChemMedChem *8*, 313–321. https://doi.org/10.1002/cmdc.201200428.

Bern, C., Kjos, S., Yabsley, M.J., and Montgomery, S.P. (2011). *Trypanosoma cruzi* and Chagas' Disease in the United States. Clin. Microbiol. Rev. *24*, 655–681. https://doi.org/10.1128/CMR.00005-11.

Bettiol, E., Samanovic, M., Murkin, A.S., Raper, J., Buckner, F., and Rodriguez, A. (2009). Identification of three classes of heteroaromatic compounds with activity against intracellular *Trypanosoma cruzi* by chemical library screening. PLOS Negl. Trop. Dis. *3*, e384. https://doi.org/10.1371/journal.pntd.0000384.

Buckner, F.S., Griffin, J.H., Wilson, A.J., and Van Voorhis, W.C. (2001). Potent anti-*Trypanosoma cruzi* activities of oxidosqualene cyclase inhibitors. Antimicrob. Agents Chemother. *45*, 1210–1215. https://doi.org/10.1128/AAC.45.4.1210-1215.2001.

Buckner, F.S., Verlinde, C.L., La Flamme, A.C., and Van Voorhis, W.C. (1996). Efficient technique for screening drugs for activity against *Trypanosoma cruzi* using parasites expressing beta-galactosidase. Antimicrob. Agents Chemother. *40*, 2592–2597.

Burgos, J.M., Diez, M., Vigliano, C., Bisio, M., Risso, M., Duffy, T., Cura, C., Brusses, B., Favaloro, L., Leguizamon, M.S., *et al.* (2010). Molecular identification of *Trypanosoma cruzi* discrete typing units in end-stage chronic Chagas heart disease and reactivation after heart transplantation. Clin. Infect. Dis. *51*, 485–495. https://doi.org/10.1086/655680.

Cal, M., Ioset, J.R., Fügi, M.A., Mäser, P., and Kaiser, M. (2016). Assessing anti-*T. cruzi* candidates in vitro for sterile cidality. Int J Parasitol Drugs Drug Resist *6*, 165–170.

Canavaci, A.M., Bustamante, J.M., Padilla, A.M., Perez Brandan, C.M., Simpson, L.J., Xu, D., Boehlke, C.L., and Tarleton, R.L. (2010). In vitro and in vivo high-throughput assays for the testing of anti-*Trypanosoma cruzi* compounds. PLOS Negl. Trop. Dis. *4*, e740. https://doi.org/10.1371/journal.pntd.0000740.

Carmody, L.C., Germain, A., Barker, D., Galan-Rodriguez, C., Bettiol, E., Rodriguez, A., MacPherson, L., Palmer, M., and Schreiber, S.L. (2010a). Identification of small-molecule inhibitors of *Trypansoma cruzi* infection. In Probe Reports from the NIH Molecular Libraries Program (National Center for Biotechnology Information, Bethesda, MD).

Carmody, L.C., Germain, A., Barker, D., Galan-Rodriguez, C., Bettiol, E., Rodriguez, A., MacPherson, L., Palmer, M., and Schreiber, S.L. (2010b). Identification of small-molecule inhibitors of *Trypansoma cruzi* infection – Probe 1. In Probe Reports from the NIH Molecular Libraries Program (National Center for Biotechnology Information, Bethesda, MD).

Carmody, L.C., Germain, A., Barker, D., Galan-Rodriguez, C., Bettiol, E., Rodriguez, A., MacPherson, L., Palmer, M., and Schreiber, S.L. (2010c). Identification of small-molecule inhibitors of *Trypansoma cruzi* infection – Probe 3. In Probe Reports from the NIH Molecular Libraries Program (National Center for Biotechnology Information, Bethesda, MD).

Carmody, L.C., Germain, A.R., Engel, J.C., Gut, J., Kaiser, M., Jewett, I., LeQuemen, S., Marie, J.C., Dandapani, S., Rodriguez, A., *et al.* (2010d). Identification of diversity-oriented synthesis derived small molecule, ML341, with cidal activity against *Trypanosoma cruzi*. In Probe Reports from the NIH Molecular Libraries Program (National Center for Biotechnology Information, Bethesda, MD).

Chatelain, E. (2015). Chagas disease drug discovery: toward a new era. J. Biomol. Screen. *20*, 22–35. https://doi.org/10.1177/1087057114550585.

De Rycker, M., Thomas, J., Riley, J., Brough, S.J., Miles, T.J., and Gray, D.W. (2016). Identification of trypanocidal activity for known clinical compounds using a new *Trypanosoma cruzi* hit-discovery screening cascade. PLOS Negl. Trop. Dis. *10*, e0004584. https://doi.org/10.1371/journal.pntd.0004584.

Ekins, S., de Siqueira-Neto, J.L., McCall, L.I., Sarker, M., Yadav, M., Ponder, E.L., Kallel, E.A., Kellar, D., Chen, S., Arkin, M., *et al.* (2015). Machine learning models and pathway genome data base for *Trypanosoma cruzi* drug discovery. PLOS Negl. Trop. Dis. *9*, e0003878. https://doi.org/10.1371/journal.pntd.0003878.

El-Sayed, N.M., Myler, P.J., Bartholomeu, D.C., Nilsson, D., Aggarwal, G., Tran, A.N., Ghedin, E., Worthey, E.A., Delcher, A.L., Blandin, G., *et al.* (2005). The genome sequence of *Trypanosoma cruzi*, etiologic agent of Chagas disease. Science *309*, 409–415.

Engel, J.C., Ang, K.K., Chen, S., Arkin, M.R., McKerrow, J.H., and Doyle, P.S. (2010). Image-based high-throughput drug screening targeting the intracellular stage of *Trypanosoma cruzi*, the agent of Chagas' disease. Antimicrob. Agents Chemother. *54*, 3326–3334. https://doi.org/10.1128/AAC.01777-09.

Francisco, A.F., Lewis, M.D., Jayawardhana, S., Taylor, M.C., Chatelain, E., and Kelly, J.M. (2015). Limited ability of posaconazole to cure both acute and chronic *Trypanosoma cruzi* infections revealed by highly sensitive in vivo imaging. Antimicrob. Agents Chemother. *59*, 4653–4661. https://doi.org/10.1128/AAC.00520-15.

Gamo, F.J., Sanz, L.M., Vidal, J., de Cozar, C., Alvarez, E., Lavandera, J.L., Vanderwall, D.E., Green, D.V., Kumar, V., Hasan, S., *et al.* (2010). Thousands of chemical starting points for antimalarial lead identification. Nature *465*, 305–310. https://doi.org/10.1038/nature09107.

Gascon, J., and Pinazo, M.-J. (2015). Chagas disease: from Latin America to the world. Reports in Parasitology. *4*, 7–14.

Germain, A.R., Carmody, L.C., Dockendorff, C., Galan-Rodriguez, C., Rodriguez, A., Johnston, S., Bittker, J.A., MacPherson, L., Dandapani, S., Palmer, M., *et al.* (2011). Identification of small-molecule inhibitors of *Trypansoma cruzi* replication. Bioorg. Med. Chem. Lett. *21*, 7197–7200. https://doi.org/10.1016/j.bmcl.2011.09.057.

Gregori-Puigjané, E., Setola, V., Hert, J., Crews, B.A., Irwin, J.J., Lounkine, E., Marnett, L., Roth, B.L., and Shoichet, B.K. (2012). Identifying mechanism-of-action targets for drugs and probes. Proc. Natl. Acad. Sci. U.S.A. *109*, 11178–11183. https://doi.org/10.1073/pnas.1204524109.

Hay, M., Thomas, D.W., Craighead, J.L., Economides, C., and Rosenthal, J. (2014). Clinical development success rates for investigational drugs. Nat. Biotechnol. *32*, 40–51. https://doi.org/10.1038/nbt.2786.

Herrera, L.J., Brand, S., Santos, A., Nohara, L.L., Harrison, J., Norcross, N.R., Thompson, S., Smith, V., Lema, C., Varela-Ramirez, A., *et al.* (2016). Validation of N-myristoyltransferase as potential chemotherapeutic target in mammal-dwelling stages of *Trypanosoma cruzi*. PLOS Negl. Trop. Dis. *10*, e0004540. https://doi.org/10.1371/journal.pntd.0004540.

Holloway, G.A., Charman, W.N., Fairlamb, A.H., Brun, R., Kaiser, M., Kostewicz, E., Novello, P.M., Parisot, J.P., Richardson, J., Street, I.P., *et al.* (2009). Trypanothione reductase high-throughput screening campaign identifies novel classes of inhibitors with antiparasitic activity. Antimicrob. Agents Chemother. *53*, 2824–2833. https://doi.org/10.1128/AAC.01568-08.

Hotez, P.J., Pecoul, B., Rijal, S., Boehme, C., Aksoy, S., Malecela, M., Tapia-Conyer, R., and Reeder, J.C. (2016). Eliminating the neglected tropical diseases: translational science and new technologies. PLOS Negl. Trop. Dis. *10*, e0003895. https://doi.org/10.1371/journal.pntd.0003895.

Hyland, K.V., Asfaw, S.H., Olson, C.L., Daniels, M.D., and Engman, D.M. (2008). Bioluminescent imaging of *Trypanosoma cruzi* infection. Int. J. Parasitol. *38*, 1391–1400. https://doi.org/10.1016/j.ijpara.2008.04.002.

Iversen, P.W., Beck, B., Chen, Y.F., Dere, W., Devanarayan, V., Eastwood, B.J., Farmen, M.W., Iturria, S.J., Montrose, C., Moore, R.A., Weidner, J.R., and Sittampalam, G.S. (2012). HTS assay validation. In Assay Guidance Manual [Internet]: Sittampalam, G.S., Coussens, N.P., Brimacombe, K., Grossman, A., Arkin, M., Auld, D., Austin, C., Bejcek, B., Glicksman, M., Inglese, J., Iversen, P.W., Li, Z., McGee, J., McManus, O., Minor, L., Napper, A., Peltier, J.M., Riss, T., Trask, O.J. Jr., Weidner, J., editors. (Eli Lilly & Company and the National Center for Advancing Translational Sciences, Bethesda, MD).

Katsuno, K., Burrows, J.N., Duncan, K., Hooft van Huijsduijnen, R., Kaneko, T., Kita, K., Mowbray, C.E., Schmatz, D., Warner, P., and Slingsby, B.T. (2015). Hit and lead criteria in drug discovery for infectious diseases of the developing world. Nat Rev Drug Discov *14*, 751–758. https://doi.org/10.1038/nrd4683.

Khare, S., Liu, X., Stinson, M., Rivera, I., Groessl, T., Tuntland, T., Yeh, V., Wen, B., Molteni, V., Glynne, R., *et al.* (2015a). Antitrypanosomal treatment with benznidazole is superior to posaconazole regimens in mouse models of Chagas disease. Antimicrob. Agents Chemother. *59*, 6385–6394. https://doi.org/10.1128/AAC.00689-15.

Khare, S., Nagle, A.S., Biggart, A., Lai, Y.H., Liang, F., Davis, L.C., Barnes, S.W., Mathison, C.J., Myburgh, E., Gao, M.Y., *et al.* (2016). Proteasome inhibition for treatment of leishmaniasis, Chagas disease and sleeping sickness. Nature *537*, 229–233. https://doi.org/10.1038/nature19339.

Khare, S., Roach, S.L., Barnes, S.W., Hoepfner, D., Walker, J.R., Chatterjee, A.K., Neitz, R.J., Arkin, M.R., McNamara, C.W., Ballard, J., *et al.* (2015b). Utilizing chemical genomics to identify cytochrome b as a novel drug target for Chagas disease. PLOS Pathog. *11*, e1005058. https://doi.org/10.1371/journal.ppat.1005058.

Lee, J.A., Uhlik, M.T., Moxham, C.M., Tomandl, D., and Sall, D.J. (2012). Modern phenotypic drug discovery is a viable, neoclassic pharma strategy. J. Med. Chem. *55*, 4527–4538. https://doi.org/10.1021/jm201649s.

Lewis, M.D., Fortes Francisco, A., Taylor, M.C., Burrell-Saward, H., McLatchie, A.P., Miles, M.A., and Kelly, J.M. (2014). Bioluminescence imaging of chronic *Trypanosoma cruzi* infections reveals tissue-specific parasite dynamics and heart disease in the absence of locally persistent infection. Cell. Microbiol. *16*, 1285–1300. https://doi.org/10.1111/cmi.12297.

Lewis, M.D., Francisco, A.F., Taylor, M.C., Jayawardhana, S., and Kelly, J.M. (2016). Host and parasite genetics shape a link between *Trypanosoma cruzi* infection dynamics and chronic cardiomyopathy. Cell. Microbiol. *18*, 1429–1443. https://doi.org/10.1111/cmi.12584.

Maccari, G., Jaeger, T., Moraca, F., Biava, M., Flohé, L., and Botta, M. (2011). A fast virtual screening approach to identify structurally diverse inhibitors of trypanothione reductase. Bioorg. Med. Chem. Lett. *21*, 5255–5258. https://doi.org/10.1016/j.bmcl.2011.07.036.

Magarinos, M.P., Carmona, S.J., Crowther, G.J., Ralph, S.A., Roos, D.S., Shanmugam, D., Van Voorhis, W.C., and Aguero, F. (2012). TDR Targets: a chemogenomics resource for neglected diseases. Nucleic Acids Res. *40*, D1118–1127.

Martinez-Perez, A., Poveda, C., Ramírez, J.D., Norman, F., Gironés, N., Guhl, F., Monge-Maillo, B., Fresno, M., and López-Vélez, R. (2016). Prevalence of *Trypanosoma* cruzi's Discrete Typing Units in a cohort of Latin American migrants in Spain. Acta Trop. *157*, 145–150. https://doi.org/10.1016/j.actatropica.2016.01.032.

Minning, T.A., Weatherly, D.B., Atwood, J., Orlando, R., and Tarleton, R.L. (2009). The steady-state transcriptome of the four major life-cycle stages of *Trypanosoma cruzi*. BMC Genomics *10*, 370. https://doi.org/10.1186/1471-2164-10-370.

Moeller, T.A., Shukla, S.J., and Xia, M. (2012). Assessment of compound hepatotoxicity using human plateable cryopreserved hepatocytes in a 1536-well-plate format. Assay Drug Dev Technol *10*, 78–87. https://doi.org/10.1089/adt.2010.0365.

Moraes, C.B., Giardini, M.A., Kim, H., Franco, C.H., Araujo-Junior, A.M., Schenkman, S., Chatelain, E., and Freitas-Junior, L.H. (2014). Nitroheterocyclic compounds are more efficacious than CYP51 inhibitors against *Trypanosoma cruzi*: implications for Chagas disease drug discovery and development. Sci Rep *4*, 4703. https://doi.org/10.1038/srep04703.

Neitz, R.J., Chen, S., Supek, F., Yeh, V., Kellar, D., Gut, J., Bryant, C., Gallardo-Godoy, A., Molteni, V., Roach, S.L., *et al.* (2015). Lead identification to clinical candidate selection: drugs for Chagas disease. J Biomol Screen *20*, 101–111. https://doi.org/10.1177/1087057114553103.

Olliaro, P.L., Kuesel, A.C., and Reeder, J.C. (2015). A changing model for developing health products for poverty-related infectious diseases. PLOS Negl. Trop. Dis. *9*, e3379. https://doi.org/10.1371/journal.pntd.0003379.

Ottilie, S., Goldgof, G.M., Calvet, C.M., Jennings, G.K., LaMonte, G., Schenken, J., Vigil, E., Kumar, P., McCall, L.I., Lopes, E.S., *et al.* (2017). Rapid Chagas Disease Drug Target Discovery Using Directed Evolution in Drug-Sensitive Yeast. ACS Chem. Biol. *12*, 422–434. https://doi.org/10.1021/acschembio.6b01037.

Pedrique, B., Strub-Wourgaft, N., Some, C., Olliaro, P., Trouiller, P., Ford, N., Pécoul, B., and Bradol, J.H. (2013). The drug and vaccine landscape for neglected diseases (2000-11): a systematic assessment. Lancet Glob Health *1*, e371–9. https://doi.org/10.1016/S2214-109X(13)70078-0.

Peña, I., Pilar Manzano, M., Cantizani, J., Kessler, A., Alonso-Padilla, J., Bardera, A.I., Alvarez, E., Colmenarejo, G., Cotillo, I., Roquero, I., *et al.* (2015). New compound sets identified from high throughput phenotypic screening against three kinetoplastid parasites: an open resource. Sci Rep *5*, 8771. https://doi.org/10.1038/srep08771.

Perez-Pineiro, R., Burgos, A., Jones, D.C., Andrew, L.C., Rodriguez, H., Suarez, M., Fairlamb, A.H., and Wishart, D.S. (2009). Development of a novel virtual screening cascade protocol to identify potential trypanothione reductase inhibitors. J. Med. Chem. *52*, 1670–1680. https://doi.org/10.1021/jm801306g.

Pinazo, M.J., Guerrero, L., Posada, E., Rodríguez, E., Soy, D., and Gascon, J. (2013). Benznidazole-related adverse drug reactions and their relationship to serum drug concentrations in patients with chronic chagas disease. Antimicrob. Agents Chemother. *57*, 390–395. https://doi.org/10.1128/AAC.01401-12.

Riley, J., Brand, S., Voice, M., Caballero, I., Calvo, D., and Read, K.D. (2015). Development of a fluorescence-based *Trypanosoma cruzi* CYP51 Inhibition Assay for Effective Compound Triaging in Drug Discovery Programmes for Chagas Disease. PLOS Negl. Trop. Dis. *9*, e0004014. https://doi.org/10.1371/journal.pntd.0004014.

Robertson, L.J., Devleesschauwer, B., Alarcón de Noya, B., Noya González, O., and Torgerson, P.R. (2016). *Trypanosoma cruzi*: Time for International Recognition as a Foodborne Parasite. PLOS Negl. Trop. Dis. *10*, e0004656. https://doi.org/10.1371/journal.pntd.0004656.

Rodriguez, A., and Tarleton, R.L. (2012). Transgenic parasites accelerate drug discovery. Trends Parasitol. *28*, 90–92. https://doi.org/10.1016/j.pt.2011.12.003.

Sykes, M.L., and Avery, V.M. (2013). Approaches to protozoan drug discovery: phenotypic screening. J. Med. Chem. *56*, 7727–7740. https://doi.org/10.1021/jm4004279.

Sykes, M.L., and Avery, V.M. (2015). Development and application of a sensitive, phenotypic, high-throughput image-based assay to identify compound activity against *Trypanosoma cruzi* amastigotes. Int J Parasitol Drugs Drug Resist 5, 215–228. https://doi.org/10.1016/j.ijpddr.2015.10.001.

Trouiller, P., Olliaro, P., Torreele, E., Orbinski, J., Laing, R., and Ford, N. (2002). Drug development for neglected diseases: a deficient market and a public-health policy failure. Lancet 359, 2188–2194.

Van Voorhis, W.C., Adams, J.H., Adelfio, R., Ahyong, V., Akabas, M.H., Alano, P., Alday, A., Alemán Resto, Y., Alsibaee, A., Alzualde, A., *et al.* (2016). Open Source Drug Discovery with the Malaria Box Compound Collection for Neglected Diseases and Beyond. PLOS Pathog. *12*, e1005763. https://doi.org/10.1371/journal.ppat.1005763.

Wiggers, H.J., Rocha, J.R., Fernandes, W.B., Sesti-Costa, R., Carneiro, Z.A., Cheleski, J., da Silva, A.B., Juliano, L., Cezari, M.H., Silva, J.S., *et al.* (2013). Non-peptidic cruzain inhibitors with trypanocidal activity discovered by virtual screening and in vitro assay. PLOS Negl. Trop. Dis. *7*, e2370. https://doi.org/10.1371/journal.pntd.0002370.

Wilkinson, S.R., and Kelly, J.M. (2009). Trypanocidal drugs: mechanisms, resistance and new targets. Expert Rev Mol Med *11*, e31. https://doi.org/10.1017/S1462399409001252.

Young, R.J., Green, D.V., Luscombe, C.N., and Hill, A.P. (2011). Getting physical in drug discovery II: the impact of chromatographic hydrophobicity measurements and aromaticity. Drug Discov. Today *16*, 822–830. https://doi.org/10.1016/j.drudis.2011.06.001.

Zanella, F., Lorens, J.B., and Link, W. (2010). High content screening: seeing is believing. Trends Biotechnol. *28*, 237–245. https://doi.org/10.1016/j.tibtech.2010.02.005.

Zhang, J.H., Chung, T.D., and Oldenburg, K.R. (1999). A Simple Statistical Parameter for Use in Evaluation and Validation of High Throughput Screening Assays. J Biomol Screen *4*, 67–73. https://doi.org/10.1177/108705719900400206.

Zingales, B., Miles, M.A., Moraes, C.B., Luquetti, A., Guhl, F., Schijman, A.G., and Ribeiro, I. (2014). Drug discovery for Chagas disease should consider *Trypanosoma cruzi* strain diversity. Mem. Inst. Oswaldo Cruz. 109, 828–833.

Protein Glycosylation in *Trypanosoma cruzi* and Mass Spectrometry-Based Strategies for Glycan and Glycoprotein Characterization

3

Rebeca Kawahara[1], Joyce Saad[1], Walter Colli[2], Maria JM Alves[2] and Giuseppe Palmisano[1]*

[1]Departamento de Parasitologia, Universidade de São Paulo, São Paulo, Brazil.
[2]Departamento de Bioquímica, Universidade de São Paulo, São Paulo, Brazil.

*Correspondence: palmisano.gp@gmail.com

https://doi.org/10.21775/9781910190838.03

Abstract

Trypanosoma cruzi is the protozoan parasite that causes Chagas disease. This parasite is genetically highly diverse and has a complex life cycle involving two different hosts and distinct developmental stages. The surface of the parasite is covered by a dense array of glycoproteins and glycolipids, which play important function in the differentiation processes and in the host–parasite interaction. The N- and O-glycan biosynthesis in *T. cruzi* has important features that involve unusual carbohydrates and glycosylation enzymes. The characterization of protein glycosylation in *T. cruzi* has unprecedented promise for the discovery of novel targets for drug design and therapeutic intervention in Chagas disease as well as to characterize the different *T. cruzi* strains. However, analytical methodologies for large-scale protein glycosylation analysis have faced challenges due to the extensive glycan micro- and macroheterogeneity. This chapter aims to provide an historical overview of key findings in *T. cruzi* protein glycosylation pathway, including the intermediates and enzymes involved especially in N-linked glycan biosynthesis and the analytical tools used to uncover the composition or structure of N and O-linked glycans. Moreover, we described how mass spectrometry is a powerful tool for large-scale characterizing the set of glycans and glycoproteins in *T. cruzi*.

Introduction

Trypanosoma cruzi is a protozoan parasite that causes Chagas disease, a debilitating condition that affects millions of people especially in South and Central America and is considered one of the world's 13 most important neglected tropical diseases (Bern, 2015). The main route of transmission is vectorial, in which the parasite is transmitted by haematophagous vectors from the Triatominae family. However, secondary modes of transmission have been described, such as congenital (Howard *et al.*, 2014), oral (Shikanai-Yasuda and Carvalho, 2012) and iatrogenic (blood transfusion or organ transplantation) (Angheben *et al.*, 2015; Coura, 2015). The disease starts with an acute phase, which is often asymptomatic and lasts a few weeks, followed by a chronic phase with diverse clinical manifestations, ranging from symptomless to severe chronic disease (Bern, 2015). The clinical outcome may be determined by a variety of factors such as parasite genetics, host genetics, mixed infections as well as cultural and geographical factors (Campbell *et al.*, 2004). Chronic cardiomyopathy is the most common manifestation in *T. cruzi* infected humans affecting approximately 30% of the patients, representing the main cause of disability and mortality (Cunha-Neto and Chevillard, 2014).

Although the genome sequence of *T. cruzi* has been described (El-Sayed *et al.*, 2005), no vaccines or new drugs are yet available to control the disease. The chemotherapy currently used is limited to benznidazole and nifurtimox, which have been shown to be effective during the acute phase of the disease; however, in chronic stages the efficacy of both drugs are low in addition to requiring prolonged treatments with severe adverse effects (Dumonteil *et al.*, 2012). Therefore, there is an urgent need for new therapeutics for use in chronic Chagas disease, which have been recently discussed in reviews by Chatelain (2017) and Paucar *et al.* (2016).

T. cruzi is a parasite with a wide genetic diversity, which strains were subdivided in six discrete typing units (DTUs), named TcI–TcVI and a recently proposed seventh, Tcbat, based on genetic, molecular or immunological markers (Brisse *et al.*, 2000; Marcili *et al.*, 2009; Zingales *et al.*, 2009). Each DTU show characteristics in terms of geographic distribution, eco-epidemiological associations, and clinical manifestations of the disease (reviewed by Zingales *et al.*, 2012) (Brenière *et al.*, 2016; Zingales *et al.*, 2012). In addition, *T. cruzi*'s genetic diversity seems to confer different drug susceptibility (Zingales *et al.*, 2014). Currently, several methods have been reported for characterizing and genotyping *T. cruzi* populations. However, they present a series of drawbacks, for example the need for multiple amplification and sequencing procedures that lead to increased costs and difficulty to achieve a reproducible and standardized protocols (Higuera *et al.*, 2013). Therefore, novel strategies are required for characterizing *T. cruzi* DTUs to understand better the association of the parasite genetic diversity with clinical presentation and outcomes, leading to a more complete diagnosis and better treatment protocols (Quijano-Hernandez and Dumonteil, 2011).

The parasite traverses a complex life cycle that alternates between haematophagous Triatominae vectors and a wide range of susceptible mammals, including humans, and it undergoes two different developmental stages. Within the insect, *T. cruzi* differentiates in two main forms: replicative epimastigote and non-replicative metacyclic trypomastigote. In mammalian hosts, non-replicative bloodstream trypomastigotes and replicative intracellular amastigotes are the typical forms of *T. cruzi* (Rassi *et al.*, 2010).

The interactions of *T. cruzi* with their different hosts are critical for the completion of each life stage and carbohydrates are important molecules in this process. Some interesting reviews discussed previously the importance of protein glycosylation in host–parasite interactions (de Souza *et al.*, 2010; Rodrigues *et al.*, 2015), as well as in the different strain-dependent characteristics (Fonseca *et al.*, 2016), which may determine the diversity of outcome of infection (Mendonça-Previato *et al.*, 2013).

The surfaces of all *T. cruzi* life stages are covered by a large number of glycosylated glycoproteins and glycolipids, which are attached to the membrane via glycosylphosphatidylinositol (GPI) anchors (de Lederkremer and Agustí, 2009). These molecules are important in the activation of the innate immune system and initiation of the acquired host immune response in the vertebrate host (Nardy *et al.*, 2016). Moreover, the remodelling of the parasite surface coat is one key aspect underlying differentiation processes (Engel and Parodi, 1985) and recently was pointed to be associated with different interaction with host galectins across *T. cruzi* lineages (Pineda *et al.*, 2015).

The study of protein glycosylation in *T. cruzi* has unprecedented promise for the discovery of vaccine candidates and for the development of novel therapeutic and diagnostic tools (Rodrigues *et al.*, 2015). However, analytical tools for in depth characterization of site specific glycan composition and structure linked to proteins still face challenges due to the extensive glycan micro- and macroheterogeneity (Thaysen-Andersen and Packer, 2014). This review aims to provide a brief history of the main findings of glycan structures and enzymes involved in *T. cruzi* N- and O-linked protein glycosylation pathways as well as to discuss how analytical methodologies based on mass spectrometry (MS) can be useful as a tool for large-scale characterizing the set of glycans and glycoproteins in parasites.

Landmarks of protein glycosylation in *T. cruzi*

Protein glycosylation is a post-translational modification that involves the covalent attachment of complex carbohydrates or oligosaccharides to specific amino acid residues such as asparagine (N-glycosylation) and serine/threonine (O-glycosylation). The N-linked glycans are linked through the amide bond of asparagine residue within the Asn–X–Ser/Thr/Cys motif (X is any amino acid except proline), and the O-linked glycans are linked through the oxygen atom of serine and threonine side chains (Moremen *et al.*, 2012). The biosynthesis of glycans is primarily determined by the glycosyltransferases, which transfer monosaccharide moieties into linear and branched glycan chains (Rini, 2009). Glycosylation of N-linked glycoproteins is initiated by the transfer of a high mannose-type oligosaccharide from a dolichol-P-P (pyrophosphate linkage) derivative to apolypeptide chains. The glycoproteins pass through the endoplasmic reticulum and the Golgi apparatus until reaching the plasma membrane or extracellular media (Aebi, 2013). The oligosaccharides present in mature glycoproteins of mammalian cells may be composed by different structures: the high mannose-type, hybrid-type and the complex-type. The last type is characterized by the two N-acetylglucosamine and three mannose units of the transferred oligosaccharide and other monosaccharides (N-acetylglucosamine, galactose, sialic acid, fucose, IV-acetylgalactosamine) that are added in the processing pathway.

One of the first evidences that *T. cruzi* is covered by carbohydrates was reported by Alves and Colli(1974) in which they demonstrated that low concentrations of concanavalin A (Con A) promoted agglutination of epimastigotes from *T. cruzi*, but not of trypomastigotes. By that time, differences in membrane structure between blood forms and epimastigotes of *T. cruzi* were already suggested and correlated to the different pathogenic properties of parasite stages. Later, the same group demonstrated glycoconjugates in epimastigote forms, called bands A, B, C and D, using phenolic extraction followed by Schiff staining (Alves and Colli, 1975).

The cell surface carbohydrates of the various *T. cruzi* stages were further analysed by Pereira *et al.* (Pereira *et al.*, 1980), using agglutination and lectin-binding assays, showing that some lectins exhibit unique selectivity in interacting with epimastigote and trypomastigote forms and describing for the first time the presence of sialic acid residues on *T. cruzi* surface by using wheat germ agglutinin (WGA) (Pereira *et al.*, 1980).

Some years later, Parodi and Allue (1982) and Parodi and Cazzulo (1982) showed that *T. cruzi* is able to form both high-mannose and hybrid-type glycoproteins. However, differently from animals, *T. cruzi* synthesizes mannose-P-dolichol instead of glucose-P-dolichol as intermediate in the formation of the other saccharides. Besides, they showed that the largest Dol-P-P-linked oligosaccharide was Man9GlcNAc2 with the dolichol moiety having about 13 isoprene residues (Parodi *et al.*, 1983). Interestingly, although typical glucosyltransferases were until today not found in *T. cruzi*, glucosidase activities degrading both Glc1Man9GlcNAc2 and Glc2Man9GlcNAc2 were described, but no glucosidase I-like enzyme acting on Glc3Man9GlcNAc2 was detected (Bosch *et al.*, 1988).

In 1985, Engel and Parodi demonstrated that the high mannose-type oligosaccharides had different structures depending on the parasite life stage. Epimastigote and intracellular amastigote forms of *T. cruzi* showed only Man9GlcNAc2 linked to dolichol-P-P, whereas infective bloodstream trypomastigotes and amastigote-like parasites isolated from an axenic medium showed both Man7GlcNAc2 and Man9GlcNAc2, confirming the earlier evidences provided by Alves and Colli (1974) and Pereira *et al.* (1980) concerning the changes in protein glycosylation during *T. cruzi* development. The difference in protein glycosylation during *T. cruzi* life stages was later functionally explored by Souto-Padrón and de Souza (1989) in the parasite–macrophage interaction context. They demonstrated that the modulation of N-glycosylation by tunicamycin promoted an increase in approximately 70% and decreased by about 27% the infection of epimastigote and trypomastigote forms, respectively. However, in macrophages, tunicamycin promoted an increase by about 65% and decrease by about 45% the ingestion of epimastigote and trypomastigote forms, respectively (Souto-Padrón and de Souza, 1989).

Although Pereira *et al.* (1980) had shown that epimastigote was able to bind to sialic-acid-affinity lectins, Previato *et al.* provided some of the first evidences on a novel pathway for the incorporation of sialic acid into *T. cruzi* glycoproteins through an unusual *trans*-glycosylase activity (Previato *et al.*, 1985). They found, employing binding of WGA, that the non-infective epimastigote forms of *T. cruzi* incorporated sialic acid onto their surface directly from exogenous sialylated glycoconjugates. In 1987, Zingales *et al.* (1987) demonstrated the existence of a sialic acid trans-glycosylase activity in trypomastigotes by the *in vivo* and *in vitro* labelling of endogenous parasite sialoglycolipids with fetuin-bound sialic acid. Later, Schenkman *et al.* (1991) rebaptized the *trans*-glycosylase activity from infective trypomastigotes as *trans*-sialidase (TcTS) identifying and characterizing the enzyme as important for attachment of trypomastigotes to host cells. The 3D structure and catalytic mechanism of this enzyme were revealed by Buschiazzo *et al.* (Buschiazzo *et al.*, 2002), thus providing a framework for the design of inhibitors with potential therapeutic applications. In fact, many TcTS inhibitors were developed and explored as therapeutic target against the parasite,

Box 3.1 Unusual features of protein glycosylation in *T. cruzi*

- Gal*f*, not found in humans, is present in *T. cruzi* glycoprotein oligosaccharides and glycoinositolphospholipids and it is important for virulence. Its biosynthetic pathway is a potential target for novel drugs against *T. cruzi*.
- *T. cruzi* expresses a unique enzyme, the TcTS, which transfers sialic acid from host glycoconjugates to cell surface proteins of the parasite.
- *T. cruzi* has UDP-GlcNAc:polypeptide α-N-acetylglucosaminyltransferase (pp-α-GlcNAcT), which transfers N-acetylglucosamine (GlcNAc) units to Ser and Thr residues through α-glycosidic O-linkage.

although no effective TcTS inhibitor has been reported yet (Kashif *et al.*, 2017) (Box 3.1). The TcTS has remarkable specificity, by transferring in a α2–3 linkage configuration, the sialic acid from Neu5Acα2→3Galpβ-containing exogenous donor molecules to terminal βGalp-but not αGalp-containing acceptors, (reviewed in Mendoca-Previato *et al.*, 2013; Kashif *et al.*, 2017; Agusti *et al.*, 2007). Besides, Agustí *et al.* reported that TcTS efficiently transfers α-(2–3) linked N-glycolyl neuraminic acid (NeuGc) to terminal β-galactopyranosyl groups (Agustí *et al.*, 2007). Although the sialic acid acceptors on the surface of *T. cruzi* are mainly a family of highly O-glycosylated, threonine-rich mucin-like glycoproteins, other proteins were shown to be N-linked modified by sialic acid containing glycans, such as T-85 (Couto *et al.*, 1987) and cruzipain (Parodi *et al.*, 1995). Therefore, both N and O-linked oligosaccharides served as acceptors of sialic acid units (Ferrero-García *et al.*, 1993). The importance of TcTS function in cell invasion, parasite survival, evasion of host's immune system and host–parasite interaction were extensively reviewed elsewhere (Dc-Rubin and Schenkman, 2012; Frasch, 2000; Freire-de-Lima *et al.*, 2015; Giorgi and de Lederkremer, 2011; Nardy *et al.*, 2016; Schenkman and Eichinger, 1993). Interestingly, Lantos *et al.* (2016) demonstrated recently that the transference of sialic acid to mucins, the main acceptors of this sugar at the cell surface of *T. cruzi*, is catalysed by the *trans*-sialidase present in vesicles released by the parasite into the medium.

Although glycosyltransferases responsible for complex-type glycan biosynthesis, such as fucosyltransferases, acetylglucosaminyltransferase, galactosyltransferase were not identified in *T. cruzi*, our group recently reported complex structures in *T. cruzi* trypomastigotes and epimastigotes using mass spectrometry-based approaches (Alves *et al.*, 2017). In addition, the occurrence of the following nucleotide sugars in *T. cruzi* epimastigote form: α-D-Manp, β-D-Manp, α-D-GlcNAcp, α-D-Glcp, α-D-Galp, α-D-Galf and α-L-Fucp has been shown by Turnock and Ferguson (2007), which provides indirect evidence for the existence of glycosyltransferases belonging to complex-glycan biosynthetic steps.

Galactofuranose is absent in mammalian, but it was detected in N-linked glycans of 80–90 kDa glycoproteins of the infective trypomastigote stage of *T. cruzi* (De Arruda *et al.*, 1989). In the latter study, terminal β-D-galactofuranosyl epitopes were recognized by antibodies, which were able to inhibit internalization of the parasite into mammalian cells (De Arruda *et al.*, 1989). Gal*f*, a sugar that had originally been described in epimastigotes lipopeptidophosphoglycan (LPPG) (de Lederkremer *et al.*, 1980) (later shown to be a GIPL), is present in different molecular weight epitopes among glycoconjugates of epimastigotes and trypomastigotes (Golgher *et al.*, 1993). It was also shown that β-Gal*f* is present in *T.*

cruzi strains G, Tulahuen (Jones *et al.*, 2004), and Dm28c (Agrellos *et al.*, 2003); however, β-Gal*f* was not found in membrane-anchored glycoprotein of *T. cruzi* strains CL-Brener (Todeschini *et al.*, 2001) and Y (Previato *et al.*, 1995). The biosynthetic pathway of Galf involves two enzymes: UDP-galactopyranose mutase (UGM), which is responsible for converting UDP-Galp to UDP-Galf and galactofuranosyl transferases (GalfTs) (Oppenheimer *et al.*, 2012). UGMs and GalfTs are not found in humans, thus they have been explored as potential drug targets. de Lederkremer and Colli (1995) and Oppenheimer *et al.* (2011) reviewed the galactofuranose-containing glycoconjugates in trypanosomatids and pointed the importance of the Galf biosynthesis-related enzymes as good target for the development of antiparasitic drugs (Box 3.1).

Using KEGG pathway for N-glycan biosynthesis, we summarized the N-glycan biosynthesis pathways in *T. cruzi*, highlighting the enzymes that were identified by orthology in *T. cruzi* database and the glycans experimentally evidenced by mass spectrometry or other approaches (Fig. 3.1).

Glycoinositol phospholipids (GIPLs) and a glycosylphosphatidylinositol (GPI)-anchored mucin-like glycoproteins are recognized as the most abundant cell-surface molecules on the epimastigote and metacyclic trypomastigote forms of *T. cruzi* (Colli and Alves, 1999). The importance and function of these major expressed cell surface components of *T. cruzi* were discussed in excellent reviews (Acosta-Serrano *et al.*, 2001; Almeida *et al.*, 1999; Buscaglia *et al.*, 2006; de Lederkremer and Agustí, 2009; Ferguson, 1999; Frasch, 2000; Giorgi and de Lederkremer, 2011).

The O-linked glycosylation of mucins is another unusual features found in *T. cruzi* when compared to that of mammalian cells. In the mammalian, N-acetylgalactosamine (GalNAc) units are linked to Ser and Thr residues of mucins, whereas, in *T. cruzi*, N-acetylglucosamine (GlcNAc) are attached to the peptide,, through the activity of a unique: UDP-GlcNAc:polypeptide α-N-acetylglucosaminyltransferase (pp-α-GlcNAcT), which is encoded by two genes TcOGNT1 and TcOGNT2 (Heise *et al.*, 2009; Previato *et al.*, 1998). Koeller *et al.* (2014) showed that TcOGNT2 is involved in other functions besides O-glycan biosynthesis, especially in the differentiation of metacyclic trypomastigotes and functioning of culture-derived trypomastigotes. Since Previato *et al.* provided a complete review about O-linked glycosylation in *T. cruzi* (Mendonça-Previato *et al.*, 2013) we will not discuss in details the O-glycan biosynthesis in *T. cruzi*. It is important to highlight that the unusual features of O-glycosylation in *T. cruzi*, for example, the selective expression of enzymes, which are not present in the parasite's hosts, such as the pp-α-GlcNAcT and *trans*-sialidases, have been explored as targets for the development of effective treatments against Chagas disease (Box 3.1). Besides, Previato *et al.* discussed the different surface variant forms of O-glycan identified across *T. cruzi* strains, reinforcing the importance of protein glycosylation as a tool to understand the parasite heterogeneity in the development of different clinical outcomes (Mendonça-Previato *et al.*, 2013).

Mass spectrometry as a tool to study protein glycosylation in *T. cruzi*

The analysis of the glycoproteome has faced analytical challenges due to the enormous complexity of glycoforms attached to proteins and the large dynamic range of protein concentration in the sample (Cummings and Pierce, 2014). The structural diversity of the

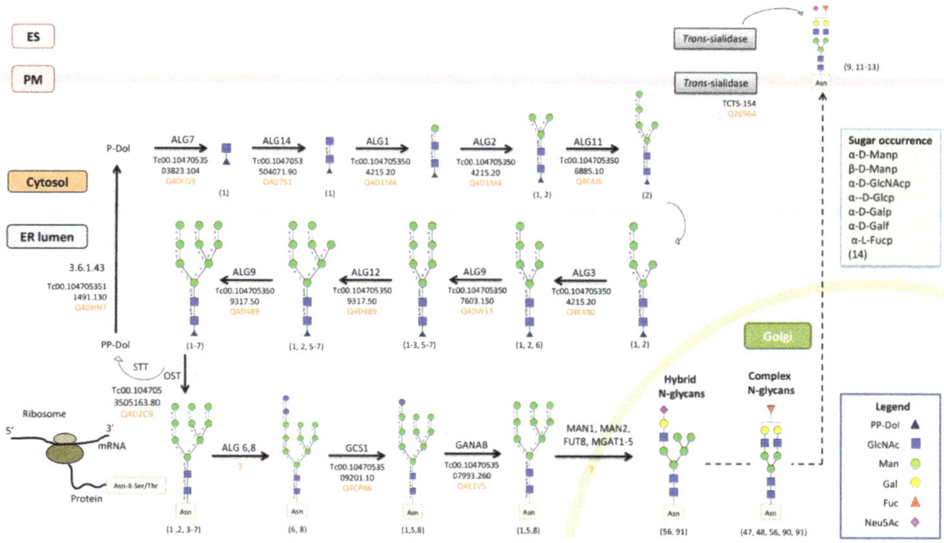

Figure 3.1 Protein N-glycosylation biosynthesis in *T. cruzi*. ES: extracellular space; PM: plasma membrane, ER: endoplasmatic reticulum. Numbers between parentheses are the references: (1) Alves *et al*. (2017); (2) Barboza *et al*. (2005); (3) Engel *et al*. (1985); (4) Parodi and Quesada-Allue *et al* (1982); (5) Parodi and Cazzulo *et al*. (1982); (6) Parodi *et al*. (1983); (7) Doyle *et al*. (1986); (8) Bosch *et al*. (1988); (9) Couto *et al*. (1987); (10) Parodi *et al*. (1995); (11) Couto *et al*. (1990); (12) Agustí *et al*. (2007); (13) Todeschini *et al*. (2009); (14) Turnock *et al*. (2007). Symbols legends: 3.6.1.43 dolichyldiphosphatase; ALG7 UDP-N-acetylglucosamine--dolichyl-phosphate N-acetylglucosaminephosphotransferase; ALG14 beta-1,4-N-acetylglucosaminyltransferase; ALG1 beta-1,4-mannosyltransferase; ALG2 alpha-1,3/alpha-1,6-mannosyltransferase; ALG11 alpha-1,2-mannosyltransferase; ALG3 alpha-1,3-mannosyltransferase; ALG9 alpha-1,2-mannosyltransferase; ALG12 alpha-1,6-mannosyltransferase; STT dolichyl-diphosphooligosaccharide--protein glycosyltransferase; OST oligosaccharyltransferase complex subunit alpha; ALG6 alpha-1,3-glucosyltransferase; ALG8 alpha-1,3-glucosyltransferas;, GCS1 mannosyl-oligosaccharide glucosidase; GANAB alpha 1,3-glucosidase; FUT8 glycoprotein 6-alpha-L-fucosyltransferase; MGAT2 alpha-1,6-mannosyl-glycoprotein beta-1,2-N-acetylglucosaminyltransferase; MGAT3 beta-1,4-mannosyl-glycoprotein beta-1,4-N-acetylglucosaminyltransferase; MGAT5 alpha-1,3(6)-mannosylglycoprotein beta-1,6-N-acetyl-glucosaminyltransferase.

glycan linked to proteins occurs due to the high combinatory possibilities of monosaccharide compositions, branching patterns and linkage types (Thaysen-Andersen and Packer, 2014). The structural complexity of glycoconjugates requires the isolation of specific population such as the N-linked, Ser/Thr-linked, and glycolipid subfractions (Rakus and Mahal, 2011).

The study of the complete set of glycans that are made by a cell or organism under specific conditions is termed 'glycomics' (Hart and Copeland, 2010). Yet 'glycoproteomics' was recently defined by the site-specific analysis of the glycoproteome at the systems level (Thaysen-Andersen *et al.*, 2016). The term glycomics often involves the analysis of the N- and/or O-glycomes, which can be released enzymatically or chemically, a process called deglycosylation (Rakus and Mahal, 2011). On the other hand, glycoproteomics yields information about the protein sequence, the glycosylation site and the structure and occupancy of the glycan (Thaysen-Andersen *et al.*, 2016).

The N-glycosidic bond can be cleaved using the enzyme peptide N-glycanase F (PNGase F), which catalyses the conversion of asparagine to aspartic acid in the glycopeptides sequence. The O-linked glycome is usually released via a reductive β-elimination reaction through the use of a strong base in the presence of sodium borohydride (Rakus and Mahal, 2011; Zaia, 2010).

Several methods have been described for identification and quantification of glycans. For detection of glycosylated proteins, two types of methods are generally used: staining and affinity-based methods. The first method is based on a separation of proteins/glycoproteins in a SDS-PAGE gel followed by a periodic acid–Schiff (PAS) reaction. In this reaction, periodic acid oxidizes two vicinal diol groups to form an aldehyde, which reacts with the Schiff reagent, producing a magenta colour. The affinity-based methods allow the determination of specific glycosylation types. For examples, lectins are proteins that selectively bind to different oligosaccharide epitopes and can be used to enrich or detect specific set of glycoproteins or glycopeptides (Mariño *et al.*, 2010; Roth, 2012).

Following glycan release and purification, the structure and composition of N- and O-linked glycans can be determined mainly by chromatography, nuclear magnetic resonance (NMR) and mass spectrometry (MS). Glycan analyses by chromatography are usually based on fluorescent labelling, which is conjugated to the glycan reducing end via reductive amination (chemical derivatization of monosaccharides). The most widely used chromatographic methods for glycan separation were reviewed by Roth *et al.* (2012) and Mariño *et al.* (2010). For the determination of the glycan structure, a combination of experimental techniques and molecular modelling is required. NMR spectroscopy can provide the entire primary structure of a glycan, including site of attachment and the stereochemistry of the saccharide linkage; however, it requires a higher amount and less complex sample than mass spectrometry (Sweeley and Nunez, 1985).

Mass spectrometry is a powerful tool for the analysis of glycans because it provides a link between mass and composition. Fast atom bombardment ionization (FAB), matrix-assisted laser desorption-ionization (MALDI) and electrospray ionization (ESI) are the main ionization techniques used for glycan ionization and further analysis by mass spectrometry. More details about each ionization techniques and their application for oligosaccharide analysis is provided by Zaia (2004). If glycan and/or peptide sequencing are the goal of the analysis, fragment ions are required. Commonly fragmentation methods used in glycan/(glyco)peptides analysis are the collision-induced dissociation (CID) and electron transfer dissociation (ETD). Each fragmentation method has particular characteristics, for example CID fragmentation predominantly causes cleavages of the glycosidic linkages of the glycopeptides generating information of the monosaccharide composition and the general topology of the N-glycan. ETD fragmentation causes predominantly peptide backbone fragmentation whereas the integrity of the glycan–peptide linkage is maintained. Therefore, complementary structural data of glycopeptides can be obtained using different fragmentation types (Thaysen-Andersen and Packer, 2014; Thaysen-Andersen *et al.*, 2016).

Glycomic and glycoproteomic studies on *T. cruzi*

In the 1980s, the composition of the intermediates, Glc1Man9GlcNac, Glc1Man8GlcNac, Glc1Man7GlcNac, Man8GlcNAc, Man7GlcNAc of N-glycan biosynthesis in *T. cruzi* (Fig. 3.1) was determined by paper chromatography (Doyle *et al.*, 1986; Engel and

Parodi, 1985; Parodi and Cazzulo, 1982; Parodi and Quesada-Allue, 1982) identifying mainly high-mannose structures. In the 1990s, several analytical technologies were used in combination, including NMR and FAB-MS aiming to characterize the composition and structures of *T. cruzi* glycoconjugates. This strategy uncovered the O-linked glycans of 38/43 kDa glycoconjugates from epimastigote forms of the G-strain (Previato *et al.*, 1994), 40/45 kDa glycoconjugates from epimastigote forms of the Y-strain (Previato *et al.*, 1995) and 35/50 kDa glycoconjugates from metacyclic forms of G-strain (Serrano *et al.*, 1995). Some of the glycan structures identified in these studies were: Gal*f*β1–4(Gal*p*β1–6)GlcNAc-ol; Gal*p*β1–3Gal*p*β1–6(Gal*f*β1–4)GlcNAc-ol; [(Gal*p*β1–3)(Gal*p*β1–2)Gal*p*β1–6](Gal*f*β1–4)GlcNAc-ol; [(Gal*p*β1–3)(Gal*p*β1–2)Gal*p*β1–6](Gal*p*β1–2Galfβ1–4)GlcNAc-ol; Gal*p*β1–3GlcNAc-ol; Gal*p*β1–6(Gal*p*β1–3)GlcNAc-ol; and Gal*p*β1–2Gal*p*β1–6(Gal*p*β1–3)GlcNAc-ol. In 2000, Jones *et al.*, reported NMR assignments for 13 gluco- or galacto-pyranosylated derivatives of GlcNAc-ol, GalNAc-ol or ManNAc-ol from the *T. cruzi* mucins (Jones *et al.*, 2000). Later, using NMR combined with FAB-MS, Todeschini *et al.* (2001) characterized the structures of the O-oligosaccharide chains from mucins of *T. cruzi* CL-Brener and *T. cruzi* Colombiana (Todeschini *et al.*, 2009) and recently, using the same strategy, O-linked chains from mucins of the Silvio X10 clone 1 strain were also characterized by Fonseca *et al.* (). In the latter study, they demonstrated the presence of a β-galactofuranose (β-Galf) unity linked to the O-4 position of the α-N-acetylglucosamine (α-GlcNAc)O-4 in mucins from *T. cruzi.*

The composition of N-linked glycoconjugates from Tc-85 kDa glycoprotein was studied by Couto *et al.* (1990) using sequential enzymatic treatment and paper chromatography. In 2005, using a combination of different techniques such as UV-MALDI-TOF mass spectrometry and anion exchange chromatography analysis, Barbosa *et al.* (2005) reported the N-glycan structure of cruzipain, which is the major cysteine proteinase of *T. cruzi.* MacRae *et al.* (2005) also used MALDI-TOF to characterize the N-glycan composition of NETNES glycoprotein.

Turnock and Ferguson (2007) identified and quantified parasites sugar nucleotides, including the epimastigote form of *T. cruzi* using LC-MSMS analysis in multiple reaction monitoring (MRM) mode. This mass spectrometry-based method is highly sensitive and selective for the detection of only a set of compounds of interest in complex mixtures. In summary, a compound is detected based upon both the m/z of the precursor ion and the m/z of one or more of its fragments. Thus, only when a molecular ion gives rise to a specific fragment ion is the ion recorded and assigned (Shi *et al.*, 2012).

The mentioned studies are some examples of analytical technologies used for protein glycosylation characterization in *T. cruzi,* but they were focused on specific glycoproteins or isolated mass-range of glycoproteins or sugars. Recently, advances in mass spectrometry and computational software have allowed deeper coverage of protein glycosylation in complex mixtures.

A glycoproteomics pipeline typically consists in the following steps: (1) extraction of proteins from biological samples, denaturation, and protease digestion ('bottom up' strategy) (2) glycopeptides enrichment or pre-fractionation (3) glycan removal (chemically or enzymatically), generating deglycosylated peptides (4) separation and detection of glycopeptides or deglycosylated peptides using LC coupled with tandem mass spectrometric analysis, and (5) bioinformatics data interpretation. Recent reviews provided detailed information of the advances in LC–MS/MS-based glycomics and glycoproteomics, including the methods used in all steps for the glycomics/glycoproteomics pipeline (Thaysen-Andersen and Packer, 2014; Thaysen-Andersen *et al.*, 2016)

In this review, we highlight the mass spectrometry-based glycoproteomic workflow applied for large-scale characterization of protein glycosylation in *T. cruzi*. In this respect, Atwood *et al.* (2006) provided the first large-scale glycosylation site profile in *T. cruzi* trypomastigotes. They used a workflow based on subcellular fractionation and lectin-based glycopeptides enrichment, followed by N-linked deglycosylation with PNGase F and analysis by LC-MS/MS. The analysis resulted in the identification of 36 N-linked glycosylation sites on 29 glycoproteins, including the first evidence for N-linked glycosylation on the newly annotated mucin associated surface protein family (MASP) and dispersed gene family (DGF) (Atwood *et al.*, 2006). By that time, only the glycosylation sites were characterized, but site-specific glycan-peptide micro-heterogeneity information were not provided (Fig. 3.2).

In 2009, Nakayasu *et al.* applied mass spectrometry to characterize in large-scale *T. cruzi* glycosylphosphatidylinositol (GPI)-anchored molecules, which is a posttranslational modification of eukaryotic surface proteins, containing a lipid and glycan portion (Nakayasu *et al.*, 2009) (Fig. 3.2). Although this review will not cover this modification, more information is provided by Nett *et al.* (2010).

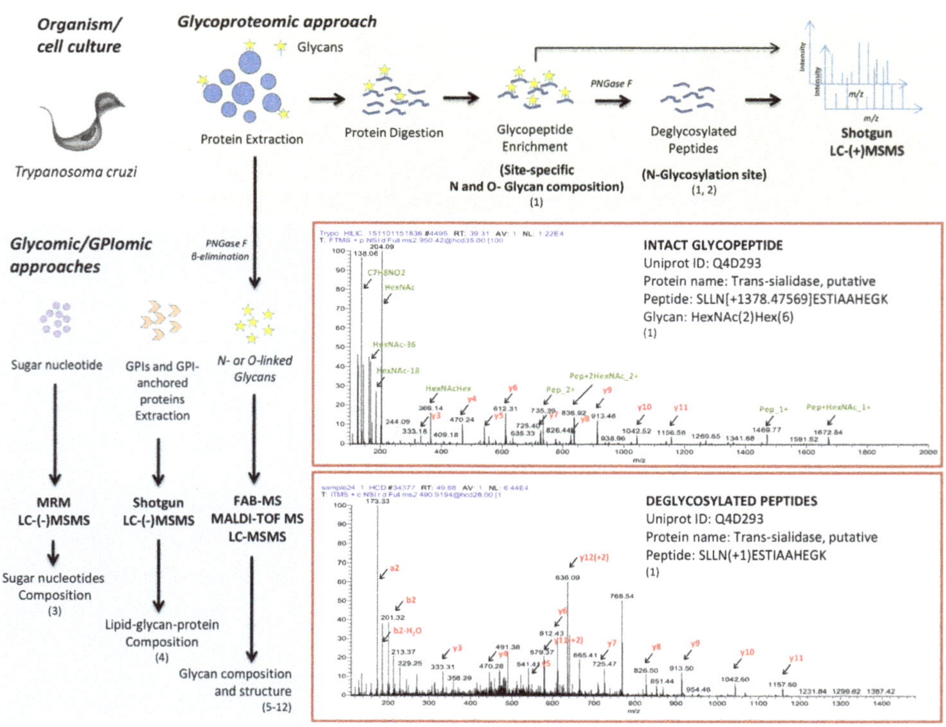

Figure 3.2 Mass spectrometry-based workflows applied for protein glycosylation analysis in *T. cruzi*. Numbers between parentheses are the references: (1) Alves *et al.* (2017); (2) Atwood *et al.* (2006); (3) Turnock *et al.* (2007); (4) Nakayasu *et al.* (2009); (5) Fonseca *et al.* (2016); (6) Todeschini *et al.* (2001); (7) Previato *et al.* (1995); (8) Previato *et al.* (1994); (9) Serrano *et al.* (1995); (10) Todeschini *et al.* (2009); (11) Barboza *et al.* (2005); (12) Macrae *et al.* (2005).

In 2017, our group reported an in-depth glycoproteomic comparison of two *T. cruzi* life stages: epimastigotes and trypomastigotes (Alves *et al.*, 2017). We conducted a glycoproteomic workflow based on bottom-up proteomics approach, followed by two glycopeptides enrichment methods: lectin affinity chromatography using concanavalin A-Sepharose and hydrophilic interaction liquid chromatography (HILIC) and analysis by high resolution LC-MS/MS of N-deglycosylated peptides generated after treatment with PNGase F and as alternative approach we also analysed intact glycopeptides (Alves *et al.*, 2017) (Fig. 3.2). It is important to highlight that the analysis of intact glycopeptides provides information of the glycan micro-heterogeneity attached to the peptide backbone, which was recently reviewed by Thaysen-Andersen (Thaysen-Andersen *et al.*, 2016). Noteworthy, this was possible due to key advances in LC-MS/MS-based detection and fragmentation as well as automated, glycopeptide identification and quantitation (Thaysen-Andersen *et al.*, 2016). In our study, a total of 1306 N-glycosylation sites in NxS/T/C motifs were identified from 690 *T. cruzi* glycoproteins and 144/466 unique N-linked and 10/97 unique O-linked intact glycopeptides in epimastigotes/trypomastigotes, respectively, were identified. The data revealed not only glycosylation sites but also site-specific N- and O-linked glycan heterogeneity information in the two life stages of *T. cruzi* (Alves *et al.*, 2017), including high mannose, hybrid and complex glycan structures attached to specific proteins which were not previously reported in the literature (Figs. 3.1 and 3.2). Besides, using the ratio of the signal intensity of the oxonium ions m/z 126 and 138 upon HCD fragmentation of O-linked glycopeptides, we could also differentiate the nature of the GalNAc or GlcNAc identity of the isobaric HexNAc variants. As mentioned before GlcNAc linked to Ser/Thr is an important feature already reported in *T. cruzi* protein glycosylation, which is catalysed by an α-GlcNAcT enzyme. In summary, using mass-spectrometry based approaches we demonstrated that trypomastigotes had increased protein glycosylation and expression of gp85/*trans*-sialidase, MASP, mucins and GP63 family proteins in trypomastigotes compared to epimastigotes. In addition, several protein pathways were found to be regulated in the investigated life cycle stages of *T. cruzi*. For example, galactose metabolism was overrepresented in trypomastigotes whereas galactose metabolism was overrepresented in epimastigotes. Important advances in the *T. cruzi* glycoproteome characterization were achieved, increasing the repertoire of potential therapeutical targets towards this pathogenic parasite (Alves *et al.*, 2017).

Conclusion

The glycobiology studies in *Trypanosoma cruzi* provided remarkable discoveries of important and unusual glycosylation enzymes and glycan structures involved in the differentiation and infection process as well as disclosed potential targets that have been explored for strain characterization and treatment of Chagas disease. Noteworthy, by reorganizing the N-glycan biosynthesis pathway using literature and orthologue-based annotations we observed that there are still some *T. cruzi* glycosylation enzymes that need to be uncovered. The development of analytical technologies for protein glycosylation analysis is indispensable and required for expanding the knowledge and characterization of the glycobiology of *T. cruzi*. In this scenario, recent advances in mass spectrometry-based glycoproteomics have allowed in depth characterization of N-glycosylation sites and site-specific N- and O-glycan composition in *T. cruzi* glycoproteins, opening exciting opportunities to increase our understanding of protein glycosylation in the parasite biology and host-interaction and to discover novel drug targets and molecular markers for *T. cruzi* strains differentiation.

Acknowledgements

The work described in the literature mentioned in this review and authored by WC, MJMA and GP has been partly supported by Fundação de Amparo à Pesquisa do Estado de São Paulo (FAPESP). GP supported by CNPQ (441878/2014-8) and FAPESP (2014/06863-3). R.K. is supported by FAPESP (2015/02866-0). J.S. is supported by CAPES.

References

Acosta-Serrano, A., Almeida, I.C., Freitas-Junior, L.H., Yoshida, N., and Schenkman, S. (2001). The mucin-like glycoprotein super-family of *Trypanosoma cruzi*: structure and biological roles. Mol. Biochem. Parasitol. *114*, 143–150.

Aebi, M. (2013). N-linked protein glycosylation in the ER. Biochim. Biophys. Acta *1833*, 2430–2437. https://doi.org/10.1016/j.bbamcr.2013.04.001.

Agrellos, O.A., Jones, C., Todeschini, A.R., Previato, J.O., and Mendonça-Previato, L. (2003). A novel sialylated and galactofuranose-containing O-linked glycan, Neu5Acalpha2 – >3Galpbeta1 – >6(Galfbeta1 – >4)GlcNAc, is expressed on the sialoglycoprotein of *Trypanosoma cruzi* Dm28c. Mol. Biochem. Parasitol. *126*, 93–96.

Agustí, R., Giorgi, M.E., and de Lederkremer, R.M. (2007). The trans-sialidase from *Trypanosoma cruzi* efficiently transfers alpha-(2 – >3)-linked N-glycolylneuraminic acid to terminal beta-galactosyl units. Carbohydr. Res. *342*, 2465–2469.

Almeida, I.C., Gazzinelli, R., Ferguson, M.A., and Travassos, L.R. (1999). *Trypanosoma cruzi* mucins: potential functions of a complex structure. Mem. Inst. Oswaldo Cruz 94 (Suppl. 1), 173–176.

Alves, M.J., and Colli, W. (1974). Agglutination of *Trypanosoma cruzi* by concanavalin A. J. Protozool. *21*, 575–578.

Alves, M.J., and Colli, W. (1975). Glycoproteins from trypanosoma cruzi: partial purification by gel chromatography. FEBS Lett. *52*, 188–190.

Alves, M.J., Kawahara, R., Viner, R., Colli, W., Mattos, E.C., Thaysen-Andersen, M., Larsen, M.R., and Palmisano, G. (2017). Comprehensive glycoprofiling of the epimastigote and trypomastigote stages of *Trypanosoma cruzi*. J. Proteomics *151*, 182–192.

Angheben, A., Boix, L., Buonfrate, D., Gobbi, F., Bisoffi, Z., Pupella, S., Gandini, G., and Aprili, G. (2015). Chagas disease and transfusion medicine: a perspective from non-endemic countries. Blood Transfus *13*, 540–550. https://doi.org/10.2450/2015.0040-15.

Atwood, J.A., Minning, T., Ludolf, F., Nuccio, A., Weatherly, D.B., Alvarez-Manilla, G., Tarleton, R., and Orlando, R. (2006). Glycoproteomics of *Trypanosoma cruzi* trypomastigotes using subcellular fractionation, lectin affinity, and stable isotope labeling. J. Proteome Res. *5*, 3376–3384. https://doi.org/10.1021/pr060364b.

Barboza, M., Duschak, V.G., Fukuyama, Y., Nonami, H., Erra-Balsells, R., Cazzulo, J.J., and Couto, A.S. (2005). Structural analysis of the N-glycans of the major cysteine proteinase of *Trypanosoma cruzi*. Identification of sulfated high-mannose type oligosaccharides. FEBS J. *272*, 3803–3815.

Bern, C. (2015). Chagas' disease. N. Engl. J. Med. *373*, 456–466. https://doi.org/10.1056/NEJMra1410150.

Bosch, M., Trombetta, S., Engstrom, U., and Parodi, A.J. (1988). Characterization of dolichol diphosphate oligosaccharide: protein oligosaccharyltransferase and glycoprotein-processing glucosidases occurring in trypanosomatid protozoa. J. Biol. Chem. *263*, 17360–17365.

Brenière, S.F., Waleckx, E., and Barnabé, C. (2016). Over Six Thousand *Trypanosoma cruzi* Strains Classified into Discrete Typing Units (DTUs): Attempt at an Inventory. PLOS Negl. Trop. Dis. *10*, e0004792. https://doi.org/10.1371/journal.pntd.0004792.

Brisse, S., Dujardin, J.C., and Tibayrenc, M. (2000). Identification of six *Trypanosoma cruzi* lineages by sequence-characterised amplified region markers. Mol. Biochem. Parasitol. *111*, 95–105.

Buscaglia, C.A., Campo, V.A., Frasch, A.C., and Di Noia, J.M. (2006). *Trypanosoma cruzi* surface mucins: host-dependent coat diversity. Nat. Rev. Microbiol. *4*, 229–236.

Buschiazzo, A., Amaya, M.F., Cremona, M.L., Frasch, A.C., and Alzari, P.M. (2002). The crystal structure and mode of action of trans-sialidase, a key enzyme in *Trypanosoma cruzi* pathogenesis. Mol. Cell *10*, 757–768.

Campbell, D.A., Westenberger, S.J., and Sturm, N.R. (2004). The determinants of Chagas disease: connecting parasite and host genetics. Curr. Mol. Med. *4*, 549–562.
Chatelain, E. (2017). Chagas disease research and development: Is there light at the end of the tunnel? Comput. Struct. Biotechnol. J. *15*, 98–103. https://doi.org/10.1016/j.csbj.2016.12.002.
Colli, W., and Alves, M.J. (1999). Relevant glycoconjugates on the surface of *Trypanosoma cruzi*. Mem. Inst. Oswaldo Cruz 94 (Suppl. 1), 37–49.
Coura, J.R. (2015). The main sceneries of Chagas disease transmission. The vectors, blood and oral transmissions – a comprehensive review. Mem. Inst. Oswaldo Cruz *110*, 277–282. https://doi.org/10.1590/0074-0276140362.
Couto, A.S., Gonçalves, M.F., Colli, W., and de Lederkremer, R.M. (1990). The N-linked carbohydrate chain of the 85-kilodalton glycoprotein from *Trypanosoma cruzi* trypomastigotes contains sialyl, fucosyl and galactosyl (alpha 1-3)galactose units. Mol. Biochem. Parasitol. *39*, 101–107.
Couto, A.S., Katzin, A.M., Colli, W., and de Lederkremer, R.M. (1987). Sialic acid in a complex oligosaccharide chain of the Tc-85 surface glycoprotein from the trypomastigote stage of *Trypanosoma cruzi*. Mol. Biochem. Parasitol. *26*, 145–153.
Cummings, R.D., and Pierce, J.M. (2014). The challenge and promise of glycomics. Chem. Biol. *21*, 1–15. https://doi.org/10.1016/j.chembiol.2013.12.010.
Cunha-Neto, E., and Chevillard, C. (2014). Chagas disease cardiomyopathy: immunopathology and genetics. Mediators Inflamm. *2014*, 683230. https://doi.org/10.1155/2014/683230.
Dc-Rubin, S.S., and Schenkman, S. (2012). T rypanosoma cruzi trans-sialidase as a multifunctional enzyme in Chagas' disease. Cell. Microbiol. *14*, 1522–1530. https://doi.org/10.1111/j.1462-5822.2012.01831.x.
De Arruda, M.V., Colli, W., and Zingales, B. (1989). Terminal beta-D-galactofuranosyl epitopes recognized by antibodies that inhibit *Trypanosoma cruzi* internalization into mammalian cells. Eur. J. Biochem. *182*, 413–421.
de Lederkremer, R.M., and Agustí, R. (2009). Glycobiology of *Trypanosoma cruzi*. Adv. Carbohydr. Chem. Biochem. *62*, 311–366. https://doi.org/10.1016/S0065-2318(09)00007-9.
de Lederkremer, R.M., Casal, O.L., Alves, M.J., and Colli, W. (1980). Evidence for the presence of D-galactofuranose in the lipopeptidophosphoglycan from Trypanosome cruzi. Modification and tritium labeling. FEBS Lett. *116*, 25–29.
de Lederkremer, R.M., and Colli, W. (1995). Galactofuranose-containing glycoconjugates in trypanosomatids. Glycobiology 5, 547–552.
de Souza, W., de Carvalho, T.M., and Barrias, E.S. (2010). Review on *Trypanosoma cruzi*: Host Cell Interaction. Int. J. Cell Biol. *2010*, 295394. https://doi.org/10.1155/2010/295394.
Doyle, P., de la Canal, L., Engel, J.C., and Parodi, A.J. (1986). Characterization of the mechanism of protein glycosylation and the structure of glycoconjugates in tissue culture trypomastigotes and intracellular amastigotes of *Trypanosoma cruzi*. Mol. Biochem. Parasitol. *21*, 93–101.
Dumonteil, E., Bottazzi, M.E., Zhan, B., Heffernan, M.J., Jones, K., Valenzuela, J.G., Kamhawi, S., Ortega, J., de Leon Rosales, S.P., Lee, B.Y., *et al.* (2012). Accelerating the development of a therapeutic vaccine for human Chagas disease: rationale and prospects. Expert Rev. Vaccines *11*, 1043–1055. https://doi.org/10.1586/erv.12.85.
El-Sayed, N.M., Myler, P.J., Bartholomeu, D.C., Nilsson, D., Aggarwal, G., Tran, A.N., Ghedin, E., Worthey, E.A., Delcher, A.L., Blandin, G., *et al.* (2005). The genome sequence of *Trypanosoma cruzi*, etiologic agent of Chagas disease. Science *309*, 409–415.
Engel, J.C., and Parodi, A.J. (1985). *Trypanosoma cruzi* cells undergo an alteration in protein N-glycosylation upon differentiation. J. Biol. Chem. *260*, 10105–10110.
Ferguson, M.A. (1999). The structure, biosynthesis and functions of glycosylphosphatidylinositol anchors, and the contributions of trypanosome research. J. Cell. Sci. *112*, 2799–2809.
Ferrero-García, M.A., Trombetta, S.E., Sánchez, D.O., Reglero, A., Frasch, A.C., and Parodi, A.J. (1993). The action of *Trypanosoma cruzi* trans-sialidase on glycolipids and glycoproteins. Eur. J. Biochem. *213*, 765–771.
Fonseca, L.M., Garcez, T.C., Penha, L., Freire-DE-Lima, L., Maes, E., Costa, K.M., Mendonça-Previato, L., and Previato, J.O. (2016). Expanding the knowledge of the chemical structure of glycoconjugates from *Trypanosoma cruzi* TcI genotype. Contribution to taxonomic studies. An. Acad. Bras. Cienc. *88*, 1519–1529. https://doi.org/10.1590/0001-3765201620160386.
Frasch, A.C. (2000). Functional diversity in the trans-sialidase and mucin families in *Trypanosoma cruzi*. Parasitol. Today *16*, 282–286.

Freire-de-Lima, L., Fonseca, L.M., Oeltmann, T., Mendonça-Previato, L., and Previato, J.O. (2015). The trans-sialidase, the major *Trypanosoma cruzi* virulence factor: Three decades of studies. Glycobiology *25*, 1142–1149. https://doi.org/10.1093/glycob/cwv057.

Giorgi, M.E., and de Lederkremer, R.M. (2011). Trans-sialidase and mucins of *Trypanosoma cruzi*: an important interplay for the parasite. Carbohydr. Res. *346*, 1389–1393. https://doi.org/10.1016/j.carres.2011.04.006.

Golgher, D.B., Colli, W., Souto-Padrón, T., and Zingales, B. (1993). Galactofuranose-containing glycoconjugates of epimastigote and trypomastigote forms of *Trypanosoma cruzi*. Mol. Biochem. Parasitol. *60*, 249–264.

Hart, G.W., and Copeland, R.J. (2010). Glycomics hits the big time. Cell *143*, 672–676. https://doi.org/10.1016/j.cell.2010.11.008.

Heise, N., Singh, D., van der Wel, H., Sassi, S.O., Johnson, J.M., Feasley, C.L., Koeller, C.M., Previato, J.O., Mendonça-Previato, L., and West, C.M. (2009). Molecular analysis of a UDP-GlcNAc:polypeptide alpha-N-acetylglucosaminyltransferase implicated in the initiation of mucin-type O-glycosylation in *Trypanosoma cruzi*. Glycobiology *19*, 918–933. https://doi.org/10.1093/glycob/cwp068.

Higuera, S.L., Guhl, F., and Ramírez, J.D. (2013). Identification of *Trypanosoma cruzi* discrete typing units (DTUs) through the implementation of a high-resolution melting (HRM) genotyping assay. Parasit. Vectors *6*, 112. https://doi.org/10.1186/1756-3305-6-112.

Howard, E.J., Xiong, X., Carlier, Y., Sosa-Estani, S., and Buekens, P. (2014). Frequency of the congenital transmission of *Trypanosoma cruzi*: a systematic review and meta-analysis. BJOG *121*, 22–33. https://doi.org/10.1111/1471-0528.12396.

Jones, C., Previato, J.O., and Mendonça-Previato, L. (2000). NMR assignments for glucosylated and galactosylated N-acetylhexosaminitols: oligosaccharide alditols related to O-linked glycans from the protozoan parasite *Trypanosoma cruzi*. Carbohydr. Res. *328*, 321–330.

Jones, C., Todeschini, A.R., Agrellos, O.A., Previato, J.O., and Mendonça-Previato, L. (2004). Heterogeneity in the biosynthesis of mucin O-glycans from *Trypanosoma cruzi* tulahuen strain with the expression of novel galactofuranosyl-containing oligosaccharides. Biochemistry *43*, 11889–11897. https://doi.org/10.1021/bi048942u.

Kashif, M., Moreno-Herrera, A., Lara-Ramirez, E.E., Ramirez-Moreno, E., Bocanegra-Garcia, V., Ashfaq, M., and Rivera, G. (2017). Recent developments in trans-sialidase inhibitors of *Trypanosoma cruzi*. J. Drug Target 25, 1–51. https://doi.org/10.1080/1061186X.2017.1289539.

Koeller, C.M., van der Wel, H., Feasley, C.L., Abreu, F., da Rocha, J.D., Montalvão, F., Fampa, P., Dos Reis, F.C., Atella, G.C., Souto-Padrón, T., *et al.* (2014). Golgi UDP-GlcNAc:polypeptide O-α-N-Acetyl-d-glucosaminyltransferase 2 (TcOGNT2) regulates trypomastigote production and function in *Trypanosoma cruzi*. Eukaryot. Cell *13*, 1312–1327. https://doi.org/10.1128/EC.00165-14.

Lantos, A.B., Carlevaro, G., Araoz, B., Ruiz Diaz, P., Camara, M.d.e.L., Buscaglia, C.A., Bossi, M., Yu, H., Chen, X., Bertozzi, C.R., *et al.* (2016). Sialic acid glycobiology unveils *Trypanosoma cruzi* trypomastigote membrane physiology. PLOS Pathog. *12*, e1005559. https://doi.org/10.1371/journal.ppat.1005559.

Macrae, J.I., Acosta-Serrano, A., Morrice, N.A., Mehlert, A., and Ferguson, M.A. (2005). Structural characterization of NETNES, a novel glycoconjugate in *Trypanosoma cruzi* epimastigotes. J. Biol. Chem. *280*, 12201–12211.

Marcili, A., Lima, L., Cavazzana, M., Junqueira, A.C., Veludo, H.H., Maia Da Silva, F., Campaner, M., Paiva, F., Nunes, V.L., and Teixeira, M.M. (2009). A new genotype of *Trypanosoma cruzi* associated with bats evidenced by phylogenetic analyses using SSU rDNA, cytochrome b and Histone H2B genes and genotyping based on ITS1 rDNA. Parasitology *136*, 641–655. https://doi.org/10.1017/S0031182009005861.

Mariño, K., Bones, J., Kattla, J.J., and Rudd, P.M. (2010). A systematic approach to protein glycosylation analysis: a path through the maze. Nat. Chem. Biol. *6*, 713–723. https://doi.org/10.1038/nchembio.437.

Mendonça-Previato, L., Penha, L., Garcez, T.C., Jones, C., and Previato, J.O. (2013). Addition of α-O-GlcNAc to threonine residues define the post-translational modification of mucin-like molecules in *Trypanosoma cruzi*. Glycoconj. J. *30*, 659–666. https://doi.org/10.1007/s10719-013-9469-7.

Moremen, K.W., Tiemeyer, M., and Nairn, A.V. (2012). Vertebrate protein glycosylation: diversity, synthesis and function. Nat. Rev. Mol. Cell Biol. *13*, 448–462. https://doi.org/10.1038/nrm3383.

Nakayasu, E.S., Yashunsky, D.V., Nohara, L.L., Torrecilhas, A.C., Nikolaev, A.V., and Almeida, I.C. (2009). GPIomics: global analysis of glycosylphosphatidylinositol-anchored molecules of *Trypanosoma cruzi*. Mol. Syst. Biol. *5*, 261. https://doi.org/10.1038/msb.2009.13.

Nardy, A.F., Freire-de-Lima, C.G., Pérez, A.R., and Morrot, A. (2016). Role of *Trypanosoma cruzi* Trans-sialidase on the Escape from Host Immune Surveillance. Front Microbiol 7, 348. https://doi.org/10.3389/fmicb.2016.00348.

Nett, I.R., Mehlert, A., Lamont, D., and Ferguson, M.A. (2010). Application of electrospray mass spectrometry to the structural determination of glycosylphosphatidylinositol membrane anchors. Glycobiology *20*, 576–585. https://doi.org/10.1093/glycob/cwq007.

Oppenheimer, M., Valenciano, A.L., Kizjakina, K., Qi, J., and Sobrado, P. (2012). Chemical mechanism of UDP-galactopyranose mutase from *Trypanosoma cruzi*: a potential drug target against Chagas' disease. PLOS ONE 7, e32918. https://doi.org/10.1371/journal.pone.0032918.

Oppenheimer, M., Valenciano, A.L., and Sobrado, P. (2011). Biosynthesis of galactofuranose in kinetoplastids: novel therapeutic targets for treating leishmaniasis and Chagas' disease. Enzyme Res. *2011*, 415976. https://doi.org/10.4061/2011/415976.

Parodi, A.J., and Cazzulo, J.J. (1982). Protein glycosylation in *Trypanosoma cruzi*. II. Partial characterization of protein-bound oligosaccharides labeled 'in vivo'. J. Biol. Chem. *257*, 7641–7645.

Parodi, A.J., Labriola, C., and Cazzulo, J.J. (1995). The presence of complex-type oligosaccharides at the C-terminal domain glycosylation site of some molecules of cruzipain. Mol. Biochem. Parasitol. *69*, 247–255.

Parodi, A.J., Lederkremer, G.Z., and Mendelzon, D.H. (1983). Protein glycosylation in *Trypanosoma cruzi*. The mechanism of glycosylation and structure of protein-bound oligosaccharides. J. Biol. Chem. *258*, 5589–5595.

Parodi, A.J., and Quesada-Allue, L.A. (1982). Protein glycosylation in *Trypanosoma cruzi*. I. Characterization of dolichol-bound monosaccharides and oligosaccharides synthesized 'in vivo'. J. Biol. Chem. *257*, 7637–7640.

Paucar, R., Moreno-Viguri, E., and Pérez-Silanes, S. (2016). Challenges in Chagas disease drug discovery: a review. Curr. Med. Chem. *23*, 3154–3170.

Pereira, M.E., Loures, M.A., Villalta, F., and Andrade, A.F. (1980). Lectin receptors as markers for *Trypanosoma cruzi*. Developmental stages and a study of the interaction of wheat germ agglutinin with sialic acid residues on epimastigote cells. J. Exp. Med. *152*, 1375–1392.

Pineda, M.A., Corvo, L., Soto, M., Fresno, M., and Bonay, P. (2015). Interactions of human galectins with *Trypanosoma cruzi*: binding profile correlate with genetic clustering of lineages. Glycobiology *25*, 197–210. https://doi.org/10.1093/glycob/cwu103.

Previato, J.O., Andrade, A.F., Pessolani, M.C., and Mendonça-Previato, L. (1985). Incorporation of sialic acid into *Trypanosoma cruzi* macromolecules. A proposal for a new metabolic route. Mol. Biochem. Parasitol. *16*, 85–96.

Previato, J.O., Jones, C., Gonçalves, L.P., Wait, R., Travassos, L.R., and Mendonça-Previato, L. (1994). O-glycosidically linked N-acetylglucosamine-bound oligosaccharides from glycoproteins of *Trypanosoma cruzi*. Biochem. J. *301*, 151–159.

Previato, J.O., Jones, C., Xavier, M.T., Wait, R., Travassos, L.R., Parodi, A.J., and Mendonça-Previato, L. (1995). Structural characterization of the major glycosylphosphatidylinositol membrane-anchored glycoprotein from epimastigote forms of *Trypanosoma cruzi* Y-strain. J. Biol. Chem. *270*, 7241–7250.

Previato, J.O., Sola-Penna, M., Agrellos, O.A., Jones, C., Oeltmann, T., Travassos, L.R., and Mendonça-Previato, L. (1998). Biosynthesis of O-N-acetylglucosamine-linked glycans in *Trypanosoma cruzi*. Characterization of the novel uridine diphospho-N-acetylglucosamine:polypeptide N-acetylglucosaminyltransferase-catalyzing formation of N-acetylglucosamine alpha1 – >O-threonine. J. Biol. Chem. *273*, 14982–14988.

Quijano-Hernandez, I., and Dumonteil, E. (2011). Advances and challenges towards a vaccine against Chagas disease. Hum Vaccin 7, 1184–1191. https://doi.org/10.4161/hv.7.11.17016.

Rakus, J.F., and Mahal, L.K. (2011). New technologies for glycomic analysis: toward a systematic understanding of the glycome. Annu Rev. Anal. Chem. *4*, 367–392. https://doi.org/10.1146/annurev-anchem-061010-113951.

Rassi, A., Jr., Rassi, A., and Marin-Neto, J.A. (2010). Chagas disease. Lancet *375*, 1388–1402. https://doi.org/10.1016/S0140-6736(10)60061-X.

Rini, J.E., Esko, J., and Varki, A. (2009). Essentials of Glycobiology, 2nd edn. (Cold Spring Harbor Laboratory Press, Cold Spring Harbor, NY).

Rodrigues, J.A., Acosta-Serrano, A., Aebi, M., Ferguson, M.A., Routier, F.H., Schiller, I., Soares, S., Spencer, D., Titz, A., Wilson, I.B., *et al.* (2015). Parasite Glycobiology: A Bittersweet Symphony. PLOS Pathog. *11*, e1005169. https://doi.org/10.1371/journal.ppat.1005169.

Roth, Z.Y., G.; Khalaila, I. (2012). Identification and quantification of protein glycosylation. Int. J. Carbo. Chem. *2012*, 10 pages. https://doi.org/10.1155/2012/640923.

Schenkman, S., and Eichinger, D. (1993). *Trypanosoma cruzi* trans-sialidase and cell invasion. Parasitol. Today *9*, 218–222.

Schenkman, S., Jiang, M.S., Hart, G.W., and Nussenzweig, V. (1991). A novel cell surface trans-sialidase of *Trypanosoma cruzi* generates a stage-specific epitope required for invasion of mammalian cells. Cell *65*, 1117–1125.

Serrano, A.A., Schenkman, S., Yoshida, N., Mehlert, A., Richardson, J.M., and Ferguson, M.A. (1995). The lipid structure of the glycosylphosphatidylinositol-anchored mucin-like sialic acid acceptors of *Trypanosoma cruzi* changes during parasite differentiation from epimastigotes to infective metacyclic trypomastigote forms. J. Biol. Chem. *270*, 27244–27253.

Shi, T., Su, D., Liu, T., Tang, K., Camp, D.G., Qian, W.J., and Smith, R.D. (2012). Advancing the sensitivity of selected reaction monitoring-based targeted quantitative proteomics. Proteomics *12*, 1074–1092. https://doi.org/10.1002/pmic.201100436.

Shikanai-Yasuda, M.A., and Carvalho, N.B. (2012). Oral transmission of Chagas disease. Clinical infectious diseases: an official publication of the Infectious Diseases Society of America *54*, 845–852. https://doi.org/10.1093/cid/cir956.

Souto-Padrón, T., and de Souza, W. (1989). The effect of tunicamycin and monensin on the association of *Trypanosoma cruzi* with resident macrophages. Parasitol. Res. *76*, 98–106.

Sweeley, C.C., and Nunez, H.A. (1985). Structural analysis of glycoconjugates by mass spectrometry and nuclear magnetic resonance spectroscopy. Annu. Rev. Biochem. *54*, 765–801. https://doi.org/10.1146/annurev.bi.54.070185.004001.

Thaysen-Andersen, M., and Packer, N.H. (2014). Advances in LC-MS/MS-based glycoproteomics: getting closer to system-wide site-specific mapping of the N- and O-glycoproteome. Biochim. Biophys. Acta *1844*, 1437–1452. https://doi.org/10.1016/j.bbapap.2014.05.002.

Thaysen-Andersen, M., Packer, N.H., and Schulz, B.L. (2016). Maturing Glycoproteomics Technologies Provide Unique Structural Insights into the N-glycoproteome and Its Regulation in Health and Disease. Mol. Cell Proteomics *15*, 1773–1790. https://doi.org/10.1074/mcp.O115.057638.

Todeschini, A.R., da Silveira, E.X., Jones, C., Wait, R., Previato, J.O., and Mendonça-Previato, L. (2001). Structure of O-glycosidically linked oligosaccharides from glycoproteins of *Trypanosoma cruzi* CL-Brener strain: evidence for the presence of O-linked sialyl-oligosaccharides. Glycobiology *11*, 47–55.

Todeschini, A.R., de Almeida, E.G., Agrellos, O.A., Jones, C., Previato, J.O., and Mendonça-Previato, L. (2009). Alpha-N-acetylglucosamine-linked O-glycans of sialoglycoproteins (Tc-mucins) from *Trypanosoma cruzi* Colombiana strain. Mem. Inst. Oswaldo Cruz 104 (Suppl. 1), 270–274.

Turnock, D.C., and Ferguson, M.A. (2007). Sugar nucleotide pools of *Trypanosoma brucei*, *Trypanosoma cruzi*, and *Leishmania major*. Eukaryotic Cell *6*, 1450–1463.

Zaia, J. (2004). Mass spectrometry of oligosaccharides. Mass Spectrom. Rev. *23*, 161–227. https://doi.org/10.1002/mas.10073.

Zaia, J. (2010). Mass spectrometry and glycomics. OMICS *14*, 401–418. https://doi.org/10.1089/omi.2009.0146.

Zingales, B., Andrade, S.G., Briones, M.R., Campbell, D.A., Chiari, E., Fernandes, O., Guhl, F., Lages-Silva, E., Macedo, A.M., Machado, C.R., *et al.* (2009). A new consensus for *Trypanosoma cruzi* intraspecific nomenclature: second revision meeting recommends TcI to TcVI. Mem. Inst. Oswaldo Cruz *104*, 1051–1054.

Zingales, B., Carniol, C., de Lederkremer, R.M., and Colli, W. (1987). Direct sialic acid transfer from a protein donor to glycolipids of trypomastigote forms of *Trypanosoma cruzi*. Mol. Biochem. Parasitol. *26*, 135–144.

Zingales, B., Miles, M.A., Campbell, D.A., Tibayrenc, M., Macedo, A.M., Teixeira, M.M., Schijman, A.G., Llewellyn, M.S., Lages-Silva, E., Machado, C.R., *et al.* (2012). The revised *Trypanosoma cruzi* subspecific nomenclature: rationale, epidemiological relevance and research applications. Infect Genet Evol. *12*, 240–253. https://doi.org/10.1016/j.meegid.2011.12.009.

Zingales, B., Miles, M.A., Moraes, C.B., Luquetti, A., Guhl, F., Schijman, A.G., and Ribeiro, I. (2014). Drug discovery for Chagas disease should consider *Trypanosoma cruzi* strain diversity. Mem. Inst. Oswaldo Cruz [Epub ahead of print].

High-throughput Genetic Engineering in *Trypanosoma cruzi*

4

Luis Miguel de Pablos Torró*

Instituto de Biotecnologíca y Departamento de Parasitología, Universidad de Granada, Granada, Spain.

*Correspondence: lpablos@ugr.es

https://doi.org/10.21775/9781910190838.04

Abstract

There are several reasons supporting the fact that kinetoplastid parasites are good models for solving biological questions: (i) these parasites are among the most ancient and divergent eukaryotes; (ii) the pathogenicity is dictated by dynamic cell/protein structures specific to this order of parasites and exposed to the host immune system; (iii) the life cycle of this parasites can be replicated *in vitro* or *in vivo* to a nice extend; (iv) their life cycles combine drastic developmental changes in response to environmental cues; and (v) kinetoplastid parasites can survive as extracellular or intracellular forms in a myriad of host including insects, plants or mammals. Thus, the versatility of scenarios faced by those cells makes them interesting to deep studies using genetic engineering. The emergency of new reverse genetic techniques such us gene editing using nucleases or new approaches for gene flanking homology DNA assembly (fusion PCR or Gibson assembly) for precise integration of different tags have been recently used and applied in the field of trypanosomatids. The aim of this chapter will be to review on advances and approaches designed for successful genetic engineering on *Trypanosoma cruzi*.

Introduction

Kinetoplastids are a widespread group of single-cell flagellated protozoa. These cells are one of the earliest eukaryotic lineages and comprise free-living species, pathogens of invertebrates, vertebrates and also plants (Maslov *et al.*, 2013). These groups of organisms include important, devastating human pathogens such as *Trypanosoma brucei*, the aetiological agent of human sleeping sickness, *Leishmania* spp., which causes visceral and cutaneous infections or *Trypanosoma cruzi*, responsible of Chagas disease. This diseases are worldwide distributed and have economic and health consequences with half a billion people at risk of contracting these diseases (Stuart *et al.*, 2008). These parasites have common core characteristics such us the presence of flagella, a kinetoplast (condensed maxi-mini mitochondrial DNA), polycistronic transcription, mitochondrial RNA editing, RNA *trans*-splicing or very few introns in their genomes. The life cycle of these parasite can be replicated *in vitro* and *in vivo*, being the culture conditions for intracellular and extracellular forms of these parasite species standardized to an extend. Therefore, the easiness of manipulating kinetoplastid parasites in the lab make these organisms suitable for integration of exogenous genetic material, gene manipulation hence genetic engineering.

In the search for essential genes responsible for specific phenotypes (responses to stress, developmental changes, drug resistance, immune evasion, etc.), forward and reverse genetics assays have been developed over the last 20 years. Forward genetics is a classical approach based on the generation of mutants, which will show different phenotypes in response to mutagens such as different chemicals, X-ray, UV or insertional (transposon based mutagenesis). Due to the limitations in space, the present chapter will be focused on the advances made in reverse genetics assays in Kinetoplastid parasites with special focus to *T. cruzi*, which has experimented a revolution with the outcome of a set of new tools and plasmids for enabling genetic engineering in those parasites. Reverse genetics is based on the modification of the activity of a target gene to search for a function or phenotype that could be further analysed. Reverse genetics could be used for developing gain-of-function or loss-of-function approaches. For instance, gain-of-function reverse genetics have been employed to find targets of a library of anti-trypanosomal compounds using a trypanosome overexpression library that could reduce drug sensitivity (Begolo *et al.*, 2014), whereas loss-of-function using genome-scale RNA interference (RNAi) approaches has been employed to identify genes related with drug resistance in *T. brucei* in loss-of-function assays (Schumann Burkard *et al.*, 2011; Baker *et al.*, 2011).

In the post-genomic era, the possibility of generating reverse genetics and high-throughput experimentation has been turned into a reality with the free access to a big repertoire of kinetoplastid genomes (up to 33 genomes uploaded to date (2017) into the kinetoplastid genome resource Tritryp (http://tritrypdb.org/tritrypdb/)), which includes different parasite species, subspecies and/or strains.

Genetic engineering for reverse genetics is an incredibly valuable avenue for the discovery and improvement of drug targets and treatments, vaccines or to increase the basic knowledge on the biology of this group of parasites.

This chapter will brought some of the knowledge acquired by researchers of this area, with special focus on the conditions required for successful experiments in *T. cruzi*.

The model system: setting the conditions for genetic engineering in kinetoplastid parasites

General considerations

As mentioned above, one of the major advantages of kinetoplastid parasites is that life cycles stages can be replicated *in vitro* or *in vivo* to some degree and success. The replicative forms of these parasites are easily grown in liquid culture media that could be expanded to high volumes depending on the needs.

However, there are some general and specific considerations that must be taken into account to generate any sort of mutant or genetically modified (GM) cell line in kinetoplastid parasites and thus high-throughput experimentation:

1 Before and after being transfected, the cells must divide to the highest possible rates for the fastest incorporation of the exogenous genetic material into the cytoplasm (episomal vectors) or nuclear DNA (integrative transfection). Although there is some exceptions, gene engineering is easier when performed over cells on their logarithmic phase of growth.
2 The media of choice must produce homogeneous populations of cells, allowing for the cloning process and should also allow for high densities of parasites.
3 Giving that kinetoplastid populations of cells rapidly adapt to culture conditions (Dumetz *et al.*, 2017), the transfection process may impact on gene expression of natural diversity of cell populations (Seco-Hidalgo *et al.*, 2015), so the parasites must be low passage cultures in order to minimize this lost.

Currently, there is a wide range of defined conditions that stimulate the transformation into different stages (including non-infectious to infectious or vice versa), which makes possible the study of how the parasites develop into the different forms. However, there are reports warning that the results taken from *in vitro* conditions do not necessarily mirror what is expected in nature. For instance, transcriptome comparisons between *L. infantum* metacyclic promastigotes isolated from stomodeal valve of infected sandflies and *in vitro* stationary-phase promastigotes or purified metacyclic promastigote showed a correlation coefficient (R^2) of 0.68 and 0.72 respectively. That means that *in vivo* and *in vitro* cells are physiologically similar but not the same, implying that any high-throughput analysis have implicit a possible discrepancy that should be analysed gene by gene (Alcolea *et al.*, 2016a,b).

Although transformations have been described for low replicative forms such us *L. infantum* axenic amastigotes (Sereno *et al.*, 2001) or *T. cruzi* amastigotes (Padmanabhan *et al.*, 2014), the kinetoplastid forms of parasites usually chosen for transformation are highly replicative stages: such as epimastigotes (*T. cruzi*), procyclic promastigotes (*Leishmania* spp.) or procyclic/bloodstream forms (*T. brucei*).

The present chapter will show the best conditions found to date to develop a successful genetic modifications mainly on *T. cruzi*, which I have divided into (i) culture media, (ii) transfection conditions, (iii) cloning and isolation of transfectans, in *T. cruzi*, *Leishmania* spp. and *T. brucei* species of kinetoplastid parasites, (iv) high-throughput genetic engeneering assays in kinetoplastid parasites.

T. cruzi culture media

As commented above *T. cruzi* is the aetiological agent of Chagas disease. The life cycle begins when the metacyclic trypomastigotes infective forms invade epithelial cells after being released from the faeces of the haematophagus triatomine insect vector (Fig. 4.1a). The metacyclic trypomastigote forms will transform into amastigote forms in approximately 24–36 h in the cytoplasm of the infected cell and divide by binary fision for up to 96–120 h approximately (Fig. 4.1b). The amastigotes will differentiate into bloodstream forms, which will be released from the infected cell to the bloodstream, where they will infect a high number of tissues. While taking blood meal, the triatomine bug will ingest circulating bloodstream forms (Fig. 4.1b). The ingested bloodstream trypomastigotes then

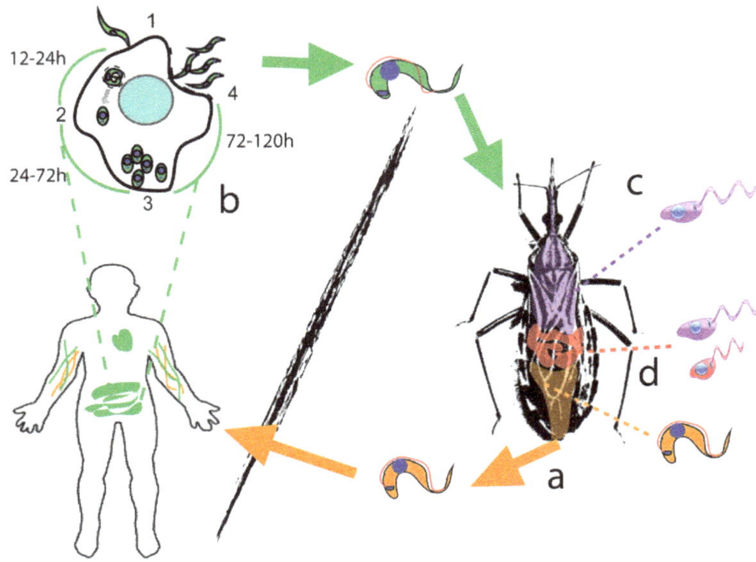

Figure 4.1 Life cycle of *T. cruzi*. The different parasite forms are code-coloured. Orange: metacyclic trypomastigotes; green: amastigotes (round shaped) and bloodstream trypomastigotes; purple: epimastigotes; red: spheromastigotes.

will differentiate into epimastigotes (Fig. 4.1c), a highly replicative form that divides inside the insect vector midgut. Finally, the epimastigote forms will finally differentiate into the infective metacyclic trypomastigote forms in the insect rectum, after adhesion of the epimastigotes to the intestinal wall (Fig. 4.1a and d).

As commented above, defining a culture media allowing for the lowest replicative times is of special importance for the success for the subsequent transfection experiment. In *T. cruzi*, the stage of choice for genetic manipulation is the highly replicative epimastigote stage that can be cultured in liquid, solid or semi-solid media. Although, there is a broad range of media where epimastigotes could be grown (RPMI, MTL, MEM, SDM-79, etc.), Liver Infusion Tryptose (LIT) (first described in the 1960s (Camargo, 1964)) and derivatives such us supplemented liver digest-neutralized tryptose (LDNT) (Tyler and Engman, 2000) are the most widely accepted media for growing epimastigote cells (Box 4.1). Due to the variability of strains, lab stocks, number of passages from isolation or even clones of cells, there is a current lack of consensus for defining epimastigote doubling times during the logarithmic phase ranging from reports of 58 ± 13 h for CL-Brener (DTU VI) strain cultured in LIT medium (Zingales *et al.*, 1997), 18–22 h SilvioX10/7 clone A (DTU I) (Roberts and Fairlamb, 2016) in SDM-78 medium, 36h-67h for a set of *T. cruzi* stocks derived from patients cultured in LIT medium (Lauria-Pires *et al.*, 1997) or 36 h for CL-strain cultIsres in supplemented liver digested neutralized tryptose (LDNT) (Peng *et al.*, 2014). In our hands, the optimization of low passage *T. cruzi* CL-brener (DTUVI) or Dm28c (DTU I) cultures showed doubling times of 21.5 ± 1.2 h and 16.6 ± 0.9 h respectively (unpublished results). In addition, those cultures can reach densities of ≈ 4–5×10^7 cells during the stationary phase in 6–7 days from a starting concentration of 1×10^6 cells/ml. Independently of the strain or doubling time obtained, all subsequent transfection experiments are typically performed

Box 4.1 LIT and LDNT media composition for *T. cruzi* culture

1029 LIT medium: liver infusion broth 9 g/l (BD 226920), tryptose 5 g/l (BD 211713), NaCl 1 g/l, Na2HPO4 8.0 g/l, KCl 0.4 g/l, glucose 1 g/l, 10% (v/v) heat-inactivated fetal calf serum, haemin 10.0 mg/l, distilled water up to 1.0 l. Adjust pH to 7.2 and filter-sterilize: (www.atcc.org/~/media/A3C1C35EC052430E858006C6E92ECFFD.ashx).

B. LDNT media (4 g/l NaCl, 0.4 g/l KCl, 8 g/l Na2HPO4, 2 g/l D-glucose, 5 g/l neutralized liver digest, 5 g/l tryptone, 10% (v/v) heat inactivated fetal calf serum, 100 mg/l haemin) Distilled water up to 1.0 l. Adjust pH to 7.2 and filter-sterilize (Tyler and Engman, 2000).

using a total of 1×10^7 to 4×10^7 cells/ml mid-log phase epimastigote cells (as consensus from several publications: Hariharan *et al.* (1993), Xu *et al.* (2009), Peng *et al.* (2014), Pacheco-Lugo *et al.* (2017).

Transfection conditions

The first transfection experiments by electroporation in *T. cruzi* were achieved in the early 1990s by Hariharan *et al.* (1993) by disrupting polyubiquitin sequence through the integration on the vector pB:PUB-Neo100 or to drive the expression of a downstream gene encoding bacterial chloramphenicol acetyltransferase (CAT) (pTcLS-CAT) (Lu and Buck, 1991) in *T. cruzi*. The conditions used in those experiments: 40–500 µg of DNA where incubated with $1–6 \times 10^8$ cells and transfected using the transfector 300 BTX machine set at 350 V and 1500 µF, paved the way for further improvement of the protocols. Recent research, have improved the transfection efficiency and reduced the amount of DNA used per experiment. To date and to our knowledge, only two papers have described optimization of transfection protocols for expression of exogenous genes in *T. cruzi*.

In a first approach, Da Rocha *et al.* (2004), tried several transfection conditions using *T. cruzi*-GFP cell reporter lines and the Bio-Rad gene pulser unit set (for more details see Box 4.2 and Fig. 4.2). The authors employed two pulses (10 s between pulse) and conditions previously used by Ramirez *et al.* (2000) (300 V and 500 µF) and then incubated the cells with dilutions (25, 50 and 100 µg) of the exogenous DNA (pTREX-GFP plasmid) used for transfection. The main conclusions of this work were (i) the efficiency of transfection is improved increasing the yields of DNA per sample (up to 3.5-fold when comparing 25 to 100 µg of plasmid delivered); and (ii) the use of an linearized vector expressing cassette flanked by *β-tubulin* (pROCKGFPNeo) for homologous recombination, increased the GFP levels up to 8% of the parasites of the DNA integrated. Moreover, these levels could be increased up to 100 times (40% of the parasite population) in the presence of 200 µg/ml G418 antibiotic for several months (DaRocha *et al.*, 2004). As the authors noticed, the conditions tested in this protocol allow for detection of positive parasites as soon as 12 h after transfections. This observation supports that duplication times (DT), hence integration of the exogenous DNA usually occurs more slowly in *T. cruzi* than inother kinetoplastid parasites (e.g. *Leishmania mexicana* procyclic promatigotes, DT ≈7.1 h; and *T. brucei* Procyclic promastigotes, DT ≈9 h), thus the addition of selection drugs after the transfection protocol in *T. cruzi* should be performed at that time point (12 h) approximately (DaRocha *et al.*, 2004).

More recently, a publication by Da Rocha's group evaluated the efficiency of

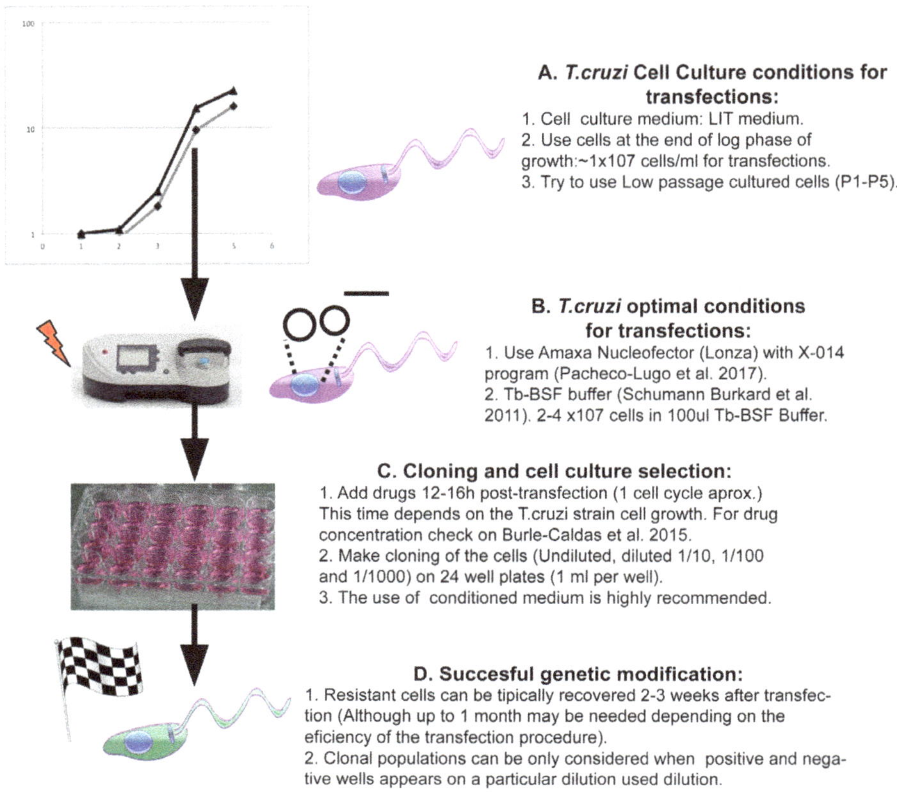

Figure 4.2 Recommended workflow for genetic modification and high-throughput experimentation in *T. cruzi*.

nucleofector^{TM} technology (Lonza, www.lonza.com) for delivering exogenous DNA in *T. cruzi,* using a range of different conditions (Pacheco-Lugo *et al.*, 2017). According to the information delivered by the manufacturer:

> Nucleofection high efficiencies can be reached using much lower substrate amounts and with only moderate impact on viability. The comprehensive way in which Nucleofector™ Programs and cell type-specific solutions are developed enables nucleic acid substrates delivery not only to the cytoplasm, but also through the nuclear membrane and into the nucleus. This allows for high efficiencies up to 99% and makes the transfection success independent from any cell proliferation.
>
> Lonza (www.lonza.com/nucleofection)

In this case the authors have used this technology to improve previous protocols using combinations of 'lower substrate' (10, 20 and 40 μg) of plasmid DNA (pTREX-GFP) with six different nucleofector programs (U-033, X-014, U-014, Z-001, X-001 and T-007). At the same time, the authors used Tb-BSF buffer (formulation provided in: Schumann Burkard *et al.*, 2011) and a range of homemade buffers (Pacheco-Lugo *et al.*, 2017) (Box 4.2). The combination of Tb-BSF buffer and X-014 program generated the best transfection efficiency

Box 4.2 *T. cruzi* optimal transfection conditions

Electroporation buffer: 90 mM sodium phosphate, 5 mM potassium chloride, 0.15 mM calcium chloride, 50 mM HEPES, pH 7.3 (Schumann Burkard *et al.*, 2011).
Transfection conditions: Amaxa Nucleofector device (Lonza), X-014 program. 2–4 × 10^7 cells in 100 µl, 2 mm gap cuvettes, 10 µg DNA (Pacheco-Lugo *et al.*, 2017).

(15% 48 h post transfection), with high survival rates (> 90% of the cells) and high fluorescence levels (Pacheco-Lugo *et al.*, 2017). In contrary to *T. cruzi*, transfection optimization in *Leishmania mexicana* using a range of nucleofector programs resulted in different results with X-001 and Z-001 to be the most effective programs for this kinetoplastid parasite (Dean *et al.*, 2015). This differences well might be due to the different plasma membrane fluidity (different lipid composition), plus different thickness of the glycocalix in each species, which clearly shows transfection conditions are sometimes somehow different among different kinetoplastid (or protozoan) parasites, thus parasite-specific protocol optimizations should be always performed for increasing the success of the DNA delivery into the cells.

Selection drugs and plasmids generated

In this regard, an excellent review by Burle-Carlas *et al.* (2015) described the technology available and to vectors for *T. cruzi* genetic manipulation (updated on 2015). Similarly, the authors also reviewed made a table with the recommended selectable markers used to generate stable *T. cruzi* transfections (Burle-Caldas *et al.*, 2015).

High-throughput genetic engineering in kinetoplastid parasites

Although several publications have described methods for overall analysis of gene function in kinetoplastid parasites, in this chapter I will focus on the available technology and platforms for genetic engineering and genome analysis in *T. cruzi and* summarized below.

A toolkit enabling efficient, scalable and reproducible gene tagging in trypanosomatids

In 2015, Gull and co-workers (Samuel Dean, Jack Sunter and Richard J. Wheerler, University of Oxford) developed a modular set of plasmids and protocols for an easy, fast and reproducible gene tagging (and knockout) in kinetoplastid parasites. The beauty of the strategy used by the authors is due to the absence of bacterial cloning, being only based on PCR reactions and thus increasing the speed of the process. As demonstrated in a series of experiments, the transfection efficiency in *Leishmania* spp. (and most probably in *T. cruzi*) is proportional to the homology length between the target gene and the tagging construct (having low chances for recombination with homologies lower than 100 bp). Therefore, the authors designed a strategy that includes a fusion PCR able to bind the cassette of interest (resistance gene + tag) with two flanking regions (FRs) of up to 500 bp. To anneal both FRs with the cassette, both FRs bearing an oligonucleotide extension of few oligonucleotides (incorporated in the primers for amplifying 5′ and 3′FRs) that are complementary to the 5′ and the 3′ ends of the cassette that can be generated by cutting with different combination of enzymes for generating cassettes with the antibiotic resistance gene up or downstream the tag depending on the strategy (N-, C- tagging (or knocking out)). This way, the three elements (cassette, 5'FR and 3'FR) are able to hybridize and further polymerize to make copies of the full cassette. For making the fusion PCR, five cycles of amplification is used for creating the long template (typically 1500–2500 bp depending on the length of the tag+ resistance gene used) that will be further re-amplified (30× cycles) in the same reaction by adding two nested primers. Therefore tagging constructs can be performed with no cloning step. Despite the fact that it is time-consuming to tag and analyse hundreds of genes (although achievable anyhow), this strategy is highly recommended for medium-throughput experiments when analysing dozens of genes. It should be pointed out that the delivery of constructs for N-tagging does not modify the endogenous 3'UTR of the gene of interest (GOI), allowing for endogenous regulation of mRNA expression, whereas C-tagging replaces the endogenous 3' UTR of the GOI by *Chritidia fasciculata* phosphoglycerate kinase A (PGKA) 3'UTR which might results in an alteration of the mRNA endogenous levels of the GOI. Finally, the same authors have designed a web tool for primer designing based in GeneDB or TritrypDB IDs for N- or C- tagging with a defined set of plasmids (www.richardwheeler.net/dnatools). Additionally, the authors have expanded the repertoire of vectors (pPOTv6, pPOTv7), to allow tagging with different fluorescent proteins or enzymes (mNeongreen, mScarlet or BirA*) (contact Samuel Dean, University of Oxford, for more info: www.sdeanresearch.com/research-interests/genetic-tools).

For the case of *T. brucei*, the same authors bypassed this step of fusion-PCR using long-primers, which amplifies tagging construct straight from the modular plasmid using primers with an overhang of 80 nucleotides for homologous recombination. This strategy could be employed in *T. brucei* because this parasite requires only a minimum of 50 Nts of identity for targeted homologous recombination (Barnes and McCulloch, 2007). To our knowledge, there are no reports testing the minimum homology length for the success of the integration of the tagging constructs in *T. cruzi*. In any case, this strategy could be also used for knocking out individual alleles with individual resistance drugs. To allow for the construct integration and knockout of multiples alleles of same gene locus, CRISPR-Cas9 is the method of choicer, being recently developed a toolkit by the group of E. Gluenz at the University of Oxford for either knock-out or knock-in of GOIs in kinetoplastid parasites (Beneke *et al.*, 2017).

CRISPR-Cas9 genome editing in *T. cruzi* and kinetoplastid parasites

The CRISPR-Cas9 system together with other endonuclease related technologies such as zinc finger nucleases (ZFNs or transcription activator like effector nucleases (TALENTs) have revolutionized the ways for efficient and precise targeted genome editing of living cells. These systems are based on the endonuclease activity of these enzymes to generate permanent mutations by introducing double-strand breaks to activate repair pathways. The subsequent repair provides opportunities for either disruption (knock-out) or integration (knock-in) with a donor DNA with homology directed repair in a targeted locus as well as small deletions repaired by microhomology-mediated end joining (MMEJ) if no donor DNA is added. Different approaches have been used for incorporating the CRISPR-Cas9 technology in trypanosomatids. However the subject of the present chapter will be to explore new efficient high-throughput approaches for research in *T. cruzi*.

CRISPR-Cas9 high-throughput genome editing toolkit for kinetoplastids

As commented above, there are several published approaches for employing this technology in kinetoplastid parasites. The beauty of the work made by Beneke *et al.* (2017) is that the authors not only generated CRISPR-Cas9 null mutants but also developed a toolkit for rapid, scalable analysis of mutant cells, creating multiple cell lines in a very efficient way. To do so, the system employ cell lines that constitutively co-express Cas9 endonuclease and T7 RNA polymerase. The expression of T7 polymerase allows for *in vivo* sgRNA transcription of PCR amplified sgRNA templates. The sgRNA template is generated by a PCR where a first oligo contains the T7 promoter followed by 20 nt of the targeted RNA and a sequence complementary to the 3′ end of the second oligo, comprising the sgRNA scaffold. Besides, the authors have designed modular vectors (pT and pPLOT) for creating donor DNA in *Leishmania* spp. (as they incorporate Leish 5′ and 3′ UTRs) by simply PCR. This technique could be adapted for generating pools of transformed cell populations where different modified cells could be tracked by incorporating barcodes to the donor or repairing cassettes, thus allowing for genome-wide mutational assays (Beneke *et al.*, 2017). For the case of *T. cruzi*, either donor DNA for tagging or knocking-out could be easily generated using any of the pPOTv1, pPOTv2, pPOTv3, pPOTv4, pPOTv6 or pPOTv7 vector generations that contain compatibles *T. cruzi* UTRs (for more info check: (www.richardwheeler.net/dnatools and www.sdeanresearch.com/research-interests/genetic-tools). However, it must be pointed out that the use of Leish 5′ and 3′ UTRs for stabilizing *T. cruzi* mRNAs has not been proved and it should be tested for ensuring correct expression. Nevertheless, another generation of vectors used for endogenous tagging (pLENT vectors: derivatives of pNUS vector which has been used for episomal expression in *C. fasciculata* and *Leishmania* spp.) contain *Crithidia fasciculata* mRNA processing signals and have been successfully applied for the endogenous tag with fluorescent proteins in *Leishmania mexicana* (Tetaud *et al.*, 2002). Although these Crispr-CAS9 toolkit strategy was tested on *L. mexicana* and *T. brucei* species, available data for PCR designing is also available for CL-Brener and Dm28c *T. cruzi* strains (www.leishgedit.net/). The system is based on the generation of Cas9-T7 RNA polymerase co-expression, so for the case of *T. cruzi*, this cell line should be first generated for employed prior to any mutational assay. One potential drawback previously reported by others, is that expression of Cas9 in *T. cruzi* could result in deleterious effect in *T. cruzi* as reflected by the slower doubling time of *T. cruzi*-Cas9 cell lines (Peng *et al.*, 2014). To overcome this potential hurdle, Peng and cols designed gRNA to mediate Cas9 disruption and 'kill' its expression. Despite that the authors were successful and recovered the WT growth phenotype after Cas9 gene disruption, a major concerns on this strategy is the time expended between first transfections and recovery of the phenotype, besides that it do not solve the problems of incorrect phenotypes due to the deleterious impact of Cas9. Another potential negative impact on expressing Cas9 and T7 RNA polymerase simultaneously is that it requires for the integration of two different constructs with two different markers for the selection of these cells. Beneke and cols solved that problem by generating a single marker plasmid pTB007 for expression of Cas9 and T7RNAP in *Leishmania* (Beneke *et al.*, 2017).

CRISPR-Cas9-mediated single-gene and gene family disruption in *Trypanosoma cruzi*

One of the major features of *T. cruzi* is that a big fraction of its genome is formed by repetitive DNA, being a significant portion (≈ 18% CL-Brener strain) composed by multigene families of genes (El-Sayed *et al.*, 2005; De Pablos and Osuna, 2012). In a paper published by Tarleton's group at the University of Georgia (USA), *T. cruzi* expressed hsPCas9 in pTREX vector, an overexpressing vector which allows high protein expression as they are under ribosomal promoter control. Despite that this cell line has deleterious effects as reflected on its slower growth phenotype, the authors were able to delete single copy genes or exogenous genes (such us the GFP) (Peng *et al.*, 2014). In our opinion, this paper adds three very important contributions for the field: (i) the micro-homology-mediated end joining (MMEJ) as the main process for repairing Cas9 double-stranded breaks in *T. cruzi* which decreases the possible mutations due to error prone joining; (ii) the authors created the Eukaryotic Pathogen gRNA Design Tool (EuPaGDT; http://grna.ctegd.uga.edu) for gRNA designing (Tarleton and Peng, 2015); and (iii) the deletion of multiple targets simultaneously using common DNA motifs found in multigene families. The potential of multiplexing the CRISPR-Cas9 action to study several targets is very useful in organisms such as *T. cruzi* which uses the multigene families to generate a great cell to cell variability on its cell surface (Seco-Hidalgo *et al.*, 2015). This feature contributes to the rapid adaptation to environmental changes including immune pressures, enzymatic attack or changes in temperature or osmotic conditions. In the work by Peng and cols, the authors successfully deleted simultaneously β-galactofuransyl glycosyltransferase (β-GalGT) family of 65 annotated genes, which were mutated via MMEJ. EuPaGDT implements an effective on-target search algorithm to identify gRNA targeting multi-gene families, which are highly represented in these pathogens and play important roles in host–pathogen interaction (Tarleton and Peng, 2015). Moreover, the vectors used in this work were kindly deposited in Addgene by the authors so they are available for the research community (www.addgene.org/Rick_Tarleton/).

Conclusion

Because this parasite is extremely diverse and variable in its clonal population structure and genetics, researchers must do an effort to define standard conditions for *T. cruzi* culture so it allows for comparative-type studies. However, as described in the chapter, many goals have been achieved in transfection protocols and genetic engineering in this parasite. As researchers move into new technologies, the conditions for successful high-throughput genetic engineering in *T. cruzi* must be continuously standardized. High-throughput genetic engineering should be integrated into common protocols for molecular and cellular biology of this parasite, so all researchers could answer complex questions with probed protocols by the community.

Acknowledgements

LMDPT would like to thank to the University of Granada (UGR) for the funding support (Assistant Professor appointment number: 16/10/PAD/1617),Department of Parasitology (UGR) for logistics provided and Professor Antonio Osuna for the technical advice and support in the *T. cruzi* molecular and cellular biology.

References

Alcolea, P.J., Alonso, A., Degayón, M.A., Moreno-Paz, M., Jiménez, M., Molina, R., and Larraga, V. (2016a). In vitro infectivity and differential gene expression of *Leishmania infantum* metacyclic promastigotes: negative selection with peanut agglutinin in culture versus isolation from the stomodeal valve of *Phlebotomus perniciosus*. BMC Genomics *17*, 375. https://doi.org/10.1186/s12864-016-2672-8.

Alcolea, P.J., Alonso, A., Domínguez, M., Parro, V., Jiménez, M., Molina, R., and Larraga, V. (2016b). Influence of the Microenvironment in the Transcriptome of *Leishmania infantum* Promastigotes: Sand Fly versus Culture. PLOS Negl. Trop. Dis. *10*, e0004693. https://doi.org/10.1371/journal.pntd.0004693.

Baker, N., Alsford, S., and Horn, D. (2011). Genome-wide RNAi screens in African trypanosomes identify the nifurtimox activator NTR *and* the eflornithine transporter AAT6. Mol. Biochem. Parasitol. *176*, 55–57. https://doi.org/10.1016/j.molbiopara.2010.11.010.

Barnes, R.L., and McCulloch, R. (2007). *Trypanosoma brucei* homologous recombination is dependent on substrate length and homology, though displays a differential dependence on mismatch repair as substrate length decreases. Nucleic Acids Res. *35*, 3478–3493.

Begolo, D., Erben, E., and Clayton, C. (2014). Drug target identification using a trypanosome overexpression library. Antimicrob. Agents Chemother. *58*, 6260–6264. https://doi.org/10.1128/AAC.03338-14.

Beneke, T., Madden, R., Makin, L., Valli, J., Sunter, J., and Gluenz, E. (2017). A CRISPR Cas9 high-throughput genome editing toolkit for kinetoplastids. R Soc Open Sci *4*, 170095. https://doi.org/10.1098/rsos.170095.

Burle-Caldas, G.d.e.A., Grazielle-Silva, V., Laibida, L.A., DaRocha, W.D., and Teixeira, S.M. (2015). Expanding the tool box for genetic manipulation of *Trypanosoma cruzi*. Mol. Biochem. Parasitol. *203*, 25–33. https://doi.org/10.1016/j.molbiopara.2015.10.004.

Camargo, E.P. (1964). Growth and differentiation in *Trypanosoma cruzi*. I. Origin of metacyclic trypanosomes in liquid media. Rev. Inst. Med. Trop. Sao Paulo *6*, 93–100.

DaRocha, W.D., Silva, R.A., Bartholomeu, D.C., Pires, S.F., Freitas, J.M., Macedo, A.M., Vazquez, M.P., Levin, M.J., and Teixeira, S.M. (2004). Expression of exogenous genes in *Trypanosoma cruzi*: improving vectors and electroporation protocols. Parasitol. Res. *92*, 113–120. https://doi.org/10.1007/s00436-003-1004-5.

Dean, S., Sunter, J., Wheeler, R.J., Hodkinson, I., Gluenz, E., and Gull, K. (2015). A toolkit enabling efficient, scalable and reproducible gene tagging in trypanosomatids. *Open Biology* *5*, 140197–140197. https://doi.org/10.1098/rsob.140197.

De Pablos, L.M., and Osuna, A. (2012). Multigene families in *Trypanosoma cruzi and* their role in infectivity. Infect. Immun. *80*, 2258–2264. https://doi.org/10.1128/IAI.06225-11.

Dumetz, F., Imamura, H., Sanders, M., Seblova, V., Myskova, J., Pescher, P., Vanaerschot, M., Meehan, C.J., Cuypers, B., De Muylder, G., *et al.* (2017). Modulation of Aneuploidy in *Leishmania donovani* during Adaptation to Different In Vitro and In Vivo Environments and Its Impact on Gene Expression. MBio *8*, e00599–17.

El-Sayed, N.M., Myler, P.J., Bartholomeu, D.C., Nilsson, D., Aggarwal, G., Tran, A.N., Ghedin, E., Worthey, E.A., Delcher, A.L., Blandin, G., *et al.* (2005). The genome sequence of *Trypanosoma cruzi*, etiologic agent of Chagas disease. Science *309*, 409–415.

Hariharan, S., Ajioka, J., and Swindle, J. (1993). Stable transformation of *Trypanosoma cruzi*: inactivation of the PUB12.5 polyubiquitin gene by targeted gene disruption. Mol. Biochem. Parasitol. 57, 15–30.

Lauria-Pires, L., Santana, J.M., Tavares, F.S., and Teixeira, A.R. (1997). Diversity of *Trypanosoma cruzi* stocks and clones derived from Chagas disease patients: I – Behavioral characterization in vitro. Rev. Soc. Bras. Med. Trop. *30*, 187–192.

Lu, H.Y., and Buck, G.A. (1991). Expression of an exogenous gene in *Trypanosoma cruzi* epimastigotes. Mol. Biochem. Parasitol. *44*, 109–114.

Maslov, D.A., Votýpka, J., Yurchenko, V., and Lukeš, J. (2013). Diversity and phylogeny of insect trypanosomatids: all that is hidden shall be revealed. Trends Parasitol. *29*, 43–52. https://doi.org/10.1016/j.pt.2012.11.001.

Pacheco-Lugo, L., Díaz-Olmos, Y., Sáenz-García, J., Probst, C.M., and DaRocha, W.D. (2017). Effective gene delivery to *Trypanosoma cruzi* epimastigotes through nucleofection. Parasitol. Int. *66*, 236–239.

Padmanabhan, P.K., Polidoro, R.B., Barteneva, N.S., Gazzinelli, R.T., and Burleigh, B.A. (2014). Transient transfection *and* expression of foreign *and* endogenous genes in the intracellular stages of *Trypanosoma cruzi*. Mol. Biochem. Parasitol. *198*, 100–103. https://doi.org/10.1016/j.molbiopara.2015.02.001.

Peng, D., Kurup, S.P., Yao, P.Y., Minning, T.A., and Tarleton, R.L. (2014). CRISPR-Cas9-mediated single-gene and gene family disruption in *Trypanosoma cruzi.* MBio *6*, e02097-14. https://doi.org/10.1128/mBio.02097-14.

Ramirez, M.I., Yamauchi, L.M., de Freitas, L.H., Uemura, H., and Schenkman, S. (2000). The use of the green fluorescent protein to monitor *and* improve transfection in *Trypanosoma cruzi.* Mol. Biochem. Parasitol. *111*, 235–240.

Roberts, A.J., and Fairlamb, A.H. (2016). The N-myristoylome of *Trypanosoma cruzi.* Sci Rep *6*, 31078. https://doi.org/10.1038/srep31078.

Schumann Burkard, G., Jutzi, P., and Roditi, I. (2011). Genome-wide RNAi screens in bloodstream form trypanosomes identify drug transporters. Mol. Biochem. Parasitol. *175*, 91–94. https://doi.org/10.1016/j.molbiopara.2010.09.002.

Seco-Hidalgo, V., Osuna, A., and Pablos, L.M. (2015). To bet or not to bet: deciphering cell to cell variation *in* protozoan infections. Trends Parasitol. *31*, 350–356. https://doi.org/10.1016/j.pt.2015.05.004.

Sereno, D., Roy, G., Lemesre, J.L., Papadopoulou, B., and Ouellette, M. (2001). DNA transformation of *Leishmania infantum* axenic amastigotes *and* their use in drug screening. Antimicrob. Agents Chemother. *45*, 1168–1173. https://doi.org/10.1128/AAC.45.4.1168-1173.2001.

Stuart, K., Brun, R., Croft, S., Fairlamb, A., Gürtler, R.E., McKerrow, J., Reed, S., and Tarleton, R. (2008). Kinetoplastids: related protozoan pathogens, different diseases. J. Clin. Invest. *118*, 1301–1310. https://doi.org/10.1172/JCI33945.

Peng, D., and Tarleton, R. (2015). EuPaGDT: a web tool tailored to design CRISPR guide RNAs for eukaryotic pathogens. Microb Genom *1*, e000033. https://doi.org/10.1099/mgen.0.000033.

Tetaud, E., Lecuix, I., Sheldrake, T., Baltz, T., and Fairlamb, A.H. (2002). A new expression vector for *Crithidia fasciculata and Leishmania*. Mol. Biochem. Parasitol. *120*, 195–204.

Tyler, K.M., and Engman, D.M. (2000). Flagellar elongation induced by glucose limitation is preadaptive for *Trypanosoma cruzi* differentiation. Cell Motil. Cytoskeleton *46*, 269–278.

Xu, D., Brandán, C.P., Basombrío, M.A., and Tarleton, R.L. (2009). Evaluation of high efficiency gene knockout strategies for *Trypanosoma cruzi.* BMC Microbiol. *9*, 90. https://doi.org/10.1186/1471-2180-9-90.

Zingales, B., Pereira, M.E., Almeida, K.A., Umezawa, E.S., Nehme, N.S., Oliveira, R.P., Macedo, A., and Souto, R.P. (1997). Biological parameters and molecular markers of clone CL Brener – the reference organism of the *Trypanosoma cruzi* genome project. Mem. Inst. Oswaldo Cruz *92*, 811–814.

Omics Approaches for Understanding Gene Expression in *Leishmania*: Clues for Tackling Leishmaniasis

5

José M Requena[1*], Pedro J Alcolea[2], Ana Alonso[2] and Vicente Larraga[2]

[1]Centro de Biologia Molecular Severo Ochoa (CSIC-UAM), Universidad Autonoma de Madrid, Madrid, Spain.
[2]Laboratorio de Parasitologia Molecular, Centro de Investigaciones Biologicas (CSIC), Madrid, Spain.

Correspondence: jmrequena@cbm.uam.es

https://doi.org/10.21775/9781910190838.05

Abstract

Leishmaniases, a group of parasitic diseases caused by species of the genus *Leishmania*, afflict millions of people across the globe and cause significant morbidity and mortality. Unfortunately, vaccine and chemotherapy options are limited. The advances in whole genome sequencing have led to renewed impetus in identifying druggable targets for future development of more effective treatments. Hence, the last decade has witnessed a revolution in our understanding of the *Leishmania* genomes through the completion of an increasing number of genome sequencing projects for several species and strains. However, the completion of a genome sequence is not the final product, rather it is just the beginning towards the objective of linking the wealth of data encoded in millions of bases to the biological processes of an organism. Moreover, the genome features (genomics) is only a part of the problem to be solved: the genome yields on transcription, the transcriptome, which in turn yields the proteome on translation, and ultimately the proteins either produce metabolites or are modulated by them. The size of genomics, transcriptomics, proteomics and metabolomics datasets has impelled a new way of analysing data together with the development of potent bioinformatics tools. During their life cycles, *Leishmania* parasites undergo significant changes in their morphology and metabolism. These changes clearly demand a developmental regulation of differential gene expression. Moreover, it is now becoming clear that epigenetic control can also regulate other aspects of the parasitic life cycle, including the control of the switch from proliferative to developmental programs, and the adaptations required for host and cellular tropisms.

Introduction

Protists of the genus *Leishmania* are aetiological agents of a spectrum of clinical diseases, known as leishmaniases, ranging from disfiguring skin lesions to life-threatening visceral infection. Depending on the *Leishmania* species, infection of humans may result in varying forms of leishmaniasis: cutaneous, diffuse cutaneous, mucocutaneous and visceral. The parasite has a worldwide distribution, both in tropical and subtropical regions, and around 1.5 million new cases occur annually, resulting in 50,000 deaths, principally from visceral leishmaniasis (Alvar *et al.*, 2012). Moreover, local socio-economical changes and world climate alterations are contributing to spreading both endemic regions and the number of affected persons (Dujardin *et al.*, 2008; Pigott *et al.*, 2014; Herrador *et al.*, 2015). Canine leishmaniasis is also a serious problem; it is estimated that 2.5 million dogs are infected in the Mediterranean basin alone (Moreno and Alvar, 2002).

Despite the existence of around 20 *Leishmania* species as human pathogens and the wide spectrum of clinical pathologies that they produce, the genus *Leishmania*, regarding morphology and life cycle features, constitutes a homogenous group (Akhoundi *et al.*, 2016). *Leishmania* parasites are transmitted between vertebrate hosts by phlebotomine sandflies of the genera *Phlebotomus* (Old World) and *Lutzomyia* (New World). In the phlebotomine insect vector, the parasite thrives as an extracellular flagellated form (named promastigote), which is attached to the alimentary tract of the insect (Dostálová and Volf, 2012). Upon transmission to the vertebrate host, the parasite is phagocytosed by macrophages, being able to survive and grow inside phagolysosomes, where it adopts a rounded form known as amastigote. These two life cycle stages are adapted to different environmental conditions. In the insect vector, promastigotes grow extracellularly, using sugars as the main nutritional source. In contrast, amastigotes grow intracellularly within the acidic environment of the phagolysosome, in which amino acids are the more abundant nutrients (McConville and Naderer, 2011). Also, drastic changes in temperature occur during transmission of promastigotes from the vector to the mammalian host. A pre-adaptation process takes place within the sandfly vector that involves the development from a noninfective, 'procyclic' stage into an infective, 'metacyclic' stage that is adapted for transmission in the fly and survival in the mammalian host (reviewed in Requena, 2012). In summary, *Leishmania* parasites must cope with many and heterogeneous stressful situations during their life cycle: (i) temperature changes, from ambient temperature in the sandfly vector to 37°C in the mammalian host; (ii) pH changes, after internalization into the macrophage, inside the phagolysosome, the environmental pH decreases to very acidic values; (iii) changes in nutrient and oxygen availability; and (iv) induction of resistance mechanisms to evade both the immune attack by complement components and the harmful effects of antimicrobial molecules produced by phagocytic cells. To cope with these hostile environmental conditions, *Leishmania* has evolved mechanisms of environmental sensing that are able to distinguish between the physiology of vector and mammalian hosts, and controlling accordingly the expression of *Leishmania* stage-specific factors. Nevertheless, how the different *Leishmania* stages are able to sense the environmental conditions, which are typical for either the insect or the mammalian, and to respond accordingly by altering gene expression are two largely unresolved issues.

Leishmania promastigotes are easily cultured in axenic media, and it is generally accepted that these forms are equivalent to the stages thriving in the sandfly midgut. In fact, the axenic promastigotes also undergo a process of programmed differentiation from a relatively noninfective stage (logarithmic-phase cultures) to a highly infective metacyclic stage, which appears when the cultures reach the stationary phase (Da Silva and Sacks, 1987). Furthermore, highly infective metacyclic promastigotes of some *Leishmania* species can be isolated from stationary-phase cultures on the basis of their failure to be agglutinated by the lectin, peanut agglutinin (Sacks *et al.*, 1985). Similarly, for some strains of particular *Leishmania* species, the amastigote forms can be obtained in axenic conditions. Thus, using conditions mimicking those encountered during insect-to-mammal transmission, i.e. temperature upshifts *in vitro* from 25°C to 37°C combined with acidification of the growth medium, promastigote-to-amastigote differentiation is readily obtained in axenic culture (Castilla *et al.*, 1995; Callahan *et al.*, 1997; Balanco *et al.*, 1998). Nevertheless, there exists some controversy regarding whether or not these axenic amastigotes must be considered *bona fide* amastigotes. On the one hand, several studies have previously shown that axenically grown amastigotes clearly resemble intracellular amastigotes with regards to their ultrastructural, biological, biochemical and immunological properties (Gupta *et al.*, 2001). On the other hand, however, it cannot be forgotten that amastigotes grow inside macrophages, where host molecules presumably modulate parasite cellular processes. In this regards, there are also several studies illustrating that axenic amastigotes (often known as amastigote-like forms) differ from intracellular amastigotes in many molecular and biochemical aspects (Gupta *et al.*, 2001; Pescher *et al.*, 2011). A conclusion emerging from these studies is that axenic growth of the *Leishmania* forms is helpful for both the characterization of parasite molecules and understanding some molecular/cellular processes, but environmental conditions are different to those encountered by the parasite along its natural life cycle, even when considering the promastigote stage. In fact, *L. infantum* promastigotes obtained from the anterior thoracic midgut of *Phlebotomus perniciosus* are considerably more infective than promastigotes in the stationary phase of axenic cultures, and important differences in their transcriptomes were reported (Alcolea *et al.*, 2016a). Moreover, it has been shown that the presence of mammalian serum in the culture media has a dramatic effect on *Leishmania* gene expression (Alcolea *et al.*, 2016d).

The mechanisms controlling gene expression in *Leishmania* substantially differ from those in other eukaryotes (Papadopoulou *et al.*, 2008; Requena, 2011; De Pablos *et al.*, 2016). Thus, regulation of transcription on a per gene basis does not exist for almost the totality of genes; instead, genes are initially transcribed as polycistronic precursor RNAs from poorly characterized promoters located at the 5′ of large directional gene clusters (DGCs). Therefore, we will dedicate the first sections of this chapter to describe our current knowledge about genome structure and the mechanisms of gene expression in *Leishmania*. Then, the readers would be ready to better appreciate the peculiarities of the analysis of Omics data derived from *Leishmania* studies. Finally, after presenting the contributions of Omics approaches to our knowledge of gene expression, we will discuss how -omics are adding new strategies for diagnosis and control of leishmaniasis.

The *Leishmania* genome

It is relevant to briefly summarize the history leading to the uncovering of the genome structure in *Leishmania* parasites. In 1992, P. Bastien and co-workers published an article entitled '*Leishmania*: Sex, Lies and Karyotype' (Bastien *et al.*, 1992); this outstanding article clearly shows the particular difficulties that scientists had to solve, many were hurdles derived from the peculiarities of the *Leishmania* genome. In fact, nowadays, questions like ploidy and genetic exchange remain unsolved. As *Leishmania* chromosomes do not condense during mitosis (Solari, 1995), their number and sizes could be studied only by pulsed field gel electrophoresis (PFGE), a technique that allows for electrophoretic separation of DNA molecules of very different size (from some kilobases (Kb) to megabase (Mb)-size molecules). The images derived from those analyses were called 'molecular karyotypes'. These images suggested that the karyotype of *Leishmania* should be extremely polymorphic, since strains of the same *Leishmania* species yielded PFGE-pictures very different with chromosomal bands of different size and intensity of staining (Blaineau *et al.*, 1991). However, this puzzling finding did not discourage to the Montpellier team in its objective of determining the *Leishmania* karyotype and chromosomal structures. For this purpose, they decided to establish physical linkage groups by hybridization of specific DNA probes to PFGE-separated chromosomes. After individual hybridization of 244 different probes, they concluded that the *L. infantum* genome comprises 36 chromosomes ranging in size from 0.35 to ≈ 3 Mb (Wincker *et al.*, 1996). Moreover, these authors demonstrated that the linkage groups were consistently conserved in other species, such as *L. major*, *L. tropica* and *L. aethiopica*. This finding contrasted with the size heterogeneity of the chromosomal bands derived from the PFGE analysis, and pointed to a high conservation of the chromosomal scaffolds within the genus *Leishmania* (Wincker *et al.*, 1996). This outstanding work paved the way for deciphering the genome sequence for this genus, first in *L. major* (Ivens *et al.*, 2005), afterwards for *L. infantum* and *L. braziliensis* (Peacock *et al.*, 2007) and, nowadays, for many other species (Cantacessi *et al.*, 2015; www.tritrypdb.org). Nevertheless, as shown in a recent article, the *L. major* (Friedlin) genome, which is the best-assembled genome to date for a *Leishmania* species, cannot be considered as set in stone (Alonso *et al.*, 2016), and users should be aware that a genome sequence is only a hypothesis.

The Old World *Leishmania* spp. have 36 chromosomes, while the New World *Leishmania* spp. have 34 or 35 chromosomes. Thus, *L. mexicana* has linkage groups of chromosomes 8 and 29 as well as of chromosomes 20 and 36, and *L. braziliensis* has a linkage group of chromosomes 20 and 34 (Britto *et al.*, 1998), regarding the *L. major* chromosomal set. Nevertheless, taking into account the evolutionary distance existing between these *Leishmania* species (see below), translocations must be considered extremely rare, and the conservation of the major linkage groups seems to be an essential feature of the genome of this parasite (Smith *et al.*, 2007).

The completion of the genome sequence in *L. major* and those of two related trypanosomatids in 2005, *Trypanosoma cruzi* (El-Sayed *et al.*, 2005) and *T. brucei* (Berriman *et al.*, 2005), revealed an extraordinary feature regarding gene organization. In the genome of these parasites, protein-coding genes are arranged into large unidirectional clusters, in which the genes present have no common nor akin function. These directional gene clusters (DGCs) are separated by short sequences termed strand-switch regions (SSRs), where the clusters converge towards or diverge from. This peculiar gene organization is dictated by another surprising mechanism existing in these parasites: polycistronic transcription (see next section). Also, this unusual way of genome organization would be linked to the very high gene density (around one gene every 3.2 kb) existing in the *Leishmania* genome (and those of related trypanosomatids); this may be an evolutionary advantage for the parasite in order to reduce cell cycle length. In this regards, recent results point to the existence of a large number of potential replication origins in *Leishmania*, whose activation would be regulated in a concerted manner with transcriptional activity (Lombraña *et al.*, 2016).

These peculiarities in gene organization and transcriptional mechanisms existing in *Leishmania* are related, in part, to the fact that these protists occupy a deep-branched position on the evolutionary tree of eukaryotes (Sogin *et al.*, 1986). The genus *Leishmania* belongs to the order Trypanosomatida (Moreira *et al.*, 2004), in which are also included *T. brucei* and *T. cruzi*, causative agents of other important infectious diseases for humans, sleeping sickness and Chagas disease, respectively. In fact, the genera *Leishmania* and *Trypanosoma* share mechanisms of gene expression, and even a remarkable synteny regarding gene organization.

According to recent estimations, the common ancestor for *Trypanosoma* and *Leishmania* lasted around 400 or 600 millions of years (Stevens *et al.*, 2001). However, in spite of the high level of divergence existing at the level of protein-coding sequences, species like *T. brucei, T. cruzi* and *L. major* exhibit a striking conservation in the gene arrangement, suggesting that selection has maintained gene order in this parasitic species over hundreds of millions of years of evolution (Ghedin *et al.*, 2004). Furthermore, when comparisons of genome sequences of *L. major, L. infantum* and *L. braziliensis* are done, the synteny across the genus is almost absolute, only a few differences regarding gene content exist (Peacock *et al.*, 2007). This is really surprising taking into account that the divergence of these species took place around 10 million (between *L. infantum* and *L. major*) and 100 million (between *L. infantum* and *L. braziliensis*) years ago, and the different pathologies that each one of these species causes in humans (McGwire and Satoskar, 2014). At present, how those distinct disease pathogeneses emerge from this strong conservation in both gene content and synteny is a question yet to be solved. It is possible that just a few species-specific genes are crucial to disease outcome, but, alternatively, it is also possible that the parasite genome may play only a minor role in determining disease phenotype, which would be mainly determined by differences in the expression levels of key conserved genes.

Vanishing of transposable elements and flourishing of retroposon-derived, repeated sequences in the *Leishmania* genome

Transposable elements (TEs) and related sequences constitute a substantial proportion of eukaryotic genomes, composing, for instance, more than 50% of the human genome (Richard *et al.*, 2008). TEs are classified into two major groups based on the mechanisms used for their transposition: retroelements or retroposons, which transpose via reverse transcription of an RNA intermediate, and DNA transposons that move through a DNA intermediate. Based on the presence (or its absence) of a long terminal repeat (LTR), retrotransposons are subdivided into LTR- and non-LTR-retrotransposons. A particularly abundant group of non-LTR retrotransposons are the long interspersed nuclear elements (LINEs), which in the human genome have a copy number of 850,000, and 660,000 LINEs were found in the mouse genome (Richard *et al.*, 2008). LINEs encode a reverse transcriptase and other enzymes required for retrotransposition, and some of these elements are indeed functional (Belancio *et al.*, 2008). However, the tendency to accumulate LINEs in the genomes has not been a general rule in eukaryotes, and even some organisms, as many species of *Leishmania*, have eliminated them from their genomes (Bringaud *et al.*, 2006).

In contrast, the related trypanosomatids *T. brucei* and *T. cruzi* contain a significant amount of retrotransposons, which are very similar in both species, suggesting a common evolutionary history. Thus, they contain VIPER, an LTR retrotransposon originally characterized in the *T. cruzi* genome (Vazquez *et al.*, 2000). Also, they contain non-LTR retrotransposons, which are divided into site-specific retroelements (*T. brucei* SLACS, *T. cruzi* CZAR and *Crithidia fasciculata* CRE1/CRE2 elements), which are always inserted at the same relative position in the spliced leader (SL) RNA genes (Aksoy, 1991) and non-site-specific retroposons (Ingi and L1Tc) (Bringaud *et al.*, 2008; Thomas *et al.*, 2010). The genomes of those trypanosomatids also contain small non-autonomous retroposons derived from the autonomous retroelements (Ingi and L1Tc): NARTc in the *T. cruzi* genome and RIME in the *T. brucei* one (Bringaud *et al.*, 2008; Thomas *et al.*, 2010). These retroelements and their derivatives constitute up to 5% of the genomic content of both species of trypanosomes. In contrast to trypanosomes, the genomes of the Old world species *L. major* and *L. infantum* do not contain any complete retroelement. However, a few highly degenerated Ingi/L1Tc-related elements (DIREs) have been identified in the genomes of these species, indicating that the genome of their ancestor contained active retroposons but they have been lost over time to become vestigial retroelements in the genome of the present day *L. major* and *L. infantum* (Bringaud *et al.*, 2006).

Another remarkable feature of Ingi and L1Tc retroelements is the presence at their 5′-end of a conserved stretch of 77–79 bp (Bringaud *et al.*, 2002), which was also present in the short non-autonomous non-LTR retrotransposons identified in *T. brucei* (RIME; Hasan *et al.*, 1984) and *T. cruzi* (NARTc; Bringaud *et al.*, 2002). This sequence, named 'non-LTR retrotransposon signature' (Bringaud *et al.*, 2006) or Pr77 (Heras *et al.*, 2007), was found to be present in the *L. major* genome at the time of NARTc characterization in *T. cruzi* (Bringaud *et al.*, 2002), and afterwards in some of the *L. major* DIREs (Bringaud *et al.*, 2006). Subsequently, using the 79-bp signature (the hallmark of trypanosomatid retroposons) as bait, Bringaud and co-workers identified in the *L. major* genome two large families of small repeated elements that were named LmSIDER1 (785 copies) and LmSIDER2 (1,073 copies); LmSIDER stands for *L. major* Short Interspersed DEgenerated Retroposon (Bringaud *et al.*, 2007). A detailed analysis of LmSIDER2 elements in the *L. major* indicated that most LmSIDER2 would be located within the 3′UTR of mRNAs, suggesting a role of these elements in modulating mRNA stability (Bringaud *et al.*, 2007). Moreover, it was shown that, unlike most eukaryotic transcripts, some of the SIDER2-bearing mRNAs do not undergo poly(A) tail shortening prior to rapid turnover, but instead, they are targeted for degradation by a site-specific endonucleolytic cleavage (Müller *et al.*, 2010a; Müller *et al.*, 2010b). In this regard, the auto-catalytic ribozyme activity, demonstrated in the Pr77 sequence (retroposon-hallmark, see above) of L1Tc (Sánchez-Luque *et al.*, 2011), might be responsible for the endonucleolytic cleavage observed in those SIDER2-bearing mRNAs (Sánchez-Luque *et al.*, 2014). Another salient feature of SIDER2 elements is that they can group, based on sequence similarity, into subfamilies, whose members are restricted to particular regions of a sole chromosome (Requena *et al.*, 2008; Requena *et al.*, 2017). This peculiar pattern of distribution would be related to the genomic plasticity of this parasite (see below). Also, members of the family SIDER1 have been involved in regulatory mechanisms of gene expression as stage-specific translational regulators (McNicoll *et al.*, 2005). In summary, *Leishmania* would be an outstanding example in eukaryotes of domestication and expansion processes in which TEs have evolved to fulfil essential roles related to gene expression, genome plasticity and perhaps transcriptional initiation (Bringaud *et al.*, 2008; Requena *et al.*, 2017).

In this context, it was surprising to detect potentially active retroposons in the genome of *L. braziliensis* (Peacock *et al.*, 2007). Thus, non-LTR retrotransposons, equivalent to *T. brucei* SLACS and *T. cruzi* CZAR, are associated also with the SL gene array in *L. braziliensis*. In addition, the telomeres in the *L. braziliensis* genome accommodate DNA transposable elements, called TATEs (telomeric associated transposable elements), encoding putative reverse transcriptase, phage integrase (site-specific recombinase) and DNA/RNA polymerase domains. Another remarkably distinct feature found in *L. braziliensis*, but absent from the genome of Old World *Leishmania* species, is the existence of an active mechanism of RNA interference (RNAi) (Lye *et al.*, 2010). In trypanosomatids, an RNAi mechanism was initially identified in *T. brucei* (Ngô *et al.*, 1998), but could not be demonstrated in other species including *L. major* and *T. cruzi* (Lye *et al.*, 2010). An open question is whether RNAi machinery has been lost or acquired across the *Leishmania* genus. Examination of the syntenic regions in *L. major* and *L. infantum* identified remnants of Argonaute in both species, an observation that favours the proposal that RNAi was lost in *L. major*/*L. infantum* after their divergence from the *L. braziliensis* evolutionary branch (Smith *et al.*, 2007). It can be hypothesized that the presence of retrotransposons in *L. braziliensis* parasites forced them to maintain the RNAi pathway, taking into account that it functions as a genome immune defence mechanism to limit the potentially deleterious consequences of transposon/retroposon mobilization (Carthew and Sontheimer, 2009).

Plasticity of the *Leishmania* genome: gene copy number alterations, repeated sequences and aneuploidy

It may result paradoxical to describe the *Leishmania* genome as unstable or plastic, after having indicated in the previous section that gene organization has been maintained static among evolutionarily distant *Leishmania* species, and even relatively few changes have occurred after the separation of the genera *Leishmania* and *Trypanosoma*. However, the *Leishmania* genome, when analysed as a whole, exhibits a remarkable degree of inter- and intraspecific variability. Initial gene mapping studies suggested that *Leishmania* is mostly diploid (Iovannisci and Beverley, 1989). However, it was later shown during gene knock-out experiments that *Leishmania* parasites can undergo changes in ploidy. Currently, the standard approach for deletion of a given gene in *Leishmania* consists of the sequential deletion of both chromosomal alleles by homologous gene replacement using constructs carrying different drug resistance markers flanked with the boundary regions of the gene to be deleted (Cruz *et al.*, 1991). Nevertheless, attempts to delete essential genes in *Leishmania* more often yield parasites bearing the planned allelic replacements but containing additional gene copies that arise via aneuploidy or polyploidy (Cruz *et al.*, 1993; Dumas *et al.*, 1997; Genest *et al.*, 2005). In fact, this observation is considered as a clue about the essentiality of the targeted gene, especially when accompanied by successful replacement in the presence of an ectopic gene (Ilgoutz *et al.*, 1999; Vergnes *et al.*, 2005; Murta *et al.*, 2009).

Other manifestations of the genomic plasticity of *Leishmania* are found in the DNA amplification events that occur after incubation of this parasite with a variety of anti-leishmanial drugs, and even amplicons have been found in unselected laboratory isolates (Beverley, 1991; Grondin *et al.*, 1996). In fact, an idea widely accepted is that *Leishmania* parasites can re-shape their genome rapidly *in vitro* to survive stressful environments (Leprohon *et al.*, 2009). Homologous recombination events between direct repeats (mostly involving SIDER elements) would lead to the generation of the circular amplicons, whereas recombination between inverted repeats would generate linear or circular, inverted amplicons (Ubeda *et al.*, 2014). Gene amplification through homologous recombination between repeated sequences is a common mechanism of drug resistance in *Leishmania* (Grondin *et al.*, 1993, 1996), and it is probably the only possible strategy for altering gene expressions in organisms like trypanosomatids lacking transcriptional control (see below). Clues about the enzymes and molecular mechanisms responsible for these amplification events are being obtained. Extrachromosomal circular amplicons are dependent on the activity of RAD51 and RAD51–4 (Ubeda *et al.*, 2014; Genois *et al.*, 2015) while linear amplicons depend on MRE11 (Laffitte *et al.*, 2014). In summary, genomic rearrangements in *Leishmania* occur genome-wide (Ubeda *et al.*, 2014), they can lead to the formation of extrachromosomal elements (Leprohon *et al.*, 2009; Downing *et al.*, 2011), to supernumerary chromosomes and to mosaic aneuploidy (Sterkers *et al.*, 2011).

A special comment merits aneuploidy as a genetic feature of *Leishmania* parasites. Thus, different sets of data support the idea that chromosomal dosage in *Leishmania* genomes is aneuploid rather than strictly diploid (Ravel *et al.*, 1998; Sunkin *et al.*, 2000; Dubessay *et al.*, 2002b; Martínez-Calvillo *et al.*, 2005; Rogers *et al.*, 2011; Sterkers *et al.*, 2011). More strikingly, the genome of *Leishmania,* when considered at the population level, exhibits an unusual feature termed 'mosaic aneuploidy' (Sterkers *et al.*, 2011). This concept serves to describe that in a given *Leishmania* population, there are cells that are mono-, di-, trisomic (and so on) in proportions that vary from one chromosome to another, from one strain to another, and even from cell to cell (Lachaud *et al.*, 2014; Seco-Hidalgo *et al.*, 2015). How this remarkable genome plasticity is generated and tolerated is poorly understood at present.

Regulation of gene expression in *Leishmania*

Initial studies dealing with the analysis of the transcriptional activity of particular *Leishmania* genes showed somewhat unexpected results: transcription of individual genes remains constant irrespective of the abundance of the mRNA products. The heat shock protein (HSP) genes of *Leishmania* were among the first to be analysed, because it was easy to find them due to their remarkable sequence conservation along the evolutionary scale and, on the other hand, because they seemed to be ideal candidates for identification of inducible RNA polymerase II promoters (reviewed in Folgueira and Requena, 2007). Such analyses were based on the technically demanding nuclear-run-on procedure, which consists of the isolation of nuclei from promastigotes and their incubation with labelled ribonucleotide triphosphates. Thus, polymerases engaged to DNA in the act of transcriptional elongation continue to transcribe (or 'run-on') for several hundred additional nucleotides. Labelled nuclear RNA is then purified and hybridized to blots containing specific DNA probes. The hybridization signals equate directly to the density of PolII and, consequently, the rate of transcription of a given gene. Run-on analyses of the expression of the *HSP70* and *HSP83/90* genes in several *Leishmania* species concluded that there was not transcriptional activation of these genes when the parasites are exposed to a heat shock, in spite that accumulation of the respective mRNAs occurred (Brandau *et al.*, 1995; Quijada *et al.*, 1997). Subsequent studies analysing the transcriptional regulation of additional genes fostered the idea that most protein-coding genes in *Leishmania,* and related trypanosomatids, do not have their own promoters and are transcribed as part of polycistronic transcripts (Clayton, 2002). In an outstanding article, Martínez-Calvillo and co-workers (2003), by using strand-specific nuclear run-on assays, showed that most of the transcriptional activity occurs on the sense DNA strand, and that RNA polymerase II-mediated transcription initiates within the strand-switch regions (SSRs) and radiates bidirectionally from them. The idea that transcription initiation is occurring at the SSRs was strengthened by the finding that acetylated histones are concentrated at divergent SSRs (putative sites of transcription initiation) but are absent from convergent SSRs (putative sites of transcription termination) in *T. cruzi* (Respuela *et al.*, 2008). In *T. brucei*, two histone modifications associated with open chromatin (H4K10ac and H3K4me3) and two histone variants (H2AZ and H2BV) concentrate in divergent SSRs, whereas variants of H3 and H4 are found near the probable pol II transcription termination sites of DGCs (Siegel *et al.*, 2009; Wright *et al.*, 2010). In *L. major*, acetylated histones H3 accumulate at all divergent SSRs (as well as other internal locations that are in turn predicted to be transcription start sites) in a manner that is more pronounced in dividing cells than in stationary-phase cells. Moreover, a remarkable enrichment of transcription factors TATA-binding protein (TBP) and SNAP50 has been also detected at these putative transcription start sites (Thomas *et al.*, 2009). In *Leishmania*, a modified thymidine, base J (β-D-glucosyl-hydroxymethyluracil), is located at chromosome-internal RNA polymerase II termination sites and promotes termination of transcription (van Luenen *et al.*, 2012; Reynolds *et al.*, 2016). This DNA modification, which has been found in the nuclear DNA of trypanosomatids and in the genera *Diplonema* and *Euglena* (van Leeuwen *et al.*, 1998; Dooijes *et al.*, 2000), is introduced enzymatically through the hydroxylation of thymidine, forming hydroxymethyluridine, followed by the transfer of a glucose by a specific glucosyltransferase (Bullard *et al.*, 2015).

However, currently, it is not clear where transcription precisely starts or how RNA pol II is recruited to the DNA. Furthermore, the principle that SSRs are the sole sites of transcription initiation is not absolute, since resistance genes introduced into a copy of *L. major* chromosome 1 lacking the divergent SSR can still be transcribed at levels sufficient for selection (Dubessay *et al.*, 2002a).

A polycistronic mode of transcription, as occurs in *Leishmania* and related trypanosomatids, does not leave much scope for regulation at the level of transcription initiation and explains why gene expression is mainly controlled post-transcriptionally in these organisms. The first task after polycistronic transcription is to separate the individual open reading frames into translatable mRNAs. In eukaryotes, crucial are the addition of a 5′-cap structure, which is made of modified bases and is usually put in place by the transcribing polymerase, RNA polymerase II, and the addition at the 3′-end of a poly(A) tail. Both modifications are required in an mRNA to be recognized as properly processed in order to be escorted to the cytosol; otherwise, the RNA may be held back in the nucleus and degraded (Moore, 2005).

In *Leishmania*, and related trypanosomatids, processing of the polycistronic transcripts to monocistronic RNAs and capping are accomplished by transcriptional *trans*-splicing of a capped, 39-nt RNA exon (named spliced leader -SL- or miniexon) to the 5′ untranslated region (5′UTR) and polyadenylation of the 3′UTR (Liang *et al.*, 2003). The mini-exon is transcribed by RNA polymerase II (pol II), yielding a ≈ 110–140-nt RNA donor (SL medRNA) for the *trans*-splicing reaction (Campbell *et al.*, 2000). The SL medRNA is encoded in an array of monocistronic tandem repeats located on chromosome 2; in contrast to the case for pol II-transcribed protein-coding genes, the SL genes are transcribed from bona fide pol II promoters (Gilinger and Bellofatto, 2001).The signal for *trans*-splicing consists of an invariant dinucleotide AG (3′-splice acceptor site) and an upstream polypyrimidine tract of variable length (Requena *et al.*, 2003). Interestingly, polyadenylation is inextricably coupled to *trans*-splicing, and contrary to most eukaryotes, polyadenylation signals have not been found in trypanosomatids. Instead, poly(A) site selection is conditioned by the position of the downstream splice acceptor site (LeBowitz *et al.*, 1993).

Taking into account that the two principal roles of promoters in other eukaryotes (capping and mRNA start site selection) are instead handled in trypanosomatids by the *trans*-splicing machinery, it is conceivable that promoters may be not necessary and possibly might not exist for pol II-transcribed protein-encoding genes. Nevertheless, the lack of transcriptional regulation in gene expression does not mean that these parasites have renounced to regulatory purposes; quite the contrary, the complexity of their life cycles makes differential gene expression of central importance to complete their biphasic developmental programmes. Therefore, their needs of controlling gene expression must be accomplished through downstream mechanisms that involve RNA and protein regulatory processes.

The requirement for the addition of the SL sequence to the polycistronic RNAs represents a potential point for modulation of mRNA levels between transcripts, since differences in the efficiency of *trans*-splicing would contribute to producing differences in steady state levels of specific mRNAs. Similarly, the polyadenylation process may also contribute to generating differences in expression for transcripts derived from the same polycistronic precursor. In fact, mutagenesis of the polypyrimidine tracts leads to the utilization of cryptic 3′-splice and polyadenylation sites (Matthews *et al.*, 1994). Additionally, for a given transcript, alternative *trans*-splicing events may occur, generating differences in the length of 5′-UTRs, but also in the protein-coding regions. The relevance of these processes for regulation of gene expression is currently unknown, but answers are likely to be forthcoming soon.

The heat shock response in *Leishmania*: a prototypical model for studying gene expression

For all living organisms, a moderate increase of just a few degrees above the optimum growth temperature represents a challenging problem for survival (Richter *et al.*, 2010). Temperature affects many physiological processes in the cell, as protein conformation is highly dependent on this physiochemical factor, and proteins tend to denature in response to temperature increases. As a survival mechanism, in response to heat stress, all the cells and organisms activate the so called heat shock response, which consists of several signalling pathways aimed to increase the expression of the heat shock proteins (HSPs; de Nadal *et al.*, 2011). The main role of HSPs (also known as molecular chaperones) is alleviating the deleterious effects of heat increase on the cellular homeostasis, by facilitating protein folding and preventing (or reversing) protein misfolding. In response to heat shock, gene expression changes quickly; hence, the heat shock response is being used as an amenable system to investigate gene expression mechanisms in different cells and organisms.

Changes in temperature occur regularly in the life cycle of *Leishmania*, when the parasite is transmitted between the insect vector and mammalian hosts, and accordingly temperature shift is considered one of the factors triggering differentiation from the promastigote form to the amastigote stage (Requena, 2012). However, the parasite also experiences abrupt changes in temperature in the poikilothermic vector, for example when it is being exposed directly to the sunlight. Therefore, it is conceivable that heat shock response is being used as a suitable model for analysing gene expression in *Leishmania* (Schwede *et al.*, 2012; Requena, 2012). In most eukaryotes, the regulation of HSP synthesis is mediated by heat shock transcription factors (HSFs), which act as master regulators of stress-induced gene expression through binding to defined cis-acting heat shock elements (HSE) inside promoters or enhancers of HSP genes (Santoro, 2000). In *Leishmania*, there is no evidence for the existence of either HSE promoter elements or HSFs. This is not so surprising taking into account that, indeed, the entire gene expression in *Leishmania* is regulated mainly at the post-transcriptional level (Requena, 2011). Transcript stabilization and preferential translation of HSP mRNAs are the main mechanisms by which these parasites control HSP synthesis (Quijada *et al.*, 1997, 2000; Zilka *et al.*, 2001; Larreta *et al.*, 2004; Folgueira *et al.*, 2005; David *et al.*, 2010). By way of example, the regulation of HSP70 expression is illustrated in Fig. 5.1, and presented briefly. The genomic organization and expression of *HSP70* genes have extensively been studied in *L. infantum*. The *L. infantum* genome contains a single locus with six *HSP70* genes arranged tandemly in a head-to-tail manner (Quijada *et al.*, 1997). This gene organization was found to be conserved in other *Leishmania* species (Folgueira *et al.*, 2007; Ramírez *et al.*, 2011). All six *HSP70* genes have identical 5′-UTRs and coding sequences, the sole difference is the 3′-UTR of gene 6 (located at the 3′ end of the cluster; Fig. 5.1A) that is absolutely divergent in sequence relative to the 3′-UTR of the other five genes. Hence, for simplicity, genes 1 to 5 are referred as *HSP70-I* genes and gene 6 as *HSP70-II* gene (Folgueira *et al.*, 2005). Regarding transcripts levels, it is remarkable that nearly all the mRNAs come from gene 6, but only expression levels from *HSP70-I* genes are increased by heat shock (Fig. 5.1B; Quijada *et al.*, 1997). Subsequently, it was shown that the temperature-dependent accumulation of *HSP70-I* transcripts is associated with the presence of a *cis*-acting sequence in their 3′-UTR (Quijada *et al.*, 2000). Regarding translation, *HSP70-II* mRNAs are preferentially translated at heat shock temperatures, but not

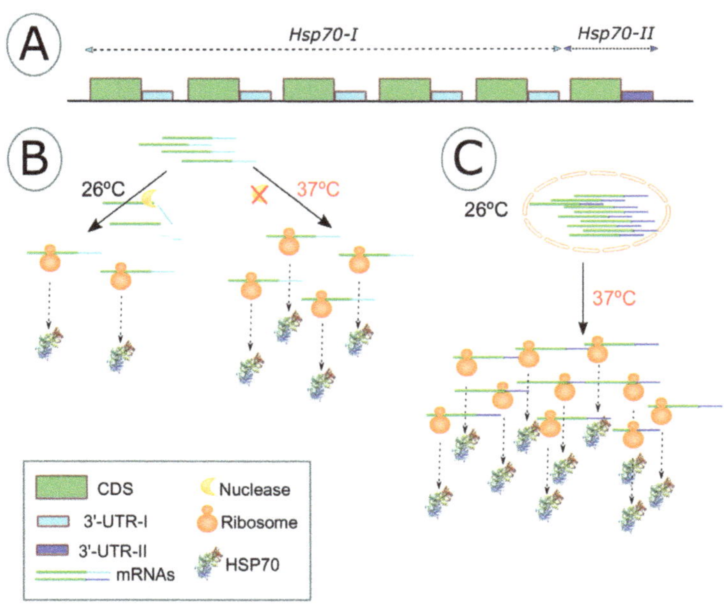

Figure 5.1 Regulation of *HSP70* gene expression in *Leishmania*. (A) Organization of the *HSP70* locus. Expression of (B) *HSP70-I* and (C) *HSP70-II* transcripts at normal (26°C) or heat shock (37°C) temperatures. See text for additional details.

at 26°C, whereas *HSP70-I* mRNAs are bound to polysomes at 26°C and 37°C (Fig. 5.1B; Folgueira *et al.*, 2005). This well-documented gene model illustrates how the expression of a gene may be controlled by variations in mRNA abundance, stability and translatability through differences in sequence elements residing in 3′-UTRs. Differential gene expression controlled by mechanisms affecting mRNA stability and translation has been documented in other *Leishmania* genes (Boucher *et al.*, 2002; Myung *et al.*, 2002; Soto *et al.*, 2004).

This post-transcriptional regulation relies on the interaction between trans-acting RNA-binding proteins and cis-acting RNA elements (Kramer and Carrington, 2011). In this regard, a putative master regulator of HSP mRNA stability has been described in *T. brucei*. Thus, the zinc finger protein ZC3H11 was identified as a specific binding factor for a common AUU repeat motif, which is found inside the 3′UTR of HSP70 and other genes coding for stress proteins (Droll *et al.*, 2013).

Another regulatory level of the stress response in *Leishmania* may operate through phosphorylation/dephosphorylation of HSPs and other proteins (Morales *et al.*, 2010). Thus, during the promastigote-to-amastigote differentiation, an increase of phosphorylated sites in HSP70 and HSP90 has been demonstrated (Morales *et al.*, 2008). These findings suggest that stress-response regulation in *Leishmania* would be done in part by protein kinases that regulate chaperone function (Späth *et al.*, 2015). Among the candidate kinases for regulating the stress response, it has been suggested the *Leishmania* PKR-homologue (named PERK), based on the observation that exposure of promastigotes to a combination of elevated temperature and acidic pH, key signals triggering amastigote differentiation, leads to a marked decrease in global translation initiation and a concomitant increase in the phosphorylation of the translation factor $eIF_{2\alpha}$ (Cloutier *et al.*, 2012). Moreover, impairing PERK leads to a defect in intracellular amastigote differentiation (Chow *et al.*, 2011), which provides a further link between *Leishmania* stress signalling and parasite differentiation. Nevertheless, other *Leishmania* kinases have also been postulated as putative master regulators of the heat shock response (Späth *et al.*, 2015).

Although important advances in the characterization of the response to stress in *Leishmania* have been achieved, little is known about the molecular details of the switch that triggers the heat shock response. In this way, a relevant finding was the observation that calcium uptake and calcineurin signalling are required for the development of parasite thermotolerance at 34–37°C (Naderer *et al.*, 2011).

Genomics and transcriptomics approaches

The Omics technologies are tools designed to gather information on the vast majority of constituents of a cell, a tissue, an organ or even the whole organism. They are aimed primarily at the universal detection of genes (genomics), mRNA (transcriptomics), proteins (proteomics) and metabolites (metabolomics) in a specific biological sample and in a non-biased manner. In this section and the following ones, brief descriptions on the foundations of the main Omics technologies, together with their contribution to increasing our knowledge on *Leishmania* biology, are provided.

Genomics is the study of the sequence and structure of the genome, paying a particular attention to gene organization. Transcriptomics deals with genome-wide identification and quantification of RNA species such as mRNAs, non-coding RNAs and small RNAs. Another Omic, closely related to genomics, is epigenomics, which study modifications and structural alterations of the chromatin, and their relationships with genome replication and transcription dynamics.

When genome-sequencing projects began to generate a wealth of information about gene content in *Leishmania*, and other organisms, DNA microarrays became a major high-throughput platform by which genome-wide analysis of gene expression could be explored at that time (Boothroyd *et al.*, 2003; Duncan, 2004). Serial Analysis of Gene Expression (SAGE) was another sequencing-based methodology used for transcriptomics studies (Velculescu *et al.*, 1995), but currently it has fallen into disuse.

Microarrays consist of ordered collections of thousands of nucleotide sequences (genomic DNA, cDNA, or oligonucleotides) set as microscopic spots on a glass slide (Boothroyd *et al.*, 2003; Duncan, 2004). With DNA microarrays, gene expression levels are measured as the amount of RNA in the sample that matches the set of probes fixed on the array. Either RNA molecules or more often their cDNA copies are fluorescently labelled and hybridized onto the array, where the intensity of the signal, measured for a given probe, is assumed to be proportional to the quantity of RNA present in the sample. The simultaneously use of two probes (labelled separately with Cy3- or Cy5-fluorescent dyes) allows for direct comparison between two environmental/physiological conditions, and the relative abundance of a given sequence under the two conditions can be deduced from the ratio of fluorescence intensity between both dyes. According to the nature of the probes, which are mainly derived from genomic DNA or cDNAs, synthesized from the RNA samples of interest, the application of this technology serves to address genomic or transcriptomic purposes, respectively.

The quality of the microarrays relies largely on the quality of the libraries, and the construction of them is the most cumbersome step. Thus, cDNA libraries have the problem that some genes are over-represented (those with a high level of expression) whereas others, corresponding to low expressed transcripts, may be completely absent. To avoid this problem of complexity, genomic libraries are a valid alternative in *Leishmania*, as the vast majority of genes lack introns and its genome has a remarkable gene density. However, a clear drawback of genomic libraries is that some segments can span two or more genes. For *Leishmania* species, whose genomes have been sequenced and annotated, these limitations were solved by using oligonucleotides that, furthermore, can be prepared and spotted directly onto the microarray slides.

In the last 10 years, an extraordinary progress has been made in sequencing technology. Currently, various sequencing platforms can generate simultaneously several millions of sequences with very affordable costs. These methodologies are often referred as next-generation sequencing (NGS); for readers interested in the technical details about the main types of NGS approaches, we suggest consulting an outstanding review published recently (Goodwin *et al.*, 2016). NGS platforms can be classified into two categories regarding the size of the sequences generated: short-reads (35–700 nucleotides) and long-reads (several hundreds of nucleotides). Within the category of short-read sequencing, NGS research is increasingly being conducted with Illumina instruments, which provide vast quantities of data; whereas the most widely used long-read platform is the single-molecule real-time (SMRT) sequencing approach (Eid *et al.*, 2009), developed by Pacific Biosciences (PacBio). Nevertheless, this is a field in continuous evolution and surely, in the next few years, novel sequencing solutions will emerge.

NGS can indeed be used for DNA or RNA sequence analyses. In particular, high-throughput sequencing of cDNA copies of mRNA (RNA-seq), because millions of sequences are generated, provides a digital read-out of mRNA levels over several orders of magnitude and allows for mapping of transcripts to the nucleotide level. Hence, RNA-seq has become particularly useful for quantitation of relative abundance of transcripts and for detection of polymorphisms, minor RNA species, and RNA editing. Currently, RNA-seq is replacing the use of microarrays for analysis of gene expression, because it has several advantages over microarrays: array design is not required, it has better resolution, and very low amounts of input material are used. Furthermore, microarrays can be limited by probe cross-reactivity, high background, and signal saturation. But depending on the array design, chips may provide rapid and high-quality data at lower cost. In fact, microarrays remain widely used in genomic research, such as allelic polymorphism determinations, pathogen identification and measuring of expression levels for selected genes. For example, DNA microarrays have been developed for simultaneous identification of 18 species of common blood protozoa affecting humans, among them *Leishmania* (Chen *et al.*, 2016).

Translation of the vast amount of genetic data into biological contexts is currently a bottleneck; mapping the vast amount of short read sequences (Martin and Wang, 2011) and assessing the significance of differences in reads coverage are two relevant questions to address. In addition, well-validated experimental samples would be a requisite, before proceeding with NGS data collection. Finally, powerful databases containing this plethora of data, continuously organized and updated, are crucial. Currently, TriTrypDB (http://tritrypdb.org) is the most relevant database aimed to provide access to genome-scale datasets for many kinetoplastid parasites (Aslett *et al.*, 2010). This database forms part of the platform Eukaryotic Pathogen Bioinformatics Resource Centre (EuPathDB; Aurrecoechea *et al.*, 2013).

Leishmania transcriptomics

Just before the completion of the genome sequence for a *Leishmania* species, DNA microarrays emerged as a powerful method to study global gene expression in terms of quantitation of mRNA levels (see previous section for details on microarrays foundations). The initial studies dealt with the use of whole-genome shotgun DNA microarrays (Saxena *et al.*, 2003; Akopyants *et al.*, 2004), but the completion of the genome sequencing of *L. major* (Ivens *et al.*, 2005), *L. infantum* and *L. braziliensis* (Peacock *et al.*, 2007) allowed the generation of high-density oligonucleotide microarrays suitable for genome-wide expression profiling analyses (Rochette *et al.*, 2008). Microarrays monitoring gene expression have been used for the study of various aspects of *Leishmania* biology: metacyclogenesis (Saxena *et al.*, 2003; Alcolea *et al.*, 2009), promastigote-to-amastigote differentiation (Holzer *et al.*, 2006; McNicoll *et al.*, 2006; Leifso *et al.*, 2007; Srividya *et al.*, 2007; Rochette *et al.*, 2009; Alcolea *et al.*, 2010; Lahav *et al.*, 2011), sandfly-derived promastigotes (Alcolea *et al.*, 2014, 2016a), species-specific gene expression (Depledge *et al.*, 2009), intraspecific heterogenicity of clinical isolates (Adaui *et al.*, 2011), resistance to antileishmanial drugs (Guimond *et al.*, 2003; Leprohon *et al.*, 2009; Singh *et al.*, 2010), and response to stress conditions imposed by heavy metals (Alcolea *et al.*, 2011a). On the other hand, microarrays have also been used for monitoring the expression of host genes being altered by *Leishmania* infection (Novais *et al.*, 2015).

The application of high-throughput RNA sequencing (RNA-seq) to the study of *Leishmania* parasites has represented a milestone in the knowledge of their transcriptome, both regarding the real number of genes and their primary structure. The first transcriptome for a species of the genus *Leishmania* was reported by Rastrojo *et al* (2013). These authors documented the existence of 10,285 different transcripts, of which 1884 were considered novel, as they did not match previously annotated genes. In addition, the annotations for around 500 ORFs were corrected based on the location of SL-addition sites. In 2015, Fiebig *et al* (2015), using also RNA-seq, defined the transcriptome for *L. mexicana*; a total of 9169 protein-coding transcripts, including 936 corresponding to previously non-annotated genes, were identified.

Apart from its usefulness for establishing the gene models, RNA-seq methodology is an excellent tool for quantitation of relative expression levels for each one of the transcripts composing the *Leishmania* genome (Rastrojo *et al.*, 2013). Also, RNA-seq has begun to be applied to the measurement of mRNA transcript abundance in some developmental stages of *Leishmania*. Thus, Dillon *et al.* (2015b) used RNA-seq to identify global changes in gene expression occurring as *L. major* undergoes metacyclogenesis, a developmental progression that is well mimicked *in vitro* using reliable axenic cultivation methods (Sacks and Perkins, 1984). Remarkably, this analysis identified 3138 genes that were expressed at significantly different levels between procyclic and metacyclic promastigotes (Dillon *et al.*, 2015b). The transcriptomes of *L. mexicana* promastigotes, axenic amastigotes and intracellular amastigotes have also been compared by RNA-seq (Fiebig *et al.*, 2015). More than one-third of all genes were found to be differentially expressed between promastigotes and intracellular amastigotes.

RNA-seq methodologies have also been used for analysing the complete repertoire of *L. major* small nucleolar RNAs (snoRNAs), which are the molecules guiding modifications by methylation and pseudouridylation on rRNAs (Eliaz *et al.*, 2015). rRNA processing events in trypanosomatids are unique, as the large subunit rRNA undergoes trypanosome-specific cleavages during rRNA maturation, yielding two large rRNA molecules and four small ones (Hernández and Cevallos, 2014).

RNA-seq has also been used to simultaneously identify global transcriptional changes in host cells and *L. major* parasites during the entry into and persistence within murine macrophages (Dillon *et al.*, 2015a). Similar studies were carried out using human macrophages, and the infection with two different *Leishmania* species, *L. amazonensis* and *L. major* (Fernandes *et al.*, 2016). No significant differences were observed between the two *Leishmania* species transcriptomes or in the transcriptional response of human macrophages infected with each species. Nevertheless, depending on the host cells (human or mouse macrophages), remarkable differences in the *L. major* transcriptome were observed (Fernandes *et al.*, 2016). More recently, host and parasite gene expression in skin biopsies from *L. braziliensis*-infected patients were simultaneously analysed using RNA-seq (Christensen *et al.*, 2016). It was evidenced a clear uniformity in the parasite gene expression in all the patient biopsies analysed. Remarkably, 4579 host genes were differentially expressed in patient's skin lesions relative to the expression occurring in normal skin. Again, the most abundant parasite transcripts expressed in lesions were distinct from transcripts expressed *in vitro* in human macrophage cultures infected with *L. amazonensis* or *L. major* (Christensen *et al.*, 2016). Additionally, RNA-seq approaches have been used to uncover microRNAs (short single-stranded ribonucleic acids that act blocking mRNA translation) that are specifically expressed in human monocyte-derived dendritic cells and macrophages after infection with *Leishmania* parasites (Geraci *et al.*, 2015).

Determination of the mRNA levels existing in a cell at a given moment does not allow for directly inferring the abundance of the encoded protein, as the latter also depends on other factors like nuclear-cytoplasmic transport, transcript-specific silencing mechanisms and mRNA translational efficiency. However, technologies for assessing the relative abundance of all proteins existing in a cell is not currently an achievable goal. A recently developed technique, ribosome profiling, has emerged as a valuable approach to filling this gap, as it allows inferring global protein production via sequencing small mRNA fragments protected by the assembled ribosomes (Ingolia *et al.*, 2012). This approach has only recently been applied to the identification of actively translated transcripts in protozoan parasites *T. brucei*, *T. cruzi*, and *Plasmodium falciparum* (reviewed in Parsons and Myler, 2016), and in *L. major* (Dillon *et al.*, 2015a).

Application of whole-genome sequencing to *Leishmania* studies

Currently, whole genome sequencing (WGS) is becoming one of the most widely used applications of NGS methodologies. Through these platforms, researchers can obtain the most comprehensive view of genomic information. Using these methodologies, an increasing number of genomic sequences for *Leishmania* species is being reported (Downing *et al.*, 2011; Rogers *et al.*, 2011; Raymond *et al.*, 2012; Real *et al.*, 2013; Gupta *et al.*, 2015; Llanes *et al.*, 2015; see also the TriTryp database at www.tritryp.org).

The relevance of WGS for epidemiological purposes has been highlighted recently in an article in which genome sequences from 204 clinical isolates have been determined and used to track the evolution and epidemiology of *L. donovani* in the Indian subcontinent (Imamura *et al.*, 2016). In addition, this approach results crucial to understand how drug resistance emerges and spreads among parasite populations. These studies have furtherly demonstrated the plasticity of *Leishmania* genomes that allows these parasites to develop resistance to drugs by increasing the copy number of particular genes.

Leishmania proteomics

Overview of proteomics methodologies

The term proteome was coined to refer to the protein complement encoded in a given genome, and the process of studying the proteome is known as proteomics. The experimental objective of proteomics approaches is to detect and identify in a reproducible and robust quantitative manner as many proteins in a single sample as possible. The development of mass spectrometry (MS), which was initially an analytical technique to determine mass to charge (m/z) ratios of chemical compounds (reviewed in Aebersold and Mann, 2003), resulted pivotal for proteomics analyses. Mass spectrometry is widely used to study proteins (proteomics) and also metabolites (metabolomics, see next section). Several methods of sample analysis and types of spectrometers exist, but their description falls out of the scope of this chapter. Detailed information can be obtained elsewhere (Dunn, 2011).

Proteomics aims at defining all of the proteins present in a cell, a tissue, or an organism (or any other biological compartment) and employs large-scale, high-throughput techniques dealing with protein content, modifications, subcellular localization, and protein interactions. Initial proteomic studies were based on a combination of protein fractionation using either one- or two-dimensional gel electrophoresis (2-DE) and MS determination of peptides derived from trypsin digestion of selected protein spots. Firstly, total protein extracts and/or subcellular fractions are separated by isoelectric focusing and afterwards in a second dimension by SDS-PAGE. For comparative purposes, images of stained gels for the different protein extracts are superimposed, spots are matched between the different gels and their intensity quantitated using specialized software. This critical and error-prone analytical step must be meticulously conducted. Many of these difficulties disappeared after the introduction of the difference gel electrophoresis (DIGE) technique (Tannu and Hemby, 2006). Although DIGE technology also relies on 2-DE electrophoresis, up to three different protein samples can be analysed in the same gel. Prior to 2-DE, protein samples are separately labelled with size-matched, charge-matched spectrally resolvable fluorescent dyes (cyanine derivatives Cy2, Cy3 and Cy5). Once spots of interest are selected and excised from 2-DE gels, proteins are fragmented into peptides for subsequent MS identification. Trypsin is the most common protease used to produce peptides, but other proteases can be also used. Peptides in the digested samples can be separated by HPLC prior to identification by tandem MS. In the first MS round, peptides are ionized and separated producing mass spectra with peaks corresponding to peptides (peptide fingerprint). Then, peptides are sequenced in the second round (fragmentation spectra).

Quantitative proteomics approaches based on 2DE-MS/MS are being replaced by shotgun proteomics, which relies exclusively on MS analysis because the former are cumbersome and expensive, especially DIGE. Currently, complex protein samples are separated by means of liquid chromatography (LC) or liquid-phase isoelectric focusing (L-IEF). These separation methods precede qualitative and quantitative analysis by MS. Liquid chromatographers are coupled to mass spectrometers in many cases, but large samples may require additional separation by L-IEF or an alternative method prior to LC-MS. This will depend on the genome size of the organism of interest and the desired coverage. Once samples have been run, a complex bioinformatics analysis is required for peptide identification, assembly of polypeptides and protein database search. Additionally, for quantification purposes, differential isotopic labelling is used; hence, different samples may be mixed and analysed together in the same run. Among the labelling methods is the Stable Isotope-Labelling by Amino acids in Culture (SILAC) approach (Ong *et al.*, 2002), which is based on the growth of cell cultures in the presence of ^{13}C6-arginine, ^{15}N4-arginine, ^{13}C6-lysine or another labelled amino acid. Then, the samples are mixed, digested and analysed by means of LC-MS/MS. Another labelling method is Isotope-coded Affinity Tag (ICAT); this technique only allows comparison of two samples because proteins are labelled on their cysteine sulfhydryl groups after protein extraction and before digestion (Smolka *et al.*, 2001). The ICAT reagent consists of an iodoacetylamide residue coupled to a linker region (light or heavy) containing a biotin residue for avidin-affinity LC coupled to MS. In this regard, Isobaric Tag for Relative and Absolute Quantitation (iTRAQ) is more advantageous because samples are labelled after digestion and up to eight samples can be pooled and simultaneously compared (Phanstiel *et al.*, 2009).

Finally, protein microarrays based on parasite proteins can be also considered a proteomic approach as they offer the potential for a high-throughput screen to determining either protein interactomes or host antibody responses when using sera from patients suffering parasitic infections (Davies *et al.*, 2005).

Proteomics studies in *Leishmania*

Although the protein-coding genes in *Leishmania* are regulated at the post-transcriptional level, mRNA abundance may not necessarily reflect protein levels, and methodologies to globally analyse gene expression at the protein level are particularly relevant for understanding the molecular dynamics in this organism. Thus, proteomics analyses are being conducted to understand different aspects of *Leishmania* biology (Cuervo *et al.*, 2010; Paape and Aebischer, 2011; Tsigankov *et al.*, 2012), such as stage differentiation, but also to study aspects of the host that result altered during the infection and disease processes (Veras and Bezerra de Menezes, 2016). Initial proteomics approaches were based on visual comparison of two-dimensional gel electrophoresis (2-DE) of proteins, followed by liquid chromatography – tandem mass spectrometry (LC-MS/MS) of the selected protein spots (Acestor *et al.*, 2002; El Fakhry *et al.*, 2002; Bente *et al.*, 2003; Drummelsmith *et al.*, 2003). When genome annotations for *Leishmania* species became available, larger numbers of proteins could be identified; conversely, proteomic data can be used as a complementary approach (referred as proteogenomics) to conventional informatics annotation (Pawar *et al.*, 2014). Additionally, technological improvements like protein fractionation by isoelectric point using free-flow electrophoresis were introduced for better resolving in the 2-DE gels of the basic proteins, which tend to be poorly represented on classical 2-DE gels compared to the acidic ones (Brotherton *et al.*, 2010). In many of these studies, it was often observed that some particular proteins were present in multiple spots, suggesting, therefore, that post-translational modifications might be extensive in *Leishmania*. Remarkably, for several proteins, different isoforms appeared to be stage-specific (McNicoll *et al.*, 2006; Dea-Ayuela *et al.*, 2006; Gupta *et al.*, 2007; Vergnes *et al.*, 2007; Alcolea *et al.*, 2011b, 2016c; Coelho *et al.*, 2012).

The use of conventional 2-DE electrophoresis for analysing relative levels of proteins has a number of limitations, mainly derived from the need of densitometry analysis of the stained gels and superimposition of images from different 2-DE runs. Some methodological approaches are being used for solving these drawbacks (see the previous section). Thus, quantitative proteomic approaches based on Isobaric Tags for Relative and Absolute quantification (iTRAQ) coupled to high-resolution mass spectrometry are being used to identify differentially expressed proteins. This method is based on the tagging of primary amines present in tryptic-digested proteins; by using different tags, it is possible multiplexing up to eight distinct samples in a single experiment. Samples are then combined at an equal ratio and subjected to LC-MS/MS and, finally, relative abundance values of all peptides attributed to a specific protein are averaged to represent the relative abundance of the entire protein. This methodology has been used for quantification of protein changes during the promastigote-to-amastigote differentiation (Rosenzweig *et al.*, 2008b; Biyani and Madhubala, 2012; Lynn *et al.*, 2013), and also for detecting protein modifications (phosphorylation, methylation, acetylation, and glycosylation sites) throughout the differentiation process (Rosenzweig *et al.*, 2008a; Tsigankov *et al.*, 2014). Also, Sardar *et al.* (2013) used iTRAQ proteome analyses to assess the effect of oxidative and nitrosative stresses on protein abundance in *L. donovani* promastigotes.

Quantitative proteomics, using the isotope-coded affinity tag (ICAT) technology and mass spectrometry (MS) were used to identify *Leishmania* stage-specific, differentially expressed proteins (Leifso *et al.*, 2007). Nevertheless, new technical improvements like label-free quantification of proteins are greatly contributing to improve comparative proteomics (Wong and Cagney, 2010). This technique is relatively simple as it relies on computational procedures for recovering the quantitative information derived from the mass spectrometry data of peptides, obtained after trypsin enzymatic cleavage of protein extracts and subsequent separation through LC. This methodology has been recently used to identify differentially modulated proteins after hypericin treatment, an inhibitor of spermidine synthase (Singh and Dubey, 2016).

Additionally, studies aimed to identify phosphorylated proteins (phosphoproteomics) in *Leishmania* have been done by immobilized metal affinity chromatography (IMAC), separation by 2-DE, fluorescent staining and peptide identification by MS (Morales *et al.*, 2008). Analysis of the phosphoproteome has revealed that the overall phosphorylation pattern in *Leishmania* substantially changes during the differentiation process (Tsigankov *et al.*, 2014). In this regards, it should be noticed that *Leishmania*, and related trypanosomatids, possess a large set of protein kinases, comprising approximately 2% of the proteome (Parsons *et al.*, 2005). Overall, these findings suggest a key role of phosphorylation in regulating protein function and signalling pathways in these parasites. In fact, a few number of studies have been published, in which the relevance of some protein kinases begins to be envisaged. For example, temperature increase and pH decrease trigger activation of three mitogen-activated protein kinases (MAP4, 7 and 10), suggesting their involvement in signalling events during the *Leishmania* life cycle (Morales *et al.*, 2007). In another study, *L. donovani* promastigotes were engineered to express a mammalian protein tyrosine phosphatase, or were treated with inhibitors of protein tyrosine kinases. As a result, following both approaches, a partial differentiation from promastigotes to amastigotes was elicited (Nascimento *et al.*, 2003). However, we are still at the initial stages to understand the regulatory pathways involved. In fact, regulation of phosphorylation patterns and intracellular signalling cascades have not been characterized in these parasites to date, and there is no guarantee that any of the regulation pathways described in model organisms are present in these protists.

Other relevant phenotypic features of *Leishmania* that have been studied through proteomic approaches are: virulence factors (Magalhães *et al.*, 2014), antigenic proteins (Forgber *et al.*, 2006; Costa *et al.*, 2011; Coelho *et al.*, 2012), resistance to oxidative and nitrosative stresses (Alcolea *et al.*, 2016b,e). Also, proteomics techniques have been used to analyse composition of exosomes (Hassani and Olivier, 2013), glycosomes (Jamdhade *et al.*, 2015) and plasma membrane proteins (Yao *et al.*, 2010).

Metabolomics applied to the study of *Leishmania*

In order to fully understand the phenotypic characteristics of an organism, it is essential to complement genomic, transcriptomic and proteomic approaches with metabolomics studies. Metabolomics is the high-throughput characterization of the mixture of all metabolites in a biological system, both endogenous and exogenous small molecules. Metabolites are lipids, peptides, and amino, nucleic and organic acids. Metabolite levels are not easily predicted by protein or transcript levels due to the continuous flux and interconversions of metabolites occurring between the different metabolic pathways. Furthermore, particular aspects of the cellular physiology are likely impacting on both the steady state levels of metabolites and the metabolic dynamics, and consequently, measurements of the metabolome can be also used to detect subtle changes in a biological system. Metabolomics is now widely used in microbiology, nutrition, agriculture and environmental sciences, and in the clinical and pharmaceutical fields. Also, metabolomics has emerged as a key area in the study of *Leishmania* and related trypanosomatids (t'Kindt *et al.*, 2010b; Creek *et al.*, 2012; Saunders *et al.*, 2014).

The analysis of cellular metabolomes is complicated by the chemical diversity and wide range of concentrations of different metabolite classes and by the fact that there is no single analytical platform that provides complete coverage of all cellular metabolites. However, recent advances in both NMR and hyphenated mass spectrometry, in which liquid chromatography, gas chromatography and capillary electrophoresis (CE) are interfaced with mass spectrometry, are allowing to detect and quantitate a significant proportion of a given cellular metabolome in broad untargeted analyses, at least in relative terms (Scheltema *et al.*, 2010; Berg *et al.*, 2013b). The LTQ-orbitrap mass spectrometry is also used for metabolomics purposes (t'Kindt *et al.*, 2010a). The CE-ESI-TOF-MS metabolic fingerprinting has also been used as a complementary technique to LC in order to separate metabolites (Canuto *et al.*, 2012). Additionally, stable-isotope labelling approaches result appropriate for identification of metabolite fluxes and mapping metabolic pathways. This method, for instance, has been used to characterize the changes in metabolism occurring during the promastigote-to-amastigote differentiation in *L. mexicana*, and the results showed an increased dependence of the amastigote stage on hexose and mitochondrial metabolism (Saunders *et al.*, 2014).

Metabolic pathways databases began to be established years ago, and they are an inextricable complement to metabolomics analyses. Among the many metabolic pathways databases available, several have emerged as front runners, i.e. LeishCyc (Doyle *et al.*, 2009), KEGG (Kanehisa *et al.*, 2010; www.genome.jp/kegg/), and BioCyc (Caspi *et al.*, 2010; http://metacyc.org/). Metabolic reconstructions of *L. major*, *L. infantum* and *L. braziliensis* are all available at KEGG. Network theory applied to metabolism has been the subject of a comprehensive recent review (Lacroix *et al.*, 2008). Various web servers, such as MetExplore (Cottret *et al.*, 2010) and MetaboAnalyst 2.0 (www.Metaboanalyst.ca; Xia and Wishart, 2016), are available for metabolomic data analysis and interpretation.

To date, only a few metabolomics studies of *Leishmania* have been published. These have been directed towards the analysis of changes in metabolites associated with the differentiation process in some *Leishmania* species (Silva *et al.*, 2011; Arjmand *et al.*, 2016). Interestingly, using heavy water labelling of infected BALB/c mice and metabolomics determinations, it has been possible to estimate *Leishmania* metabolic activity and growth rate *in vivo* (Kloehn *et al.*, 2015). Also, metabolomics studies have been used for the detection of metabolic differences that may correlate to drug resistance in parasite strains from clinical isolates (Scheltema *et al.*, 2010; Shaw *et al.*, 2016), for analysing the mode of drug action (Berg *et al.*, 2013a, 2015; Vincent *et al.*, 2014; Rojo *et al.*, 2015), and for the detection of possible new drug targets using *in vitro* studies (Cunha-Júnior *et al.*, 2017). Additionally, metabolomics approaches have been found useful for identification of *Leishmania* species (Westrop *et al.*, 2015). Complementary analyses based on the identification of genes coding for metabolism enzymes serve for drawing up particular biosynthetic routes (Opperdoes *et al.*, 2016). Additionally, integration of proteomics and metabolomics data has been done in order to elucidate metabolic adaptation in *Leishmania* mutant parasites (Akpunarlieva *et al.*, 2017), and integration of genomics and metabolomics analysis for elucidating differences in virulence (Alves-Ferreira *et al.*, 2015). Obviously, metabolomics approaches are also well suited for analysing relevant pathways, such as polyamine metabolism (Castilho-Martins *et al.*, 2015).

Future trends

Whole-genome sequencing is becoming every day more accessible and affordable, and soon it will be an essential tool for investigating the evolution and epidemiology of *Leishmania* and all other human infectious agents. Moreover, parallel sequencing of different strains, representative of the clinical and phenotypic diversity of a given *Leishmania* species, is revealing genome-wide sequence differences in pathogen populations (Imamura *et al.*, 2016) that might explain the phenotypic variation of the infection outcomes. Insights derived from those genomic data should aid to leishmaniasis control in disease-burdened regions as well as facilitate drug resistance diagnosis and vaccine development.

A recurrent theme emerging from some people working in the field of gene expression in *Leishmania* and related trypanosomatids is that transcriptomics data may have a marginal value for quantitation of protein expression as mRNA levels are often poorly correlated with protein or enzymatic activity. This represents a simplistic view, since every gene-encoded protein is expressed through the translation of the corresponding mRNA. Therefore, the mRNAs are central players for gene expression, which is achieved by regulation one or more of the following aspects: successful maturation of pre-mRNA (*trans*-splicing and polyadenylation, in *Leishmania*), export from the nucleus, cytoplasmic mRNA half-life, the fraction of mRNA in translation and the efficiency of that translation. In this regards, transcripts might be sequestered into ribonucleoprotein particles (or RNA granules) that ultimately control their availability to the translation machinery, storage or degradation in response to different stresses (i.e. temperature, starvation, oxidative stress) as shown in *T. brucei* (Fernández-Moya *et al.*, 2012; Kramer *et al.*, 2012, 2013). The regulation of these aspects is mediated by *cis*-elements mainly located at 5′ and 3′ untranslated region (UTR) sequences in concert with *trans*-acting protein factors. Thus, the 5′- and 3′-UTR should be considered, together with the coding regions, when defining gene models (Rastrojo *et al.*, 2013). Additionally, RNA-binding proteins (RBPs), which are responsible for the rapid gene expression remodelling that occurs during cell differentiation in *Leishmania*, and the identification of the cohorts of mRNAs they bind, must be the objects of extensive studies in the forthcoming years (Alves and Goldenberg, 2016; Kramer, 2014).

Another of the most exciting advances to look forward to is the ability to integrate data across genomics, transcriptomics, proteomics, metabolomics and perhaps other broad approaches. Pioneering studies, using this integrative approach, are being produced (Vacchina *et al.*, 2016). Current high-throughput technologies produce very large data sets and have shifted the bottleneck from data production to data analysis. Thus, great efforts have to be invested in the training of a new generation of scientists in the fields of bioinformatics and systems biology.

Although we are still faced with inadequate chemotherapy and no effective vaccines, the information that is emerging from the -omics techniques promise to alleviate this problem in the near future. For *Leishmania*, this means identifying the genes that are expressed in forms of the parasite that invade and survive in the vertebrate host. This is the case of the recently described DNA polymerase theta from *L. infantum*, which increases the infection ability of the parasite by repairing the DNA damaged by the oxidative stress that the parasite experience inside the phagolysosomal vacuole of macrophages (Fernández-Orgiler *et al.*, 2016). Similarly, metabolite profiling of both the parasite and the mammalian host during the course of infection will lead to the identification of novel diagnostic biomarkers, as well as the development of new strategies for controlling leishmaniasis.

Currently, genetic manipulation of *Leishmania* is a difficult and challenging task, mainly if essential genes are targeted (Späth and Clos, 2016). Fortunately, into the two last years, novel and very useful tools for the phenotypic analysis of essential genes have been developed. Firstly, the CRISPR/Cas9 system, which has been shown to be a powerful tool for genome editing in *Leishmania* (Sollelis *et al.*, 2015; Zhang and Matlashewski, 2015). More recently, it has been implemented an inducible gene deletion system based on a dimerized Cre recombinase (diCre) that allows the targeting of essential genes (Duncan *et al.*, 2016). Thus, these new methods will serve to directly assess whether a gene is essential to parasite viability, and therefore they will become useful tools for obtaining novel insights into the function of essential genes in *Leishmania*.

Acknowledgements

This field is already too expansive for a comprehensive, single review; thus we apologize in advance for the many omissions. Research activity is supported by grants from Proyecto del Ministerio de Economía y Competitividad (SAF2013-47556-R, co-financed with FEDER funds), and the Fondo de Investigaciones Sanitarias (ISCIII-RETIC RD12/0018/0009-FEDER). An institutional grant from the Fundación Ramón Areces is also acknowledged.

References

Acestor, N., Masina, S., Walker, J., Saravia, N.G., Fasel, N., and Quadroni, M. (2002). Establishing two-dimensional gels for the analysis of *Leishmania* proteomes. Proteomics *2*, 877–879. https://doi.org/10.1002/1615-9861(200207)2:7<877::AID-PROT877>3.0.CO;2-D.

Adaui, V., Castillo, D., Zimic, M., Gutierrez, A., Decuypere, S., Vanaerschot, M., De Doncker, S., Schnorbusch, K., Maes, I., Van der Auwera, G., *et al.* (2011). Comparative gene expression analysis throughout the life cycle of *Leishmania braziliensis*: diversity of expression profiles among clinical isolates. PLOS Negl. Trop. Dis. *5*, e1021. https://doi.org/10.1371/journal.pntd.0001021.

Aebersold, R., and Mann, M. (2003). Mass spectrometry-based proteomics. Nature *422*, 198–207. https://doi.org/10.1038/nature01511.

Akhoundi, M., Kuhls, K., Cannet, A., Votýpka, J., Marty, P., Delaunay, P., and Sereno, D. (2016). A Historical Overview of the Classification, Evolution, and Dispersion of *Leishmania* Parasites and Sandflies. PLOS Negl. Trop. Dis. *10*, e0004349. https://doi.org/10.1371/journal.pntd.0004349.

Akopyants, N.S., Matlib, R.S., Bukanova, E.N., Smeds, M.R., Brownstein, B.H., Stormo, G.D., and Beverley, S.M. (2004). Expression profiling using random genomic DNA microarrays identifies differentially expressed genes associated with three major developmental stages of the protozoan parasite *Leishmania major*. Mol. Biochem. Parasitol. *136*, 71–86. https://doi.org/10.1016/j.molbiopara.2004.03.002.

Akpunarlieva, S., Weidt, S., Lamasudin, D., Naula, C., Henderson, D., Barrett, M., Burgess, K., and Burchmore, R. (2017). Integration of proteomics and metabolomics to elucidate metabolic adaptation in *Leishmania*. J Proteomics *155*, 85–98. https://10.1016/j.jprot.2016.12.009.

Aksoy, S. (1991). Site-specific retrotransposons of the trypanosomatid protozoa. Parasitol. Today *7*, 281–285.

Alcolea, P.J., Alonso, A., Sánchez-Gorostiaga, A., Moreno-Paz, M., Gómez, M.J., Ramos, I., Parro, V., and Larraga, V. (2009). Genome-wide analysis reveals increased levels of transcripts related with infectivity in peanut lectin non-agglutinated promastigotes of *Leishmania infantum*. Genomics *93*, 551–564. https://doi.org/10.1016/j.ygeno.2009.01.007.

Alcolea, P.J., Alonso, A., Gómez, M.J., Moreno, I., Domínguez, M., Parro, V., and Larraga, V. (2010). Transcriptomics throughout the life cycle of *Leishmania infantum*: high down-regulation rate in the amastigote stage. Int. J. Parasitol. *40*, 1497–1516. https://doi.org/10.1016/j.ijpara.2010.05.013.

Alcolea, P.J., Alonso, A., and Larraga, V. (2011a). Genome-wide gene expression profile induced by exposure to cadmium acetate in *Leishmania infantum* promastigotes. Int. Microbiol. *14*, 1–11.

Alcolea, P.J., Alonso, A., and Larraga, V. (2011b). Proteome profiling of *Leishmania infantum* promastigotes. J. Eukaryot. Microbiol. *58*, 352–358. https://doi.org/10.1111/j.1550-7408.2011.00549.x.

Alcolea, P.J., Alonso, A., Gómez, M.J., Postigo, M., Molina, R., Jiménez, M., and Larraga, V. (2014). Stage-specific differential gene expression in *Leishmania infantum*: from the foregut of *Phlebotomus perniciosus* to the human phagocyte. BMC Genomics *15*, 849. https://doi.org/10.1186/1471-2164-15-849.

Alcolea, P.J., Alonso, A., Domínguez, M., Parro, V., Jiménez, M., Molina, R., and Larraga, V. (2016a). Influence of the Microenvironment in the Transcriptome of *Leishmania infantum* Promastigotes: Sand Fly versus Culture. PLOS Negl. Trop. Dis. *10*, e0004693. https://doi.org/10.1371/journal.pntd.0004693.

Alcolea, P.J., Alonso, A., García-Tabares, F., Mena, M.C., Ciordia, S., and Larraga, V. (2016b). Increased Abundance of Proteins Involved in Resistance to Oxidative and Nitrosative Stress at the Last Stages of Growth and Development of *Leishmania* amazonensis Promastigotes Revealed by Proteome Analysis. PLOS ONE *11*, e0164344. https://doi.org/10.1371/journal.pone.0164344.

Alcolea, P.J., Alonso, A., García-Tabares, F., Mena, M.C., Ciordia, S., and Larraga, V. (2016c). Proteome profiling of the growth phases of *Leishmania* pifanoi promastigotes in axenic culture reveals differential abundance of immunostimulatory proteins. Acta Trop. *158*, 240–247. https://doi.org/10.1016/j.actatropica.2016.03.015.

Alcolea, P.J., Alonso, A., Moreno-Izquierdo, M.A., Degayón, M.A., Moreno, I., and Larraga, V. (2016d). Serum Removal from Culture Induces Growth Arrest, Ploidy Alteration, Decrease in Infectivity and Differential Expression of Crucial Genes in *Leishmania infantum* Promastigotes. PLOS ONE *11*, e0150172. https://doi.org/10.1371/journal.pone.0150172.

Alcolea, P.J., Tuñón, G.I., Alonso, A., García-Tabares, F., Ciordia, S., Mena, M.C., Campos, R.N., Almeida, R.P., and Larraga, V. (2016e). Differential protein abundance in promastigotes of nitric oxide-sensitive and resistant *Leishmania chagasi* strains. Proteomics Clin Appl *10*, 1132–1146. https://doi.org/10.1002/prca.201600054.

Alonso, G., Rastrojo, A., López-Pérez, S., Requena, J.M., and Aguado, B. (2016). Resequencing and assembly of seven complex loci to improve the *Leishmania major* (Friedlin strain) reference genome. Parasit Vectors *9*, 74. https://doi.org/10.1186/s13071-016-1329-4.

Alvar, J., Vélez, I.D., Bern, C., Herrero, M., Desjeux, P., Cano, J., Jannin, J., and den Boer, M. (2012). Leishmaniasis worldwide and global estimates of its incidence. PLOS ONE *7*, e35671. https://doi.org/10.1371/journal.pone.0035671.

Alves-Ferreira, E.V., Toledo, J.S., De Oliveira, A.H., Ferreira, T.R., Ruy, P.C., Pinzan, C.F., Santos, R.F., Boaventura, V., Rojo, D., López-Gonzálvez, Á., *et al.* (2015). Differential Gene Expression and Infection Profiles of Cutaneous and Mucosal *Leishmania braziliensis* Isolates from the Same Patient. PLOS Negl. Trop. Dis. *9*, e0004018. https://doi.org/10.1371/journal.pntd.0004018.

Alves, L.R., and Goldenberg, S. (2016). RNA-binding proteins related to stress response and differentiation in protozoa. World J. Biol. Chem. *7*, 78–87. https://doi.org/10.4331/wjbc.v7.i1.78.

Arjmand, M., Madrakian, A., Khalili, G., Najafi Dastnaee, A., Zamani, Z., and Akbari, Z. (2016). Metabolomics-Based Study of Logarithmic and Stationary Phases of Promastigotes in *Leishmania major* by 1H NMR Spectroscopy. Iran. Biomed. J. *20*, 77–83.

Aslett, M., Aurrecoechea, C., Berriman, M., Brestelli, J., Brunk, B.P., Carrington, M., Depledge, D.P., Fischer, S., Gajria, B., Gao, X., *et al.* (2010). TriTrypDB: a functional genomic resource for the Trypanosomatidae. Nucleic Acids Res. *38*, D457–462. https://doi.org/10.1093/nar/gkp851.

Aurrecoechea, C., Barreto, A., Brestelli, J., Brunk, B.P., Cade, S., Doherty, R., Fischer, S., Gajria, B., Gao, X., Gingle, A., *et al.* (2013). EuPathDB: the eukaryotic pathogen database. Nucleic Acids Res. *41*, D684–691. https://doi.org/10.1093/nar/gks1113.

Balanco, J.M., Pral, E.M., da Silva, S., Bijovsky, A.T., Mortara, R.A., and Alfieri, S.C. (1998). Axenic cultivation and partial characterization of *Leishmania braziliensis* amastigote-like stages. Parasitology *116*, 103–113.

Bastien, P., Blaineau, C., and Pages, M. (1992). *Leishmania*: sex, lies and karyotype. Parasitol. Today *8*, 174–177.https://doi.org/10.1016/0169-4758(92)90016-U.

Belancio, V.P., Hedges, D.J., and Deininger, P. (2008). Mammalian non-LTR retrotransposons: for better or worse, in sickness and in health. Genome Res. *18*, 343–358. https://doi.org/10.1101/gr.5558208.

Bente, M., Harder, S., Wiesgigl, M., Heukeshoven, J., Gelhaus, C., Krause, E., Clos, J., and Bruchhaus, I. (2003). Developmentally induced changes of the proteome in the protozoan parasite *Leishmania donovani*. Proteomics 3, 1811–1829. https://doi.org/10.1002/pmic.200300462.

Berg, M., García-Hernández, R., Cuypers, B., Vanaerschot, M., Manzano, J.I., Poveda, J.A., Ferragut, J.A., Castanys, S., Dujardin, J.C., and Gamarro, F. (2015). Experimental resistance to drug combinations in *Leishmania donovani*: metabolic and phenotypic adaptations. Antimicrob. Agents Chemother. 59, 2242–2255. https://doi.org/10.1128/AAC.04231-14.

Berg, M., Mannaert, A., Vanaerschot, M., Van Der Auwera, G., and Dujardin, J.C. (2013a). (Post-) Genomic approaches to tackle drug resistance in *Leishmania*. Parasitology *140*, 1492–1505. https://doi.org/10.1017/S0031182013000140.

Berg, M., Vanaerschot, M., Jankevics, A., Cuypers, B., Maes, I., Mukherjee, S., Khanal, B., Rijal, S., Roy, S., Opperdoes, F., *et al.* (2013b). Metabolic adaptations of *Leishmania donovani* in relation to differentiation, drug resistance, and drug pressure. Mol. Microbiol. *90*, 428–442. https://doi.org/10.1111/mmi.12374.

Berriman, M., Ghedin, E., Hertz-Fowler, C., Blandin, G., Renauld, H., Bartholomeu, D.C., Lennard, N.J., Caler, E., Hamlin, N.E., Haas, B., *et al.* (2005). The genome of the African trypanosome *Trypanosoma brucei*. Science *309*, 416–422.

Beverley, S.M. (1991). Gene amplification in *Leishmania*. Annu. Rev. Microbiol. *45*, 417–444. https://doi.org/10.1146/annurev.mi.45.100191.002221.

Biyani, N., and Madhubala, R. (2012). Quantitative proteomic profiling of the promastigotes and the intracellular amastigotes of *Leishmania donovani* isolates identifies novel proteins having a role in *Leishmania* differentiation and intracellular survival. Biochim. Biophys. Acta *1824*, 1342–1350. https://doi.org/10.1016/j.bbapap.2012.07.010.

Blaineau, C., Bastien, P., Rioux, J.A., Roizès, G., and Pagès, M. (1991). Long-range restriction maps of size-variable homologous chromosomes in *Leishmania infantum*. Mol. Biochem. Parasitol. *46*, 292–302.

Boothroyd, J.C., Blader, I., Cleary, M., and Singh, U. (2003). DNA microarrays in parasitology: strengths and limitations. Trends Parasitol. *19*, 470–476.

Boucher, N., Wu, Y., Dumas, C., Dube, M., Sereno, D., Breton, M., and Papadopoulou, B. (2002). A common mechanism of stage-regulated gene expression in *Leishmania* mediated by a conserved 3'-untranslated region element. J. Biol. Chem. *277*, 19511–19520. https://doi.org/10.1074/jbc.M200500200.

Brandau, S., Dresel, A., and Clos, J. (1995). High constitutive levels of heat-shock proteins in human-pathogenic parasites of the genus *Leishmania*. Biochem. J. *310*, 225–232.

Bringaud, F., García-Pérez, J.L., Heras, S.R., Ghedin, E., El-Sayed, N.M., Andersson, B., Baltz, T., and Lopez, M.C. (2002). Identification of non-autonomous non-LTR retrotransposons in the genome of *Trypanosoma cruzi*. Mol. Biochem. Parasitol. *124*, 73–78.

Bringaud, F., Ghedin, E., Blandin, G., Bartholomeu, D.C., Caler, E., Levin, M.J., Baltz, T., and El-Sayed, N.M. (2006). Evolution of non-LTR retrotransposons in the trypanosomatid genomes: *Leishmania major* has lost the active elements. Mol. Biochem. Parasitol. *145*, 158–170.

Bringaud, F., Ghedin, E., El-Sayed, N.M., and Papadopoulou, B. (2008). Role of transposable elements in trypanosomatids. Microbes Infect. *10*, 575–581. https://doi.org/10.1016/j.micinf.2008.02.009.

Bringaud, F., Müller, M., Cerqueira, G.C., Smith, M., Rochette, A., El-Sayed, N.M., Papadopoulou, B., and Ghedin, E. (2007). Members of a large retroposon family are determinants of post-transcriptional gene expression in *Leishmania*. PLOS Pathog. *3*, 1291–1307.

Britto, C., Ravel, C., Bastien, P., Blaineau, C., Pagès, M., Dedet, J.P., and Wincker, P. (1998). Conserved linkage groups associated with large-scale chromosomal rearrangements between Old World and New World *Leishmania* genomes. Gene *222*, 107–117.

Brotherton, M.C., Racine, G., Foucher, A.L., Drummelsmith, J., Papadopoulou, B., and Ouellette, M. (2010). Analysis of stage-specific expression of basic proteins in *Leishmania infantum*. J. Proteome Res. *9*, 3842–3853. https://doi.org/10.1021/pr100048m.

Bullard, W., Cliffe, L., Wang, P., Wang, Y., and Sabatini, R. (2015). Base J glucosyltransferase does not regulate the sequence specificity of J synthesis in trypanosomatid telomeric DNA. Mol. Biochem. Parasitol. *204*, 77–80. https://doi.org/10.1016/j.molbiopara.2016.01.005.

Callahan, H.L., Portal, A.C., Devereaux, R., and Grogl, M. (1997). An axenic amastigote system for drug screening. Antimicrob. Agents Chemother. *41*, 818–822.

Campbell, D.A., Sturm, N.R., and Yu, M.C. (2000). Transcription of the kinetoplastid spliced leader RNA gene. Parasitol. Today *16*, 78–82.

Cantacessi, C., Dantas-Torres, F., Nolan, M.J., and Otranto, D. (2015). The past, present, and future of *Leishmania* genomics and transcriptomics. Trends Parasitol. *31*, 100–108. https://doi.org/10.1016/j.pt.2014.12.012.

Canuto, G.A., Castilho-Martins, E.A., Tavares, M., López-Gonzálvez, A., Rivas, L., and Barbas, C. (2012). CE-ESI-MS metabolic fingerprinting of *Leishmania* resistance to antimony treatment. Electrophoresis *33*, 1901–1910. https://doi.org/10.1002/elps.201200007.

Carthew, R.W., and Sontheimer, E.J. (2009). Origins and Mechanisms of miRNAs and siRNAs. Cell *136*, 642–655. https://doi.org/10.1016/j.cell.2009.01.035.

Caspi, R., Altman, T., Dale, J.M., Dreher, K., Fulcher, C.A., Gilham, F., Kaipa, P., Karthikeyan, A.S., Kothari, A., Krummenacker, M., *et al.* (2010). The MetaCyc database of metabolic pathways and enzymes and the BioCyc collection of pathway/genome databases. Nucleic Acids Res. *38,* D473–479. https://doi.org/10.1093/nar/gkp875.

Castilho-Martins, E.A., Canuto, G.A., Muxel, S.M., da Silva, M.F., Floeter-Winter, L.M., Del Aguila, C., Lopez-Gonzalvez, A., and Barbas, C. (2015). Capillary electrophoresis reveals polyamine metabolism modulation in *Leishmania* (*Leishmania*) *amazonensis* wild type and arginase knockout mutants under arginine starvation. Electrophoresis 36, 2314–2323. https://doi.org/10.1002/elps.201500114.

Castilla, J.J., Sanchez-Moreno, M., Mesa, C., and Osuna, A. (1995). *Leishmania donovani*: in vitro culture and [1H] NMR characterization of amastigote-like forms. Mol. Cell. Biochem. *142,* 89–97.

Chen, M.X., Ai, L., Chen, J.H., Feng, X.Y., Chen, S.H., Cai, Y.C., Lu, Y., Zhou, X.N., Chen, J.X., and Hu, W. (2016). DNA Microarray Detection of 18 Important Human Blood Protozoan Species. PLOS Negl. Trop. Dis. *10,* e0005160. https://doi.org/10.1371/journal.pntd.0005160.

Chow, C., Cloutier, S., Dumas, C., Chou, M.N., and Papadopoulou, B. (2011). Promastigote to amastigote differentiation of *Leishmania* is markedly delayed in the absence of PERK $eIF_{2\alpha}$lpha kinase-dependent $eIF_{2\alpha}$lpha phosphorylation. Cell. Microbiol. *13,* 1059–1077. https://doi.org/10.1111/j.1462-5822.2011.01602.x.

Christensen, S.M., Dillon, L.A., Carvalho, L.P., Passos, S., Novais, F.O., Hughitt, V.K., Beiting, D.P., Carvalho, E.M., Scott, P., El-Sayed, N.M., *et al.* (2016). Meta-transcriptome Profiling of the Human-*Leishmania braziliensis* Cutaneous Lesion. PLOS Negl. Trop. Dis. *10,* e0004992. https://doi.org/10.1371/journal.pntd.0004992.

Clayton, C.E. (2002). Life without transcriptional control? From fly to man and back again. EMBO J. *21,* 1881–1888. https://doi.org/10.1093/emboj/21.8.1881.

Cloutier, S., Laverdière, M., Chou, M.N., Boilard, N., Chow, C., and Papadopoulou, B. (2012). Translational control through eIF2alpha phosphorylation during the *Leishmania* differentiation process. PLOS ONE 7, e35085. https://doi.org/10.1371/journal.pone.0035085.

Coelho, V.T., Oliveira, J.S., Valadares, D.G., Chávez-Fumagalli, M.A., Duarte, M.C., Lage, P.S., Soto, M., Santoro, M.M., Tavares, C.A., Fernandes, A.P., *et al.* (2012). Identification of proteins in promastigote and amastigote-like *Leishmania* using an immunoproteomic approach. PLOS Negl. Trop. Dis. *6,* e1430. https://doi.org/10.1371/journal.pntd.0001430.

Costa, M.M., Andrade, H.M., Bartholomeu, D.C., Freitas, L.M., Pires, S.F., Chapeaurouge, A.D., Perales, J., Ferreira, A.T., Giusta, M.S., Melo, M.N., *et al.* (2011). Analysis of *Leishmania chagasi* by 2-D difference gel electrophoresis (2-D DIGE) and immunoproteomic: identification of novel candidate antigens for diagnostic tests and vaccine. J. Proteome Res. *10,* 2172–2184. https://doi.org/10.1021/pr101286y.

Cottret, L., Wildridge, D., Vinson, F., Barrett, M.P., Charles, H., Sagot, M.F., and Jourdan, F. (2010). MetExplore: a web server to link metabolomic experiments and genome-scale metabolic networks. Nucleic Acids Res. *38,* W132–137. https://doi.org/10.1093/nar/gkq312.

Creek, D.J., Anderson, J., McConville, M.J., and Barrett, M.P. (2012). Metabolomic analysis of trypanosomatid protozoa. Mol. Biochem. Parasitol. *181,* 73–84. https://doi.org/10.1016/j.molbiopara.2011.10.003.

Cruz, A., Coburn, C.M., and Beverley, S.M. (1991). Double targeted gene replacement for creating null mutants. Proc. Natl. Acad. Sci. U.S.A. *88,* 7170–7174.

Cruz, A.K., Titus, R., and Beverley, S.M. (1993). Plasticity in chromosome number and testing of essential genes in *Leishmania* by targeting. Proc. Natl. Acad. Sci. U.S.A. *90,* 1599–1603.

Cuervo, P., Domont, G.B., and De Jesus, J.B. (2010). Proteomics of trypanosomatids of human medical importance. J Proteomics 73, 845–867. https://doi.org/10.1016/j.jprot.2009.12.012.

Cunha-Júnior, E.F., Andrade-Neto, V.V., Lima, M.L., da Costa-Silva, T.A., Galisteo Junior, A.J., Abengózar, M.A., Barbas, C., Rivas, L., Almeida-Amaral, E.E., Tempone, A.G., *et al.* (2017). Cyclobenzaprine Raises ROS Levels in *Leishmania infantum* and Reduces Parasite Burden in Infected Mice. PLOS Negl. Trop. Dis. *11,* e0005281. https://doi.org/10.1371/journal.pntd.0005281.

da Silva, R., and Sacks, D.L. (1987). Metacyclogenesis is a major determinant of *Leishmania* promastigote virulence and attenuation. Infect. Immun. *55,* 2802–2806.

David, M., Gabdank, I., Ben-David, M., Zilka, A., Orr, I., Barash, D., and Shapira, M. (2010). Preferential translation of Hsp83 in *Leishmania* requires a thermosensitive polypyrimidine-rich element in the 3' UTR and involves scanning of the 5' UTR. RNA *16,* 364–374. https://doi.org/10.1261/rna.1874710.

Davies, D.H., Liang, X., Hernandez, J.E., Randall, A., Hirst, S., Mu, Y., Romero, K.M., Nguyen, T.T., Kalantari-Dehaghi, M., Crotty, S., *et al.* (2005). Profiling the humoral immune response to infection by using proteome microarrays: high-throughput vaccine and diagnostic antigen discovery. Proc. Natl. Acad. Sci. U.S.A. *102,* 547–552.

de Nadal, E., Ammerer, G., and Posas, F. (2011). Controlling gene expression in response to stress. Nat. Rev. Genet. *12*, 833–845. https://doi.org/10.1038/nrg3055.

De Pablos, L.M., Ferreira, T.R., and Walrad, P.B. (2016). Developmental differentiation in *Leishmania* lifecycle progression: post-transcriptional control conducts the orchestra. Curr. Opin. Microbiol. *34*, 82–89.

Dea-Ayuela, M.A., Rama-Iñiguez, S., and Bolás-Fernández, F. (2006). Proteomic analysis of antigens from *Leishmania infantum* promastigotes. Proteomics *6*, 4187–4194. https://doi.org/10.1002/pmic.200600101.

Depledge, D.P., Evans, K.J., Ivens, A.C., Aziz, N., Maroof, A., Kaye, P.M., and Smith, D.F. (2009). Comparative expression profiling of *Leishmania*: modulation in gene expression between species and in different host genetic backgrounds. PLOS Negl. Trop. Dis. *3*, e476. https://doi.org/10.1371/journal.pntd.0000476.

Dillon, L.A., Suresh, R., Okrah, K., Corrada Bravo, H., Mosser, D.M., and El-Sayed, N.M. (2015a). Simultaneous transcriptional profiling of *Leishmania major* and its murine macrophage host cell reveals insights into host-pathogen interactions. BMC Genomics *16*, 1108. https://doi.org/10.1186/s12864-015-2237-2.

Dillon, L.A., Okrah, K., Hughitt, V.K., Suresh, R., Li, Y., Fernandes, M.C., Belew, A.T., Corrada Bravo, H., Mosser, D.M., and El-Sayed, N.M. (2015b). Transcriptomic profiling of gene expression and RNA processing during *Leishmania major* differentiation. Nucleic Acids Res. *43*, 6799–6813. https://doi.org/10.1093/nar/gkv656.

Dooijes, D., Chaves, I., Kieft, R., Dirks-Mulder, A., Martin, W., and Borst, P. (2000). Base J originally found in kinetoplastida is also a minor constituent of nuclear DNA of *Euglena gracilis*. Nucleic Acids Res. *28*, 3017–3021.

Dostálová, A., and Volf, P. (2012). *Leishmania* development in sand flies: parasite-vector interactions overview. Parasit. Vectors *5*, 276. https://doi.org/10.1186/1756-3305-5-276.

Downing, T., Imamura, H., Decuypere, S., Clark, T.G., Coombs, G.H., Cotton, J.A., Hilley, J.D., de Doncker, S., Maes, I., Mottram, J.C., *et al.* (2011). Whole genome sequencing of multiple *Leishmania donovani* clinical isolates provides insights into population structure and mechanisms of drug resistance. Genome Res. *21*, 2143–2156. https://doi.org/10.1101/gr.123430.111.

Doyle, M.A., MacRae, J.I., De Souza, D.P., Saunders, E.C., McConville, M.J., and Likić, V.A. (2009). LeishCyc: a biochemical pathways database for *Leishmania major*. BMC Syst. Biol. *3*, 57. https://doi.org/10.1186/1752-0509-3-57.

Droll, D., Minia, I., Fadda, A., Singh, A., Stewart, M., Queiroz, R., and Clayton, C. (2013). Post-transcriptional regulation of the trypanosome heat shock response by a zinc finger protein. PLOS Pathog. *9*, e1003286. https://doi.org/10.1371/journal.ppat.1003286.

Drummelsmith, J., Brochu, V., Girard, I., Messier, N., and Ouellette, M. (2003). Proteome mapping of the protozoan parasite *Leishmania* and application to the study of drug targets and resistance mechanisms. Mol. Cell Proteomics *2*, 146–155. https://doi.org/10.1074/mcp.M200085-MCP200.

Dubessay, P., Ravel, C., Bastien, P., Crobu, L., Dedet, J.P., Pagès, M., and Blaineau, C. (2002a). The switch region on *Leishmania major* chromosome 1 is not required for mitotic stability or gene expression, but appears to be essential. Nucleic Acids Res. *30*, 3692–3697.

Dubessay, P., Ravel, C., Bastien, P., Stuart, K., Dedet, J.P., Blaineau, C., and Pagès, M. (2002b). Mitotic stability of a coding DNA sequence-free version of *Leishmania major* chromosome 1 generated by targeted chromosome fragmentation. Gene *289*, 151–159.

Dujardin, J.C., Campino, L., Cañavate, C., Dedet, J.P., Gradoni, L., Soteriadou, K., Mazeris, A., Ozbel, Y., and Boelaert, M. (2008). Spread of vector-borne diseases and neglect of Leishmaniasis, Europe. Emerging Infect. Dis. *14*, 1013–1018. https://doi.org/10.3201/eid1407.071589.

Dumas, C., Ouellette, M., Tovar, J., Cunningham, M.L., Fairlamb, A.H., Tamar, S., Olivier, M., and Papadopoulou, B. (1997). Disruption of the trypanothione reductase gene of *Leishmania* decreases its ability to survive oxidative stress in macrophages. EMBO J. *16*, 2590–2598. https://doi.org/10.1093/emboj/16.10.2590.

Duncan, R. (2004). DNA microarray analysis of protozoan parasite gene expression: outcomes correlate with mechanisms of regulation. Trends Parasitol. *20*, 211–215. https://doi.org/10.1016/j.pt.2004.02.008.

Duncan, S.M., Myburgh, E., Philipon, C., Brown, E., Meissner, M., Brewer, J., and Mottram, J.C. (2016). Conditional gene deletion with DiCre demonstrates an essential role for CRK3 in *Leishmania mexicana* cell cycle regulation. Mol. Microbiol. *100*, 931–944. https://doi.org/10.1111/mmi.13375.

Dunn, W.B. (2011). Mass spectrometry in systems biology an introduction. Meth. Enzymol. *500,* 15–35. https://doi.org/10.1016/B978-0-12-385118-5.00002-5.

Eid, J., Fehr, A., Gray, J., Luong, K., Lyle, J., Otto, G., Peluso, P., Rank, D., Baybayan, P., Bettman, B., *et al.* (2009). Real-time DNA sequencing from single polymerase molecules. Science *323,* 133–138. https://doi.org/10.1126/science.1162986.

El-Sayed, N.M., Myler, P.J., Bartholomeu, D.C., Nilsson, D., Aggarwal, G., Tran, A.N., Ghedin, E., Worthey, E.A., Delcher, A.L., Blandin, G., *et al.* (2005). The genome sequence of *Trypanosoma cruzi,* etiologic agent of Chagas disease. Science *309,* 409–415.

El Fakhry, Y., Ouellette, M., and Papadopoulou, B. (2002). A proteomic approach to identify developmentally regulated proteins in *Leishmania infantum.* Proteomics 2, 1007–1017. https://doi.org/10.1002/1615-9861(200208)2:8<1007::AID-PROT1007>3.0.CO;2-G.

Eliaz, D., Doniger, T., Tkacz, I.D., Biswas, V.K., Gupta, S.K., Kolev, N.G., Unger, R., Ullu, E., Tschudi, C., and Michaeli, S. (2015). Genome-wide analysis of small nucleolar RNAs of *Leishmania major* reveals a rich repertoire of RNAs involved in modification and processing of rRNA. RNA Biol *12,* 1222–1255. https://doi.org/10.1080/15476286.2015.1038019.

Fernandes, M.C., Dillon, L.A., Belew, A.T., Bravo, H.C., Mosser, D.M., and El-Sayed, N.M. (2016). Dual Transcriptome Profiling of *Leishmania*-Infected Human Macrophages Reveals Distinct Reprogramming Signatures. MBio 7, e00027-16. https://doi.org/10.1128/mBio.00027-16.

Fernández-Moya, S.M., García-Pérez, A., Kramer, S., Carrington, M., and Estévez, A.M. (2012). Alterations in DRBD3 ribonucleoprotein complexes in response to stress in *Trypanosoma brucei.* PLOS ONE 7, e48870. https://doi.org/10.1371/journal.pone.0048870.

Fernández-Orgiler, A., Martínez-Jiménez, M.I., Alonso, A., Alcolea, P.J., Requena, J.M., Thomas, M.C., Blanco, L., and Larraga, V. (2016). A putative *Leishmania* DNA polymerase theta protects the parasite against oxidative damage. Nucleic Acids Res. *44,* 4855–4870. https://doi.org/10.1093/nar/gkw346.

Fiebig, M., Kelly, S., and Gluenz, E. (2015). Comparative Life Cycle Transcriptomics Revises *Leishmania mexicana* Genome Annotation and Links a Chromosome Duplication with Parasitism of Vertebrates. PLOS Pathog. *11,* e1005186. https://doi.org/10.1371/journal.ppat.1005186.

Folgueira, C., Cañavate, C., Chicharro, C., and Requena, J.M. (2007). Genomic organization and expression of the HSP70 locus in New and Old World *Leishmania* species. Parasitology *134,* 369–377. https://doi.org/10.1017/S0031182006001570.

Folgueira, C., Quijada, L., Soto, M., Abanades, D.R., Alonso, C., and Requena, J.M. (2005). The translational efficiencies of the two *Leishmania infantum* HSP70 mRNAs, differing in their 3'-untranslated regions, are affected by shifts in the temperature of growth through different mechanisms. J. Biol. Chem. *280,* 35172–35183.

Folgueira, C., and Requena, J.M. (2007). A postgenomic view of the heat shock proteins in kinetoplastids. FEMS Microbiol. Rev. *31,* 359–377.

Forgber, M., Basu, R., Roychoudhury, K., Theinert, S., Roy, S., Sundar, S., and Walden, P. (2006). Mapping the antigenicity of the parasites in *Leishmania donovani* infection by proteome serology. PLOS ONE *1,* e40. https://doi.org/10.1371/journal.pone.0000040.

Genest, P.A., ter Riet, B., Dumas, C., Papadopoulou, B., van Luenen, H.G., and Borst, P. (2005). Formation of linear inverted repeat amplicons following targeting of an essential gene in *Leishmania.* Nucleic Acids Res. *33,* 1699–1709.

Genois, M.M., Plourde, M., Éthier, C., Roy, G., Poirier, G.G., Ouellette, M., and Masson, J.Y. (2015). Roles of Rad51 paralogs for promoting homologous recombination in *Leishmania infantum.* Nucleic Acids Res. *43,* 2701–2715. https://doi.org/10.1093/nar/gkv118.

Geraci, N.S., Tan, J.C., and McDowell, M.A. (2015). Characterization of microRNA expression profiles in *Leishmania*-infected human phagocytes. Parasite Immunol. *37,* 43–51. https://doi.org/10.1111/pim.12156.

Ghedin, E., Bringaud, F., Peterson, J., Myler, P., Berriman, M., Ivens, A., Andersson, B., Bontempi, E., Eisen, J., Angiuoli, S., *et al.* (2004). Gene synteny and evolution of genome architecture in trypanosomatids. Mol. Biochem. Parasitol. *134,* 183–191. https://doi.org/10.1016/j.molbiopara.2003.11.012.

Gilinger, G., and Bellofatto, V. (2001). Trypanosome spliced leader RNA genes contain the first identified RNA polymerase II gene promoter in these organisms. Nucleic Acids Res. *29,* 1556–1564.

Goodwin, S., McPherson, J.D., and McCombie, W.R. (2016). Coming of age: ten years of next-generation sequencing technologies. Nat. Rev. Genet. *17,* 333–351. https://doi.org/10.1038/nrg.2016.49.

Grondin, K., Papadopoulou, B., and Ouellette, M. (1993). Homologous recombination between direct repeat sequences yields P-glycoprotein containing amplicons in arsenite resistant *Leishmania.* Nucleic Acids Res. *21,* 1895–1901.

Grondin, K., Roy, G., and Ouellette, M. (1996). Formation of extrachromosomal circular amplicons with direct or inverted duplications in drug-resistant *Leishmania* tarentolae. Mol. Cell. Biol. *16*, 3587–3595.

Guimond, C., Trudel, N., Brochu, C., Marquis, N., El Fadili, A., Peytavi, R., Briand, G., Richard, D., Messier, N., Papadopoulou, B., *et al.* (2003). Modulation of gene expression in *Leishmania* drug resistant mutants as determined by targeted DNA microarrays. Nucleic Acids Res. *31*, 5886–5896.

Gupta, A.K., Srivastava, S., Singh, A., and Singh, S. (2015). De Novo Whole-Genome Sequence and Annotation of a *Leishmania* Strain Isolated from a Case of Post-Kala-Azar Dermal Leishmaniasis. Genome Announc. *3*, e00809–15. https://doi.org/10.1128/genomeA.00809-15.

Gupta, N., Goyal, N., and Rastogi, A.K. (2001). In vitro cultivation and characterization of axenic amastigotes of *Leishmania*. Trends Parasitol. *17*, 150–153.

Gupta, S.K., Sisodia, B.S., Sinha, S., Hajela, K., Naik, S., Shasany, A.K., and Dube, A. (2007). Proteomic approach for identification and characterization of novel immunostimulatory proteins from soluble antigens of *Leishmania donovani* promastigotes. Proteomics *7*, 816–823. https://doi.org/10.1002/pmic.200600725.

Hasan, G., Turner, M.J., and Cordingley, J.S. (1984). Complete nucleotide sequence of an unusual mobile element from *Trypanosoma brucei*. Cell *37*, 333–341.

Hassani, K., and Olivier, M. (2013). Immunomodulatory impact of leishmania-induced macrophage exosomes: a comparative proteomic and functional analysis. PLOS Negl. Trop. Dis. *7*, e2185. https://doi.org/10.1371/journal.pntd.0002185.

Heras, S.R., López, M.C., Olivares, M., and Thomas, M.C. (2007). The L1Tc non-LTR retrotransposon of *Trypanosoma cruzi* contains an internal RNA-pol II-dependent promoter that strongly activates gene transcription and generates unspliced transcripts. Nucleic Acids Res. *35*, 2199–2214.

Hernández, R., and Cevallos, A.M. (2014). Ribosomal RNA gene transcription in trypanosomes. Parasitol. Res. *113*, 2415–2424. https://doi.org/10.1007/s00436-014-3940-7.

Herrador, Z., Gherasim, A., Jimenez, B.C., Granados, M., Granados, M., San Martín, J.V., and Aparicio, P. (2015). Epidemiological changes in leishmaniasis in Spain according to hospitalization-based records, 1997-2011: raising awareness towards leishmaniasis in non-HIV patients. PLOS Negl. Trop. Dis. *9*, e0003594. https://doi.org/10.1371/journal.pntd.0003594.

Holzer, T.R., McMaster, W.R., and Forney, J.D. (2006). Expression profiling by whole-genome interspecies microarray hybridization reveals differential gene expression in procyclic promastigotes, lesion-derived amastigotes, and axenic amastigotes in *Leishmania mexicana*. Mol. Biochem. Parasitol. *146*, 198–218.

Ilgoutz, S.C., Zawadzki, J.L., Ralton, J.E., and McConville, M.J. (1999). Evidence that free GPI glycolipids are essential for growth of *Leishmania mexicana*. EMBO J. *18*, 2746–2755. https://doi.org/10.1093/emboj/18.10.2746.

Imamura, H., Downing, T., Van den Broeck, F., Sanders, M.J., Rijal, S., Sundar, S., Mannaert, A., Vanaerschot, M., Berg, M., De Muylder, G., *et al.* (2016). Evolutionary genomics of epidemic visceral leishmaniasis in the Indian subcontinent. Elife *5*, e12613. https://doi.org/10.7554/eLife.12613.

Ingolia, N.T., Brar, G.A., Rouskin, S., McGeachy, A.M., and Weissman, J.S. (2012). The ribosome profiling strategy for monitoring translation in vivo by deep sequencing of ribosome-protected mRNA fragments. Nat Protoc *7*, 1534–1550. https://doi.org/10.1038/nprot.2012.086.

Iovannisci, D.M., and Beverley, S.M. (1989). Structural alterations of chromosome 2 in *Leishmania major* as evidence for diploidy, including spontaneous amplification of the mini-exon array. Mol. Biochem. Parasitol. *34*, 177–188.

Ivens, A.C., Peacock, C.S., Worthey, E.A., Murphy, L., Aggarwal, G., Berriman, M., Sisk, E., Rajandream, M.A., Adlem, E., Aert, R., *et al.* (2005). The genome of the kinetoplastid parasite, *Leishmania major*. Science *309*, 436–442.

Jamdhade, M.D., Pawar, H., Chavan, S., Sathe, G., Umasankar, P.K., Mahale, K.N., Dixit, T., Madugundu, A.K., Prasad, T.S., Gowda, H., *et al.* (2015). Comprehensive proteomics analysis of glycosomes from *Leishmania donovani*. OMICS *19*, 157–170. https://doi.org/10.1089/omi.2014.0163.

Kanehisa, M., Goto, S., Furumichi, M., Tanabe, M., and Hirakawa, M. (2010). KEGG for representation and analysis of molecular networks involving diseases and drugs. Nucleic Acids Res. *38*, D355–360. https://doi.org/10.1093/nar/gkp896.

Kloehn, J., Saunders, E.C., O'Callaghan, S., Dagley, M.J., and McConville, M.J. (2015). Characterization of metabolically quiescent *Leishmania* parasites in murine lesions using heavy water labeling. PLOS Pathog. *11*, e1004683. https://doi.org/10.1371/journal.ppat.1004683.

Kramer, S. (2014). RNA in development: how ribonucleoprotein granules regulate the life cycles of pathogenic protozoa. Wiley Interdiscip Rev RNA *5*, 263–284. https://doi.org/10.1002/wrna.1207.

Kramer, S., Bannerman-Chukualim, B., Ellis, L., Boulden, E.A., Kelly, S., Field, M.C., and Carrington, M. (2013). Differential localization of the two T. brucei poly(A) binding proteins to the nucleus and RNP granules suggests binding to distinct mRNA pools. PLOS ONE *8*, e54004. https://doi.org/10.1371/journal.pone.0054004.

Kramer, S., and Carrington, M. (2011). Trans-acting proteins regulating mRNA maturation, stability and translation in trypanosomatids. Trends Parasitol. *27*, 23–30. https://doi.org/10.1016/j.pt.2010.06.011.

Kramer, S., Marnef, A., Standart, N., and Carrington, M. (2012). Inhibition of mRNA maturation in trypanosomes causes the formation of novel foci at the nuclear periphery containing cytoplasmic regulators of mRNA fate. J. Cell. Sci. *125*, 2896–2909. https://doi.org/10.1242/jcs.099275.

Lacroix, V., Cottret, L., Thébault, P., and Sagot, M.F. (2008). An introduction to metabolic networks and their structural analysis. IEEE/ACM Trans Comput Biol Bioinform *5*, 594–617. https://doi.org/10.1109/TCBB.2008.79.

Lachaud, L., Bourgeois, N., Kuk, N., Morelle, C., Crobu, L., Merlin, G., Bastien, P., Pagès, M., and Sterkers, Y. (2014). Constitutive mosaic aneuploidy is a unique genetic feature widespread in the *Leishmania* genus. Microbes Infect. *16*, 61–66. https://doi.org/10.1016/j.micinf.2013.09.005.

Laffitte, M.C., Genois, M.M., Mukherjee, A., Légaré, D., Masson, J.Y., and Ouellette, M. (2014). Formation of linear amplicons with inverted duplications in *Leishmania* requires the MRE11 nuclease. PLOS Genet. *10*, e1004805. https://doi.org/10.1371/journal.pgen.1004805.

Lahav, T., Sivam, D., Volpin, H., Ronen, M., Tsigankov, P., Green, A., Holland, N., Kuzyk, M., Borchers, C., Zilberstein, D., *et al.* (2011). Multiple levels of gene regulation mediate differentiation of the intracellular pathogen *Leishmania*. FASEB J. *25*, 515–525. https://doi.org/10.1096/fj.10-157529.

Larreta, R., Soto, M., Quijada, L., Folgueira, C., Abanades, D.R., Alonso, C., and Requena, J.M. (2004). The expression of HSP83 genes in *Leishmania infantum* is affected by temperature and by stage-differentiation and is regulated at the levels of mRNA stability and translation. BMC Mol. Biol. *5*, 3. https://doi.org/10.1186/1471-2199-5-3.

LeBowitz, J.H., Smith, H.Q., Rusche, L., and Beverley, S.M. (1993). Coupling of poly(A) site selection and trans-splicing in *Leishmania*. Genes Dev. *7*, 996–1007.

Leifso, K., Cohen-Freue, G., Dogra, N., Murray, A., and McMaster, W.R. (2007). Genomic and proteomic expression analysis of *Leishmania* promastigote and amastigote life stages: the *Leishmania* genome is constitutively expressed. Mol. Biochem. Parasitol. *152*, 35–46.

Leprohon, P., Légaré, D., Raymond, F., Madore, E., Hardiman, G., Corbeil, J., and Ouellette, M. (2009). Gene expression modulation is associated with gene amplification, supernumerary chromosomes and chromosome loss in antimony-resistant *Leishmania infantum*. Nucleic Acids Res. *37*, 1387–1399. https://doi.org/10.1093/nar/gkn1069.

Liang, X.H., Haritan, A., Uliel, S., and Michaeli, S. (2003). trans and cis splicing in trypanosomatids: mechanism, factors, and regulation. Eukaryotic Cell *2*, 830–840.

Llanes, A., Restrepo, C.M., Del Vecchio, G., Anguizola, F.J., and Lleonart, R. (2015). The genome of *Leishmania* panamensis: insights into genomics of the L. (Viannia) subgenus. Sci Rep *5*, 8550. https://doi.org/10.1038/srep08550.

Lombraña, R., Álvarez, A., Fernández-Justel, J.M., Almeida, R., Poza-Carrión, C., Gomes, F., Calzada, A., Requena, J.M., and Gómez, M. (2016). Transcriptionally Driven DNA Replication Program of the Human Parasite *Leishmania major*. Cell Rep *16*, 1774–1786.

Lye, L.F., Owens, K., Shi, H., Murta, S.M., Vieira, A.C., Turco, S.J., Tschudi, C., Ullu, E., and Beverley, S.M. (2010). Retention and loss of RNA interference pathways in trypanosomatid protozoans. PLOS Pathog. *6*, e1001161. https://doi.org/10.1371/journal.ppat.1001161.

Lynn, M.A., Marr, A.K., and McMaster, W.R. (2013). Differential quantitative proteomic profiling of *Leishmania infantum* and *Leishmania mexicana* density gradient separated membranous fractions. J Proteomics *82*, 179–192. https://doi.org/10.1016/j.jprot.2013.02.010.

Magalhães, R.D., Duarte, M.C., Mattos, E.C., Martins, V.T., Lage, P.S., Chávez-Fumagalli, M.A., Lage, D.P., Menezes-Souza, D., Régis, W.C., Manso Alves, M.J., *et al.* (2014). Identification of differentially expressed proteins from *Leishmania* amazonensis associated with the loss of virulence of the parasites. PLOS Negl. Trop. Dis. *8*, e2764. https://doi.org/10.1371/journal.pntd.0002764.

Martin, J.A., and Wang, Z. (2011). Next-generation transcriptome assembly. Nat. Rev. Genet. *12*, 671–682. https://doi.org/10.1038/nrg3068.

Martínez-Calvillo, S., Stuart, K., and Myler, P.J. (2005). Ploidy changes associated with disruption of two adjacent genes on *Leishmania major* chromosome 1. Int. J. Parasitol. *35*, 419–429.

Martínez-Calvillo, S., Yan, S., Nguyen, D., Fox, M., Stuart, K., and Myler, P.J. (2003). Transcription of *Leishmania major* Friedlin chromosome 1 initiates in both directions within a single region. Mol. Cell *11*, 1291–1299.

Matthews, K.R., Tschudi, C., and Ullu, E. (1994). A common pyrimidine-rich motif governs trans-splicing and polyadenylation of tubulin polycistronic pre-mRNA in trypanosomes. Genes Dev. *8*, 491–501.

McConville, M.J., and Naderer, T. (2011). Metabolic pathways required for the intracellular survival of *Leishmania*. Annu. Rev. Microbiol. *65*, 543–561. https://doi.org/10.1146/annurev-micro-090110-102913.

McGwire, B.S., and Satoskar, A.R. (2014). Leishmaniasis: clinical syndromes and treatment. QJM *107*, 7–14. https://doi.org/10.1093/qjmed/hct116.

McNicoll, F., Drummelsmith, J., Müller, M., Madore, E., Boilard, N., Ouellette, M., and Papadopoulou, B. (2006). A combined proteomic and transcriptomic approach to the study of stage differentiation in *Leishmania infantum*. Proteomics *6*, 3567–3581. https://doi.org/10.1002/pmic.200500853.

McNicoll, F., Müller, M., Cloutier, S., Boilard, N., Rochette, A., Dubé, M., and Papadopoulou, B. (2005). Distinct 3'-untranslated region elements regulate stage-specific mRNA accumulation and translation in *Leishmania*. J. Biol. Chem. *280*, 35238–35246.

Moore, M.J. (2005). From birth to death: the complex lives of eukaryotic mRNAs. Science *309*, 1514–1518.

Morales, M.A., Renaud, O., Faigle, W., Shorte, S.L., and Späth, G.F. (2007). Over-expression of *Leishmania major* MAP kinases reveals stage-specific induction of phosphotransferase activity. Int. J. Parasitol. *37*, 1187–1199.

Morales, M.A., Watanabe, R., Dacher, M., Chafey, P., Osorio y Fortéa, J., Scott, D.A., Beverley, S.M., Ommen, G., Clos, J., Hem, S., *et al.* (2010). Phosphoproteome dynamics reveal heat-shock protein complexes specific to the *Leishmania donovani* infectious stage. Proc. Natl. Acad. Sci. U.S.A. *107*, 8381–8386. https://doi.org/10.1073/pnas.0914768107.

Morales, M.A., Watanabe, R., Laurent, C., Lenormand, P., Rousselle, J.C., Namane, A., and Späth, G.F. (2008). Phosphoproteomic analysis of *Leishmania donovani* pro- and amastigote stages. Proteomics *8*, 350–363. https://doi.org/10.1002/pmic.200700697.

Moreira, D., Lopez-Garcia, P., and Vickerman, K. (2004). An updated view of kinetoplastid phylogeny using environmental sequences and a closer outgroup: proposal for a new classification of the class Kinetoplastea. Int. J. Syst. Evol. Microbiol. *54*, 1861–1875. https://doi.org/10.1099/ijs.0.63081-0.

Moreno, J., and Alvar, J. (2002). Canine leishmaniasis: epidemiological risk and the experimental model. Trends Parasitol. *18*, 399–405.

Müller, M., Padmanabhan, P.K., and Papadopoulou, B. (2010a). Selective inactivation of SIDER2 retroposon-mediated mRNA decay contributes to stage- and species-specific gene expression in *Leishmania*. Mol. Microbiol. *77*, 471–491. https://doi.org/10.1111/j.1365-2958.2010.07226.x.

Müller, M., Padmanabhan, P.K., Rochette, A., Mukherjee, D., Smith, M., Dumas, C., and Papadopoulou, B. (2010b). Rapid decay of unstable *Leishmania* mRNAs bearing a conserved retroposon signature 3'-UTR motif is initiated by a site-specific endonucleolytic cleavage without prior deadenylation. Nucleic Acids Res. *38*, 5867–5883. https://doi.org/10.1093/nar/gkq349.

Murta, S.M., Vickers, T.J., Scott, D.A., and Beverley, S.M. (2009). Methylene tetrahydrofolate dehydrogenase/cyclohydrolase and the synthesis of 10-CHO-THF are essential in *Leishmania major*. Mol. Microbiol. *71*, 1386–1401. https://doi.org/10.1111/j.1365-2958.2009.06610.x.

Myung, K.S., Beetham, J.K., Wilson, M.E., and Donelson, J.E. (2002). Comparison of the post-transcriptional regulation of the mRNAs for the surface proteins PSA (GP46) and MSP (GP63) of *Leishmania chagasi*. J. Biol. Chem. *277*, 16489–16497. https://doi.org/10.1074/jbc.M200174200.

Naderer, T., Dandash, O., and McConville, M.J. (2011). Calcineurin is required for *Leishmania major* stress response pathways and for virulence in the mammalian host. Mol. Microbiol. *80*, 471–480. https://doi.org/10.1111/j.1365-2958.2011.07584.x.

Nascimento, M., Abourjeily, N., Ghosh, A., Zhang, W.W., and Matlashewski, G. (2003). Heterologous expression of a mammalian protein tyrosine phosphatase gene in *Leishmania*: effect on differentiation. Mol. Microbiol. *50*, 1517–1526.

Ngô, H., Tschudi, C., Gull, K., and Ullu, E. (1998). Double-stranded RNA induces mRNA degradation in *Trypanosoma brucei*. Proc. Natl. Acad. Sci. U.S.A. *95*, 14687–14692.

Novais, F.O., Carvalho, L.P., Passos, S., Roos, D.S., Carvalho, E.M., Scott, P., and Beiting, D.P. (2015). Genomic profiling of human *Leishmania braziliensis* lesions identifies transcriptional modules associated with cutaneous immunopathology. J. Invest. Dermatol. *135*, 94–101.

Ong, S.E., Blagoev, B., Kratchmarova, I., Kristensen, D.B., Steen, H., Pandey, A., and Mann, M. (2002). Stable isotope labeling by amino acids in cell culture, SILAC, as a simple and accurate approach to expression proteomics. Mol. Cell Proteomics *1*, 376–386.

Opperdoes, F.R., Butenko, A., Flegontov, P., Yurchenko, V., and Lukeš, J. (2016). Comparative Metabolism of Free-living *Bodo saltans* and Parasitic Trypanosomatids. J. Eukaryot. Microbiol. *63*, 657–678. https://doi.org/10.1111/jeu.12315.

Paape, D., and Aebischer, T. (2011). Contribution of proteomics of *Leishmania* spp. to the understanding of differentiation, drug resistance mechanisms, vaccine and drug development. J Proteomics *74*, 1614–1624. https://doi.org/10.1016/j.jprot.2011.05.005.

Papadopoulou, B., Muller, M., Rochette, A., McNicoll, F., Dumas, C., and Chow, C. (2008). Regulation of gene expression in *Leishmania* throughout a complex digenetic life cycle. In: *Leishmania*, after the genome, P.J. Myler and N. Fasel, eds. (Caister Academic Press, Norfolk), pp. 29–54.

Parsons, M., and Myler, P.J. (2016). Illuminating Parasite Protein Production by Ribosome Profiling. Trends Parasitol. *32*, 446–457. https://doi.org/10.1016/j.pt.2016.03.005.

Parsons, M., Worthey, E.A., Ward, P.N., and Mottram, J.C. (2005). Comparative analysis of the kinomes of three pathogenic trypanosomatids: *Leishmania major*, *Trypanosoma brucei* and *Trypanosoma cruzi*. BMC Genomics *6*, 127.

Pawar, H., Renuse, S., Khobragade, S.N., Chavan, S., Sathe, G., Kumar, P., Mahale, K.N., Gore, K., Kulkarni, A., Dixit, T., *et al.* (2014). Neglected tropical diseases and omics science: proteogenomics analysis of the promastigote stage of *Leishmania major* parasite. OMICS *18*, 499–512. https://doi.org/10.1089/omi.2013.0159.

Peacock, C.S., Seeger, K., Harris, D., Murphy, L., Ruiz, J.C., Quail, M.A., Peters, N., Adlem, E., Tivey, A., Aslett, M., *et al.* (2007). Comparative genomic analysis of three *Leishmania* species that cause diverse human disease. Nat. Genet. *39*, 839–847.

Pescher, P., Blisnick, T., Bastin, P., and Späth, G.F. (2011). Quantitative proteome profiling informs on phenotypic traits that adapt *Leishmania donovani* for axenic and intracellular proliferation. Cell. Microbiol. *13*, 978–991. https://doi.org/10.1111/j.1462-5822.2011.01593.x.

Phanstiel, D., Unwin, R., McAlister, G.C., and Coon, J.J. (2009). Peptide quantification using 8-plex isobaric tags and electron transfer dissociation tandem mass spectrometry. Anal. Chem. *81*, 1693–1698. https://doi.org/10.1021/ac8019202.

Pigott, D.M., Bhatt, S., Golding, N., Duda, K.A., Battle, K.E., Brady, O.J., Messina, J.P., Balard, Y., Bastien, P., Pratlong, F., *et al.* (2014). Global distribution maps of the leishmaniases. Elife 3. https://doi.org/10.7554/eLife.02851.

Quijada, L., Soto, M., Alonso, C., and Requena, J.M. (1997). Analysis of post-transcriptional regulation operating on transcription products of the tandemly linked *Leishmania infantum* hsp70 genes. J. Biol. Chem. *272*, 4493–4499.

Quijada, L., Soto, M., Alonso, C., and Requena, J.M. (2000). Identification of a putative regulatory element in the 3'-untranslated region that controls expression of HSP70 in *Leishmania infantum*. Mol. Biochem. Parasitol. *110*, 79–91.

Ramírez, C.A., Requena, J.M., and Puerta, C.J. (2011). Identification of the HSP70-II gene in *Leishmania braziliensis* HSP70 locus: genomic organization and UTRs characterization. Parasit. Vectors *4*, 166. https://doi.org/10.1186/1756-3305-4-166.

Rastrojo, A., Carrasco-Ramiro, F., Martín, D., Crespillo, A., Reguera, R.M., Aguado, B., and Requena, J.M. (2013). The transcriptome of *Leishmania major* in the axenic promastigote stage: transcript annotation and relative expression levels by RNA-seq. BMC Genomics *14*, 223. https://doi.org/10.1186/1471-2164-14-223.

Ravel, C., Dubessay, P., Bastien, P., Blackwell, J.M., and Ivens, A.C. (1998). The complete chromosomal organization of the reference strain of the *Leishmania* Genome Project, *L. major* 'Friedlin'. Parasitol. Today *14*, 301–303.

Raymond, F., Boisvert, S., Roy, G., Ritt, J.F., Légaré, D., Isnard, A., Stanke, M., Olivier, M., Tremblay, M.J., Papadopoulou, B., *et al.* (2012). Genome sequencing of the lizard parasite *Leishmania* tarentolae reveals loss of genes associated to the intracellular stage of human pathogenic species. Nucleic Acids Res. *40*, 1131–1147. https://doi.org/10.1093/nar/gkr834.

Real, F., Vidal, R.O., Carazzolle, M.F., Mondego, J.M., Costa, G.G., Herai, R.H., Würtele, M., de Carvalho, L.M., Carmona e Ferreira, R., Mortara, R.A., *et al.* (2013). The genome sequence of *Leishmania* (*Leishmania*) amazonensis: functional annotation and extended analysis of gene models. DNA Res. *20*, 567–581. https://doi.org/10.1093/dnares/dst031.

Requena, J.M. (2011). Lights and shadows on gene organization and regulation of gene expression in *Leishmania*. Front. Biosci. *16*, 2069–2085.

Requena, J.M. (2012). The Stressful Life of pathogenic *Leishmania* species. In: Stress Response in Microbiology, J.M. Requena, ed. (Caister Academic Press, Norfolk), pp. 323–346.

Requena, J.M., Folgueira, C., López, M.C., and Thomas, M.C. (2008). The SIDER2 elements, interspersed repeated sequences that populate the *Leishmania* genomes, constitute subfamilies showing chromosomal proximity relationship. BMC Genomics *9*, 263. https://doi.org/10.1186/1471-2164-9-263.

Requena, J.M., Quijada, L., Soto, M., and Alonso, C. (2003). Conserved nucleotides surrounding the trans-splicing acceptor site and the translation initiation codon in *Leishmania* genes. Exp. Parasitol. *103*, 78–81.

Requena, J.M., Rastrojo, A., Garde, E., López, M.C., Thomas, M.C., and Aguado, B. (2017). Genomic cartography and proposal of nomenclature for the repeated, interspersed elements of the *Leishmania major* SIDER2 family and identification of SIDER2-containing transcripts. Mol. Biochem. Parasitol. *212*, 9–15.

Respuela, P., Ferella, M., Rada-Iglesias, A., and Aslund, L. (2008). Histone acetylation and methylation at sites initiating divergent polycistronic transcription in *Trypanosoma cruzi*. J. Biol. Chem. *283*, 15884–15892. https://doi.org/10.1074/jbc.M802081200.

Reynolds, D.L., Hofmeister, B.T., Cliffe, L., Siegel, T.N., Anderson, B.A., Beverley, S.M., Schmitz, R.J., and Sabatini, R. (2016). Base J represses genes at the end of polycistronic gene clusters in *Leishmania major* by promoting RNAP II termination. Mol. Microbiol. *101*, 559–574. https://doi.org/10.1111/mmi.13408.

Richard, G.F., Kerrest, A., and Dujon, B. (2008). Comparative genomics and molecular dynamics of DNA repeats in eukaryotes. Microbiol. Mol. Biol. Rev. *72*, 686–727. https://doi.org/10.1128/MMBR.00011-08.

Richter, K., Haslbeck, M., and Buchner, J. (2010). The heat shock response: life on the verge of death. Mol. Cell *40*, 253–266. https://doi.org/10.1016/j.molcel.2010.10.006.

Rochette, A., Raymond, F., Corbeil, J., Ouellette, M., and Papadopoulou, B. (2009). Whole-genome comparative RNA expression profiling of axenic and intracellular amastigote forms of *Leishmania infantum*. Mol. Biochem. Parasitol. *165*, 32–47. https://doi.org/10.1016/j.molbiopara.2008.12.012.

Rochette, A., Raymond, F., Ubeda, J.M., Smith, M., Messier, N., Boisvert, S., Rigault, P., Corbeil, J., Ouellette, M., and Papadopoulou, B. (2008). Genome-wide gene expression profiling analysis of *Leishmania major* and *Leishmania infantum* developmental stages reveals substantial differences between the two species. BMC Genomics *9*, 255. https://doi.org/10.1186/1471-2164-9-255.

Rogers, M.B., Hilley, J.D., Dickens, N.J., Wilkes, J., Bates, P.A., Depledge, D.P., Harris, D., Her, Y., Herzyk, P., Imamura, H., *et al.* (2011). Chromosome and gene copy number variation allow *major* structural change between species and strains of *Leishmania*. Genome Res. *21*, 2129–2142. https://doi.org/10.1101/gr.122945.111.

Rojo, D., Canuto, G.A., Castilho-Martins, E.A., Tavares, M.F., Barbas, C., López-Gonzálvez, Á., and Rivas, L. (2015). A Multiplatform Metabolomic Approach to the Basis of Antimonial Action and Resistance in *Leishmania infantum*. PLOS ONE *10*, e0130675. https://doi.org/10.1371/journal.pone.0130675.

Rosenzweig, D., Smith, D., Myler, P.J., Olafson, R.W., and Zilberstein, D. (2008a). Post-translational modification of cellular proteins during *Leishmania donovani* differentiation. Proteomics *8*, 1843–1850. https://doi.org/10.1002/pmic.200701043.

Rosenzweig, D., Smith, D., Opperdoes, F., Stern, S., Olafson, R.W., and Zilberstein, D. (2008b). Retooling *Leishmania* metabolism: from sand fly gut to human macrophage. FASEB J. *22*, 590–602.

Sacks, D.L., Hieny, S., and Sher, A. (1985). Identification of cell surface carbohydrate and antigenic changes between noninfective and infective developmental stages of *Leishmania major* promastigotes. J. Immunol. *135*, 564–569.

Sacks, D.L., and Perkins, P.V. (1984). Identification of an infective stage of *Leishmania* promastigotes. Science *223*, 1417–1419.

Sánchez-Luque, F.J., López, M.C., Carreira, P.E., Alonso, C., and Thomas, M.C. (2014). The wide expansion of hepatitis delta virus-like ribozymes throughout trypanosomatid genomes is linked to the spreading of L1Tc/ingi clade mobile elements. BMC Genomics *15*, 340. https://doi.org/10.1186/1471-2164-15-340.

Sánchez-Luque, F.J., López, M.C., Macias, F., Alonso, C., and Thomas, M.C. (2011). Identification of an hepatitis delta virus-like ribozyme at the mRNA 5'-end of the L1Tc retrotransposon from *Trypanosoma cruzi*. Nucleic Acids Res. *39*, 8065–8077. https://doi.org/10.1093/nar/gkr478.

Santoro, M.G. (2000). Heat shock factors and the control of the stress response. Biochem. Pharmacol. *59*, 55–63.

Sardar, A.H., Kumar, S., Kumar, A., Purkait, B., Das, S., Sen, A., Kumar, M., Sinha, K.K., Singh, D., Equbal, A., *et al.* (2013). Proteome changes associated with *Leishmania donovani* promastigote adaptation to oxidative and nitrosative stresses. J Proteomics *81*, 185–199. https://doi.org/10.1016/j.jprot.2013.01.011.

Saunders, E.C., Ng, W.W., Kloehn, J., Chambers, J.M., Ng, M., and McConville, M.J. (2014). Induction of a stringent metabolic response in intracellular stages of *Leishmania mexicana* leads to increased dependence on mitochondrial metabolism. PLOS Pathog. *10*, e1003888. https://doi.org/10.1371/journal.ppat.1003888.

Saxena, A., Worthey, E.A., Yan, S., Leland, A., Stuart, K.D., and Myler, P.J. (2003). Evaluation of differential gene expression in *Leishmania major* Friedlin procyclics and metacyclics using DNA microarray analysis. Mol. Biochem. Parasitol. *129*, 103–114.

Scheltema, R.A., Decuypere, S., T'kindt, R., Dujardin, J.C., Coombs, G.H., and Breitling, R. (2010). The potential of metabolomics for *Leishmania* research in the post-genomics era. Parasitology *137*, 1291–1302. https://doi.org/10.1017/S0031182009992022.

Schwede, A., Kramer, S., and Carrington, M. (2012). How do trypanosomes change gene expression in response to the environment? Protoplasma *249*, 223–238. https://doi.org/10.1007/s00709-011-0282-5.

Seco-Hidalgo, V., Osuna, A., and de Pablos, L.M. (2015). To bet or not to bet: deciphering cell to cell variation in protozoan infections. Trends Parasitol. *31*, 350–356. https://doi.org/10.1016/j.pt.2015.05.004.

Shaw, C.D., Lonchamp, J., Downing, T., Imamura, H., Freeman, T.M., Cotton, J.A., Sanders, M., Blackburn, G., Dujardin, J.C., Rijal, S., *et al.* (2016). In vitro selection of miltefosine resistance in promastigotes of *Leishmania donovani* from Nepal: genomic and metabolomic characterization. Mol. Microbiol. *99*, 1134–1148. https://doi.org/10.1111/mmi.13291.

Siegel, T.N., Hekstra, D.R., Kemp, L.E., Figueiredo, L.M., Lowell, J.E., Fenyo, D., Wang, X., Dewell, S., and Cross, G.A. (2009). Four histone variants mark the boundaries of polycistronic transcription units in *Trypanosoma brucei*. Genes Dev. *23*, 1063–1076. https://doi.org/10.1101/gad.1790409.

Silva, A.M., Cordeiro-da-Silva, A., and Coombs, G.H. (2011). Metabolic variation during development in culture of *Leishmania donovani* promastigotes. PLOS Negl. Trop. Dis. *5*, e1451. https://doi.org/10.1371/journal.pntd.0001451.

Singh, R., Kumar, D., Duncan, R.C., Nakhasi, H.L., and Salotra, P. (2010). Overexpression of histone H2A modulates drug susceptibility in *Leishmania* parasites. Int. J. Antimicrob. Agents *36*, 50–57. https://doi.org/10.1016/j.ijantimicag.2010.03.012.

Singh, S., and Dubey, V.K. (2016). Quantitative Proteome Analysis of *Leishmania donovani* under Spermidine Starvation. PLOS ONE *11*, e0154262. https://doi.org/10.1371/journal.pone.0154262.

Smith, D.F., Peacock, C.S., and Cruz, A.K. (2007). Comparative genomics: from genotype to disease phenotype in the leishmaniases. Int. J. Parasitol. *37*, 1173–1186.

Smolka, M.B., Zhou, H., Purkayastha, S., and Aebersold, R. (2001). Optimization of the isotope-coded affinity tag-labeling procedure for quantitative proteome analysis. Anal. Biochem. *297*, 25–31. https://doi.org/10.1006/abio.2001.5318.

Sogin, M.L., Elwood, H.J., and Gunderson, J.H. (1986). Evolutionary diversity of eukaryotic small-subunit rRNA genes. Proc. Natl. Acad. Sci. U.S.A. *83*, 1383–1387.

Solari, A.J. (1995). Mitosis and genome partition in trypanosomes. Biocell *19*, 65–84.

Sollelis, L., Ghorbal, M., MacPherson, C.R., Martins, R.M., Kuk, N., Crobu, L., Bastien, P., Scherf, A., Lopez-Rubio, J.J., and Sterkers, Y. (2015). First efficient CRISPR-Cas9-mediated genome editing in *Leishmania* parasites. Cell. Microbiol. *17*, 1405–1412. https://doi.org/10.1111/cmi.12456.

Soto, M., Iborra, S., Quijada, L., Folgueira, C., Alonso, C., and Requena, J.M. (2004). Cell-cycle-dependent translation of histone mRNAs is the key control point for regulation of histone biosynthesis in *Leishmania infantum*. Biochem. J. *379*, 617–625. https://doi.org/10.1042/BJ20031522.

Späth, G.F., and Clos, J. (2016). Joining forces: first application of a rapamycin-induced dimerizable Cre system for conditional null mutant analysis in *Leishmania*. Mol. Microbiol. *100*, 923–927. https://doi.org/10.1111/mmi.13374.

Späth, G.F., Drini, S., and Rachidi, N. (2015). A touch of Zen: post-translational regulation of the *Leishmania* stress response. Cell. Microbiol. *17*, 632–638. https://doi.org/10.1111/cmi.12440.

Srividya, G., Duncan, R., Sharma, P., Raju, B.V.S., Nakhasi, H.L., and Salotra, P. (2007). Transcriptome analysis during the process of in vitro differentiation of *Leishmania donovani* using genomic microarrays. Parasitology *134*, 1527–1539. https://doi.org/10.1017/S003118200700296X.
Sterkers, Y., Lachaud, L., Crobu, L., Bastien, P., and Pagès, M. (2011). FISH analysis reveals aneuploidy and continual generation of chromosomal mosaicism in *Leishmania major*. Cell. Microbiol. *13*, 274–283. https://doi.org/10.1111/j.1462-5822.2010.01534.x.
Stevens, J.R., Noyes, H.A., Schofield, C.J., and Gibson, W. (2001). The molecular evolution of Trypanosomatidae. Adv. Parasitol. *48*, 1–56.
Sunkin, S.M., Kiser, P., Myler, P.J., and Stuart, K. (2000). The size difference between leishmania *major* friedlin chromosome one homologues is localized to sub-telomeric repeats at one chromosomal end. Mol. Biochem. Parasitol. *109*, 1–15.
t'Kindt, R., Jankevics, A., Scheltema, R.A., Zheng, L., Watson, D.G., Dujardin, J.C., Breitling, R., Coombs, G.H., and Decuypere, S. (2010a). Towards an unbiased metabolic profiling of protozoan parasites: optimisation of a *Leishmania* sampling protocol for HILIC-orbitrap analysis. Anal. Bioanal. Chem. *398*, 2059–2069. https://doi.org/10.1007/s00216-010-4139-0.
t'Kindt, R., Scheltema, R.A., Jankevics, A., Brunker, K., Rijal, S., Dujardin, J.C., Breitling, R., Watson, D.G., Coombs, G.H., and Decuypere, S. (2010b). Metabolomics to unveil and understand phenotypic diversity between pathogen populations. PLOS Negl. Trop. Dis. *4*, e904. https://doi.org/10.1371/journal.pntd.0000904.
Tannu, N.S., and Hemby, S.E. (2006). Two-dimensional fluorescence difference gel electrophoresis for comparative proteomics profiling. Nat. Protoc. *1*, 1732–1742.
Thomas, M.C., Macias, F., Alonso, C., and López, M.C. (2010). The biology and evolution of transposable elements in parasites. Trends Parasitol. *26*, 350–362. https://doi.org/10.1016/j.pt.2010.04.001.
Thomas, S., Green, A., Sturm, N.R., Campbell, D.A., and Myler, P.J. (2009). Histone acetylations mark origins of polycistronic transcription in *Leishmania major*. BMC Genomics *10*, 152. https://doi.org/10.1186/1471-2164-10-152.
Tsigankov, P., Gherardini, P.F., Helmer-Citterich, M., Späth, G.F., Myler, P.J., and Zilberstein, D. (2014). Regulation dynamics of *Leishmania* differentiation: deconvoluting signals and identifying phosphorylation trends. Mol. Cell Proteomics *13*, 1787–1799. https://doi.org/10.1074/mcp.M114.037705.
Tsigankov, P., Gherardini, P.F., Helmer-Citterich, M., and Zilberstein, D. (2012). What has proteomics taught us about *Leishmania* development? Parasitology *139*, 1146–1157. https://doi.org/10.1017/S0031182012000157.
Ubeda, J.M., Raymond, F., Mukherjee, A., Plourde, M., Gingras, H., Roy, G., Lapointe, A., Leprohon, P., Papadopoulou, B., Corbeil, J., *et al.* (2014). Genome-wide stochastic adaptive DNA amplification at direct and inverted DNA repeats in the parasite *Leishmania*. PLOS Biol. *12*, e1001868. https://doi.org/10.1371/journal.pbio.1001868.
Vacchina, P., Norris-Mullins, B., Abengózar, M.A., Viamontes, C.G., Sarro, J., Stephens, M.T., Pfrender, M.E., Rivas, L., and Morales, M.A. (2016). Genomic Appraisal of the Multifactorial Basis for In Vitro Acquisition of Miltefosine Resistance in *Leishmania donovani*. Antimicrob. Agents Chemother. *60*, 4089–4100. https://doi.org/10.1128/AAC.00478-16.
van Leeuwen, F., Taylor, M.C., Mondragon, A., Moreau, H., Gibson, W., Kieft, R., and Borst, P. (1998). beta-D-glucosyl-hydroxymethyluracil is a conserved DNA modification in kinetoplastid protozoans and is abundant in their telomeres. Proc. Natl. Acad. Sci. U.S.A. *95*, 2366–2371.
van Luenen, H.G., Farris, C., Jan, S., Genest, P.A., Tripathi, P., Velds, A., Kerkhoven, R.M., Nieuwland, M., Haydock, A., Ramasamy, G., *et al.* (2012). Glucosylated hydroxymethyluracil, DNA base J, prevents transcriptional readthrough in *Leishmania*. Cell *150*, 909–921. https://doi.org/10.1016/j.cell.2012.07.030.
Vazquez, M., Ben-Dov, C., Lorenzi, H., Moore, T., Schijman, A., and Levin, M.J. (2000). The short interspersed repetitive element of *Trypanosoma cruzi*, SIRE, is part of VIPER, an unusual retroelement related to long terminal repeat retrotransposons. Proc. Natl. Acad. Sci. U.S.A. *97*, 2128–2133. https://doi.org/10.1073/pnas.050578397.
Velculescu, V.E., Zhang, L., Vogelstein, B., and Kinzler, K.W. (1995). Serial analysis of gene expression. Science *270*, 484–487. https://doi.org/10.1126/science.270.5235.484.
Veras, P.S., and Bezerra de Menezes, J.P. (2016). Using Proteomics to Understand How *Leishmania* Parasites Survive inside the Host and Establish Infection. Int J Mol Sci *17*, E1270. https://doi.org/10.3390/ijms17081270.

Vergnes, B., Gourbal, B., Girard, I., Sundar, S., Drummelsmith, J., and Ouellette, M. (2007). A proteomics screen implicates HSP83 and a small kinetoplastid calpain-related protein in drug resistance in *Leishmania donovani* clinical field isolates by modulating drug-induced programmed cell death. Mol. Cell Proteomics *6*, 88–101.

Vergnes, B., Sereno, D., Tavares, J., Cordeiro-da-Silva, A., Vanhille, L., Madjidian-Sereno, N., Depoix, D., Monte-Alegre, A., and Ouaissi, A. (2005). Targeted disruption of cytosolic SIR2 deacetylase discloses its essential role in *Leishmania* survival and proliferation. Gene *363*, 85–96.

Vincent, I.M., Weidt, S., Rivas, L., Burgess, K., Smith, T.K., and Ouellette, M. (2014). Untargeted metabolomic analysis of miltefosine action in *Leishmania infantum* reveals changes to the internal lipid metabolism. Int J Parasitol Drugs Drug Resist *4*, 20–27. https://doi.org/10.1016/j.ijpddr.2013.11.002.

Westrop, G.D., Williams, R.A., Wang, L., Zhang, T., Watson, D.G., Silva, A.M., and Coombs, G.H. (2015). Metabolomic Analyses of *Leishmania* Reveal Multiple Species Differences and Large Differences in Amino Acid Metabolism. PLOS ONE *10*, e0136891. https://doi.org/10.1371/journal.pone.0136891.

Wincker, P., Ravel, C., Blaineau, C., Pages, M., Jauffret, Y., Dedet, J.P., and Bastien, P. (1996). The *Leishmania* genome comprises 36 chromosomes conserved across widely divergent human pathogenic species. Nucleic Acids Res. *24*, 1688–1694.

Wong, J.W., and Cagney, G. (2010). An overview of label-free quantitation methods in proteomics by mass spectrometry. Methods Mol. Biol. *604*, 273–283. https://doi.org/10.1007/978-1-60761-444-9_18.

Wright, J.R., Siegel, T.N., and Cross, G.A. (2010). Histone H3 trimethylated at lysine 4 is enriched at probable transcription start sites in *Trypanosoma brucei*. Mol. Biochem. Parasitol. *172*, 141–144. https://doi.org/10.1016/j.molbiopara.2010.03.013.

Xia, J., and Wishart, D.S. (2016). Using MetaboAnalyst 3.0 for comprehensive metabolomics data analysis. Curr. Protoc. Bioinformatics *55*, 14.10.11–14.10.91.

Yao, C., Li, Y., Donelson, J.E., and Wilson, M.E. (2010). Proteomic examination of *Leishmania chagasi* plasma membrane proteins: Contrast between avirulent and virulent (metacyclic) parasite forms. Proteomics Clin Appl *4*, 4–16. https://doi.org/10.1002/prca.200900050.

Zhang, W.W., and Matlashewski, G. (2015). CRISPR-Cas9-Mediated Genome Editing in *Leishmania donovani*. MBio *6*, e00861. https://doi.org/10.1128/mBio.00861-15.

Zilka, A., Garlapati, S., Dahan, E., Yaolsky, V., and Shapira, M. (2001). Developmental regulation of heat shock protein 83 in *Leishmania*. 3' processing and mRNA stability control transcript abundance, and translation id directed by a determinant in the 3'-untranslated region. J. Biol. Chem. *276*, 47922–47929. https://doi.org/10.1074/jbc.M108271200

Natural Products in Human Leishmaniasis Therapy: Recent Advances

6

José E Piñero*, Atteneri Lopez-Arencibia, Maria Reyes-Batlle, Ines Sifaoui, Carolina Wagner, Antonio del Castillo, Elizabeth Córdoba-Lanús and Jacob Lorenzo-Morales

Instituto Universitario de Enfermedades Tropicales y Salud Pública de Canarias, Universidad de La Laguna, Tenerife, Canary Islands, Spain.

Correspondence: jpinero@ull.edu.es

https://doi.org/10.21775/9781910190838.09

Abstract

Leishmaniasis is a vector-borne neglected tropical disease caused by protozoan parasites of the *Leishmania* genus and transmitted by the female *Phlebotomus* and *Lutzomyia* sandflies. This disease affects the tropical as well as subtropical countries of the world and presents more than one clinical form, with the visceral one considered fatal. Furthermore, this disease received limited funding for discovery, development and delivery of new tools. Chemotherapy is the major treatment option considered for the control of this disease in the infected people. The currently prescribed therapies still rely on pentavalent antimonials, pentamidine, paromomycin, liposomal amphotericin B and miltefosine. However, their low efficacy, long-course treatment regimen, high toxicity, adverse side effects, induction of parasite resistance and high cost require the need for better drugs given that antileishmanial vaccines may not be available in the near future. An alternative to synthetic drugs is the search for anti-parasitic plant extracts or secondary metabolites derived from these sources. This chapter examines the later studies on natural bioactive compounds as sources of antileishmanicidal agents.

Introduction

Leishmaniasis is a vector-borne disease caused by more than 20 protozoan parasites from *Leishmania* species. These organisms produce different clinical variants of the disease: visceral leishmaniasis (VL), cutaneous leishmaniasis (CL) and mucocutaneous leishmaniasis (ML). Leishmaniasis is considered as one of the most significant among all the neglected tropical diseases (Feasey *et al.*, 2010), with an extremely high mortality and morbidity rate, standing second and fourth, respectively, among the tropical infections (Alvar *et al.*, 2006).

Current estimates show an annual global incidence of 0.2–0.4 million cases of VL and 0.7–1.2 million cases of CL. Leishmaniasis is endemic in 98 countries and three territories with a total of 350 million people at risk and death rate of 70 000 per year around the globe (Blum *et al.*, 2004; Reithinger *et al.*, 2007; WHO, 2016). The actual number of leishmaniasis affected people worldwide can be larger because most of the cases are not reported and false diagnosis is also an issue (WHO, 2016).

VL is hypoendemic in Mediterranean countries, where it is caused by *Leishmania infantum*. VL cases in this area account for 5–6% of the global burden, with an estimated annual incidence of 1200–2000 cases. Cases of *Leishmania*/HIV coinfection have been reported in the Mediterranean region, mainly in France, Italy, Portugal and Spain. In Mediterranean countries, more than 27,000 new human immunodeficiency virus (HIV) infections were diagnosed in 2012, corresponding to rates of 6.6 per 100,000 people (reviwed by Monge-Maillo *et al.*, 2014).

The digenetic life cycle of *Leishmania* consists of motile, flagellated, extracellular promastigotes (10–15 µm in length, measuring 1.5–3.5 µm at their widest part) form in the gut of sandfly vector, to the genera *Phlebotomus* in Old World (Europe, Africa and Asia) while *Lutzomyia* in New World (America), that infects mammalian host. The promastigotes are phagocytosed by host macrophages where they transform into non-motile, spherical, nonflagellated amastigotes (approximately 2–3 µm in diameter), which survive and multiply within phagolysosomal compartment of macrophages.

Infection starts after a sandfly takes a blood meal from an infected host (e.g. canines, marsupials, edentates, and rodents). Small amounts of blood, lymph, and macrophages infected with *Leishmania* amastigotes are ingested. Once ingested, the amastigotes migrate to the midgut of the sandfly, where they transform into the promastigotes by binary fission and move forward (after 4 or 5 days) to the oesophagus, and the salivary glands. Infected sandfly during the second blood meal regurgitates the infectious promastigotes into the bloodstream of the host vertebrates.

Once inside the bloodstream of reservoirs, phagocytosis of promastigotes by the host mononuclear phagocytic cells takes place. The *Leishmania* are able to resist the microbicidal action of the acid hydrolases release from the lysozymes, thus easily survive, and multiply inside the macrophages. Eventually, the host cells lyse, releasing the free parasites which spread to new cells and tissues of different organs (especially the spleen, liver and bone marrow), causing lesions and tissue destruction.

Current chemotherapy for leishmaniasis

Drugs are the most important tools in order to control different clinical manifestations of leishmaniasis. Treatment of leishmaniasis should be selected based on several factors such as patient characteristics, pharmacogenetic characteristics of the drug, causative *Leishmania* species, risk factors and geographical area (Lindoso *et al.*, 2012; Fernández *et al.*, 2014). An ideal drug for treatment of leishmaniasis should provide several requirements for patients including being effective in one or few doses, low cost, not be teratogenic, no side effects, no need for hospitalization, and no induction of resistance (Alviano *et al.*, 2012).

Pentavalent antimonials

Pentavalent antimonials have been used in the treatment regimens for VL and CL for more than 60 years (Fig. 6.1). However, an increasing incidence of resistance to pentavalent antimony (Sbv) has been reported in some parts of India (Seifer, 2011). Other issues related to treatment regimens include side-effects of low-cost generic drugs (Alvar *et al.*, 2006) and cardiotoxicity (Olliaro *et al.*, 2005).

The activity of antimony against leishmaniasis was first introduced in the early 1900s when antimony (III) potassium tartrate was found to be effective against mucocutaneous leishmaniasis. Later, efficacy was shown against the visceral form of leishmaniasis. Usage was limited, however, by the severe toxicity caused by the agent. Pentavalent antimony (V), a safer form, was developed several decades later and has been used since as the first-line treatment for VL. At present, two types of organo-antimony (V) complexes are commercially available: Glucantime® (meglumine antimoniate) and Pentostam® (sodium stibogluconate). The mechanism of pentavalent antimony action is not well elucidated. Based on extensive investigations, two major hypotheses are available. The first model is based on the reductive bioconversion of Sb(V) to Sb(III) by the parasite or by the infected host cells to create an active agent (Shaked-Mishan *et al.*, 2001). The second model posits Sb(V) as the active species. In the first model, Sb(V) enters *Leishmania*-infected host cells and is exposed to thioles in the lysosome (cysteine and cysteinylglycine) and within the parasite (trypanothione), causing its bioconversion to Sb(III) (Ferreira Cdos *et al.*, 2003). The conversion process is more efficient at lower pH *in vitro*, which reflects the environment of the intracellular amastigote rather than that of the extracellular promastigote (Ferreira Cdos *et al.*, 2003). Antimonite reductase (Zhou *et al.*, 2004) and thiol-dependent reductase 1 (Denton *et al.*, 2004) are also proposed to be involved in the conversion process. A few studies have

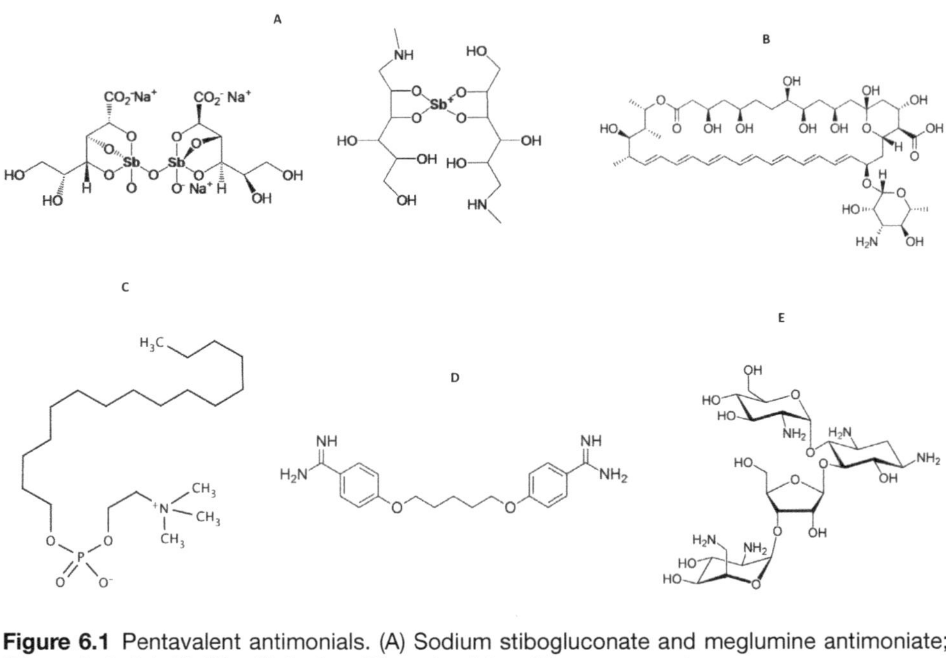

Figure 6.1 Pentavalent antimonials. (A) Sodium stibogluconate and meglumine antimoniate; (B) amphotericin B; (C) miltefosine; (D) pentamidine; (E) paramomycin.

described the downstream consequences once Sb(III), the reduced form, is generated. Sb(III) strongly binds biomolecules containing sulfhydryl moieties. Based on that characteristic, Sb(III) is reported to form conjugated pairs with trypanothione (Mukhopadhyay *et al.*, 1996), a virulence factor that helps the parasite to resist the oxidative stress of the host cell environment. Another proposed mechanism of Sb(III) action involves the inhibition of trypanothione reductase (TryR), an enzyme that recycles oxidized trypanothione to keep the trypanothione in a reducing state (Krauth-Siegel and Comini, 2008). Sb(III)-resistant clinical isolates show an up-regulation of the TryR gene (Rai *et al.*, 2013).

Another proposed scenario involves the binding of Sb(III) to the HEXBP protein, a CCHC zinc finger protein, which replaces the zinc in the active site and interferes with various downstream functions involved in DNA replication and repair (Demicheli *et al.*, 2008). In the second model, Sb(V), the pentavalent form, directly exerts activity against *Leishmania*. Using nuclear extracts containing topoisomerase I activity, inhibition by sodium stibogluconate was confirmed to be three times more selective against the topoisomerase of the promastigote than against the topoisomerase of themonocyte (Walker and Saravia, 2004). Sb(V) was also reported to bind to ribonucleosides, such as adenosine and guanosine, at low pH, an environment similar to that within lysosomes (Demicheli *et al.*, 2002). An exhaustive revision shows that the recommended dosage for antimonials is 20 mg/kg/day for 20 days in cutaneous leishmaniasis and for 28 to 30 days in mucocutaneous and visceral leishmaniasis (Sundar and Chakravarty, 2012).

Amphotericin B

Amphotericin B (AmpB) is a macrolide polyene metabolite with antifungal and antiparasitic activities (Fig. 6.1B). Currently, AmpB is widely used to treat systemic *Candida albicans* and *Aspergillus fumigatus* infections and is especially effective against fungal infections in immunodeficient patients. The use of AmpB is limited, however, by severe side-effects such as nephrotoxicity and haematotoxicity. Recently, various types of formulations, including some that lower the free AmpB concentration in the blood stream, have been investigated to reduce the toxicity. The first antileishmanial activity of AmpB was reported in the early 1960s. Currently, the liposomal formulation of AmpB (AmBisome®) is administered intravenously to treat VL and has a 95% of cure rate for a single-course therapy (Sundar *et al.*, 2010). Even with the high cure rate, the cost of AmBisome® treatment is a limiting factor for patients in developing countries. Despite negotiations between the World Health Organization (WHO) and Gilead, the producer of AmBisome®, to reduce the price of the treatment ($18 per 50-mg ampoule), the treatment is still expensive (Matheus *et al.*, 2010).

AmpB preferentially binds to the ergosterol in the fungus or the parasite rather than to human cholesterol. In general, AmpB binds to ergosterol through a hydrophobic interaction between ergosterol and the polyene stretch in the macrolide ring of AmpB.

The most extensively studied and widely accepted model for the action of AmpB is the formation of pores in the parasite membrane. The pores allow ions, water, and glucose molecules to permeate through the lipid bilayer, leading to the death of the parasite or fungus (Kamiński, 2014). The formation of the pores is not fully dependent on binding to sterols, but the ion transport is more efficient in the presence of ergosterol. Recently, a new mechanism of action was proposed in which the binding of AmpB to sequester ergosterol in the membrane is directly responsible for the therapeutic AmpB activity (Kamiński, 2014).

Miltefosine

Miltefosine (hexadecylphosphocholine, Impavido®), an anticancer drug, is orally administered as the second-line treatment in VL and PKDL patients who fail to respond to antimonial drugs, at dose of 2.5 mg/kg/day for a total of 28 days (Jain and Jain, 2013; Fernández *et al.*, 2014) (Fig. 6.1C). The mechanism of action of this drug is not well understood. Miltefosine induced an apoptosis-like death phenotype in *L. donovani* observed by DNA labelling (Paris *et al.*,2004; Verma and Dey, 2004). The programmed cell death was also found along with increased metacaspase activity in the promastigote forms of *Leishmania amazonesis* and *L. infantum* (Marinho *et al.*, 2011; Khademvatan *et al.*, 2011). Miltefosine was found to affect mitochondrial function by depolarizing the membrane potential and inhibiting cytochrome c oxidase, which is probably associated with the apoptosis-related death of the parasite (Zuo *et al.*, 2011).

Paromomycin

Paromomycin showed a 94.6% cure rate in Phase IV clinical trials in India, and its parenteral formulation is currently used to treat VL (Sundar *et al.*, 2007) (Fig. 6.1E). Orally administered paromomycin is poorly absorbed, however, which limits its usage to intramuscular injections for VL. In a phase III clinical trial conducted in Tunisia, a topical formulation of paromomycin was effective against cutaneous leishmaniasis with an index lesion cure rate of 82% for paromomycin alone and 81% for paromomycin combined with gentamycin (Ben Salah *et al.*, 2013). A similar study was conducted in Panama as a Phase II clinical trial, but the combined therapy achieved a relatively higher cure rate (Sosa *et al.*, 2013), providing a potential new approach for the treatment of cutaneous leishmaniasis.

The mechanism of paromomycin action in *Leishmania* is not precisely known, but protein synthesis has been proposed as the target, based largely on studies in bacteria. Other possible mechanisms have been proposed, including the alteration of membrane fluidity and effects on the mitochondrial membrane potential. The ribosomal complex, responsible for translating genetic information in mRNA to protein, is the usual site of action for aminoglycoside antibiotics. Paromomycin binds directly to the RNA and the binding stabilizes the ribosomal protein synthesis complex and induces the misreading of the codons in the mRNA. Based on a recent analysis of the global proteome differences between wild-type and paromomycin-resistant *L. donovani*, the up-regulation of ribosomal proteins was observed in the resistant parasite, suggesting that the protein synthesis machinery is the site of paromomycin action in *Leishmania* (Chawla *et al.*, 2011).

Pentamidine

Pentamidine is currently used as a second-line treatment for VL (Fig. 6.1D). Currently this drug is mainly used for antimony resistance and immunocompromised patients as well as in combination with other drugs but with limited success (Patel and Lockwood, 2009). Although its precise mode of action is not known but it is reported that the drug enters inside promastigote through arginine and polyamine transporters and gets accumulated in mitochondria and inhibit mitochondrial topoisomerase II (Mukherjee *et al.*, 2006; Porcheddu *et al.*, 2012). Pentamidine is highly toxic and causes hypoglycaemia, nephrotoxicity, hypotension, etc.

Natural products as promising antileishmanial agents (last two years of research)

An alternative to synthetic drugs is the search for anti-parasitic plant extracts or secondary metabolites derived from them. A number of natural products have already been screened. Among natural products many compounds have shown potential of antileishmanial lead activity: flavonoids, chalcones, lignans, coumarins, iridoids, monoterpenes, saponins, toxoids, curcumin, quinoline alkaloids, quinones, alkaloids, terpenes (Polonio and Efferth, 2008; Brito *et al.*, 2013; Singh *et al.*, 2014; Chouhan *et al*, 2014; Hussain *et al.*, 2014; Ullah *et al.*, 2016).

Alkaloids

The *in vitro* anti-*Leishmania amazonensis* activity of crude extracts of branches, leaves and fruits of *Guatteria latifolia* was evaluated by Ferreira *et al.* (2017). The branch extract (GCE) exhibited promising leishmanicidal activity against promastigotes (IC_{50} 51.7 μg/ml), and was submitted to fractionation guided by *in vitro* assays. Among the seven subfractions obtained, GF1 and GF2 were the most actives against promastigotes with IC_{50} 25.6 and 16 μg/ml, respectively. Since GCE, GF1 and GF2 were not toxic for macrophages, Effect on intracellular amastigotes was tested, and the IC_{50} values obtained were, respectively 30.5, 10.4 and 7.4 μg/ml, after 24 h treatment. The selectivity index for GCE, GF1 and GF2 were > 6.5, > 19.2 and > 27, respectively. Additionally, GCE, GF1 and GF2 affected the division pattern of the promastigotes by increasing 6.7, 9.4 and 7-fold the cells in sub-G0/G1 phase, and decreasing 1.6-, 2.5- and 1.8-fold the cells in G0/G1 phase, respectively. To assess the GCE and GFs capacity to modulate microbicide mechanisms of macrophages, nitric oxide (NO) and TNF-α production were tested. At the IC_{50}s GCE, GF1 and GF2 decreased NO production of infected macrophages stimulated with IFN-γ and LPS, besides, only GF1 decreased the production of TNF-α.

Koolen *et al.* (2017) isolated five alkaloids, in addition to hydroxytyrosol and uridine, from aerial parts of *Alternanthera littoralis P. Beauv*. Among the isolated compounds, alternamide A was an unusual tricyclic alkaloid with a bridged benzoazepine core (Fig. 6.2A). All isolated alkaloids have a catechol moiety, indicating a possible common biosynthetic route. The antiprotozoal activity of the isolated compounds was assayed against amastigotes of *Leishmania amazonensis*. Alternamine A was the most active compound, reducing markedly the viability of *Leishmania*. Antioxidant capacities evaluated by ORACFL assay showed that the isolated alkaloids (mainly alternamide B) contributed to the high activity recorded for the ethanolic crude extract; possibly, the catechol moiety present in all structures plays a central role in this result.

Martins *et al.* (2016) showed that synthetic analogues of marine sponge guanidine alkaloids showed *in vitro* antiparasitic activity against *Leishmania infantum* and *Trypanosoma cruzi*. Guanidines 10 and 11 presented the highest selectivity index when tested against *Leishmania* (Fig. 6.2B). Both compounds induced depolarization of mitochondrial membrane potential, up-regulation of reactive oxygen species levels, and increased plasma membrane permeability in *Leishmania* parasites. Immunomodulatory assays suggested an NO-independent effect of guanidines 10 and 11 on macrophages. The same compounds also promoted anti-inflammatory activity in *L. infantum*-infected macrophages cocultived with splenocytes, reducing the production of cytokines MCP-1 and IFN-γ. Guanidines

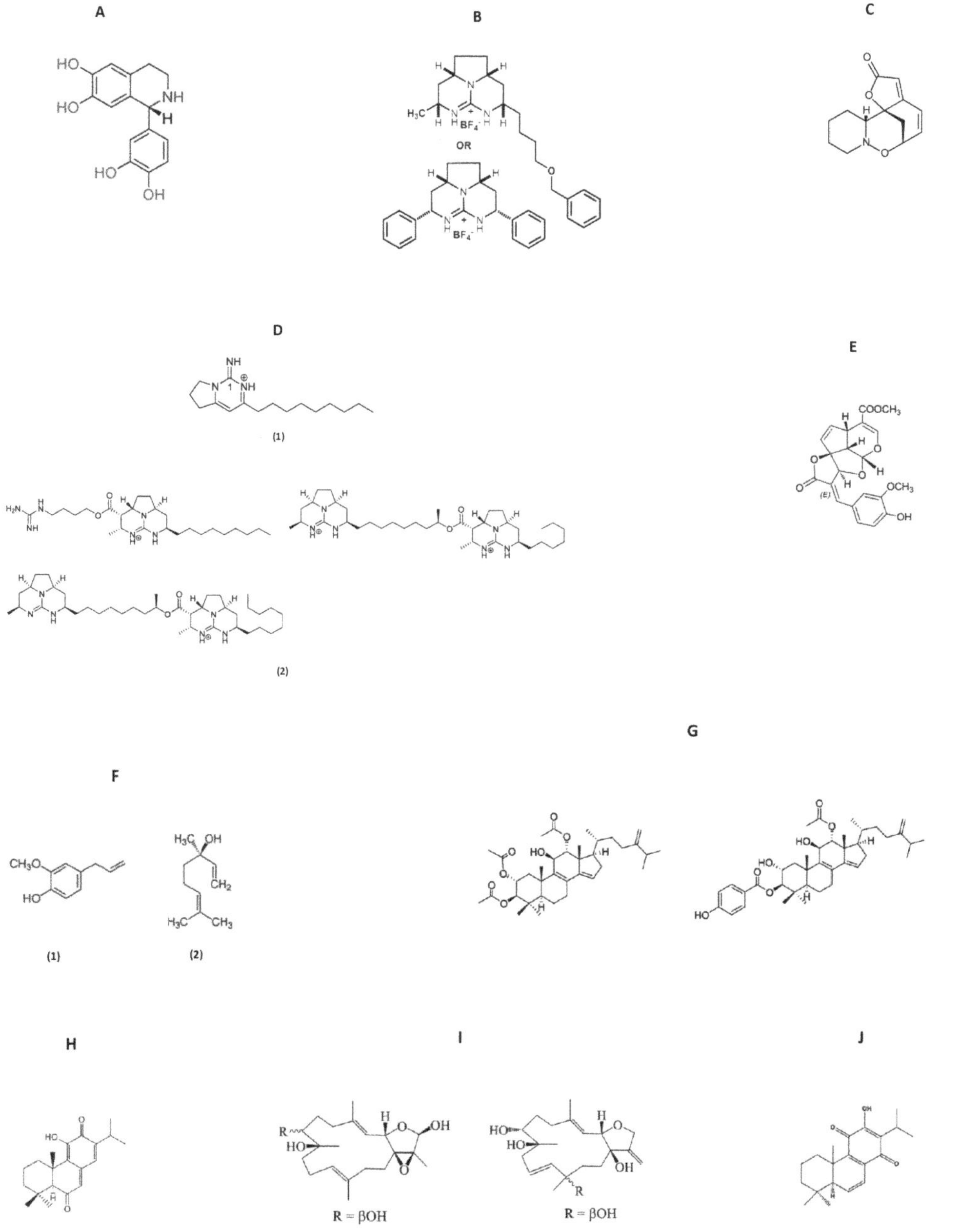

Figure 6.2 Molecules active against *Leishmania*. (A) Alternamide A (Koolen *et al.*, 2017); (B) guanidines 10 and 11 (Martins *et al.*, 2016); (C) phyllanthidine (Moraes *et al.*, 2015); (D) (1) monalidine A and (2) batzelladines D, F, L (Santos *et al.*, 2015); (E) molucidin (Amoa-Bosompem *et al.*, 2016); (F) (1) eugenol and (2) linalool (Dutra *et al*, 2016); (G) integracides H and J (Ibrahim *et al*, 2016); (H) taxodione (Búfalo *et al.*, 2016); (I) lobocrasol A and C (Thao *et al.*, 2015); (J) TrROY (Demarchi *et al.*, 2015); (K) lupane (Teles *et al.*, 2015); (L) isoquercitrin (Tracanna *et al.*, 2016); (M) pentalinonsterol (Gupta *et al.*, 2015); (N) (1 and 2) ergostanes and (3) lanostane (Leliebre-Lara *et al.*, 2016).

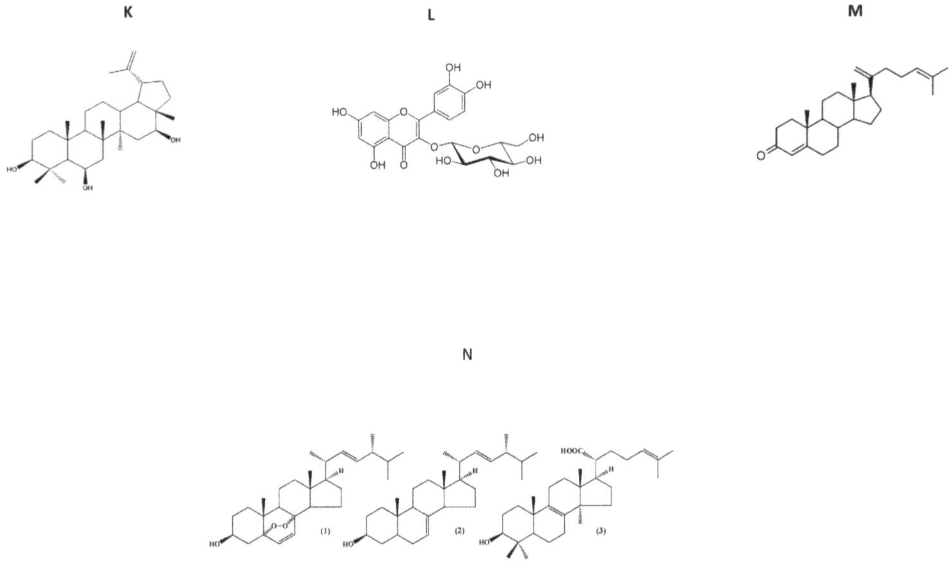

Figure 6.2 continued

10 and 11 affect the bioenergetic metabolism of *Leishmania,* with selective elimination of parasites via a host-independent mechanism.

Lorenzo *et al.* (2016) evaluated the potential activity of the azaphenanthrene alkaloids eupolaramine, imbiline 1, imbiline 4, sampangine, 3-metoxisampangine and 4-metoxisampangine, isolated from *A. dolichocarpa,* together with a homemade databank of 142 aporphynic alkaloids isolated from *Annonaceae,* through ligand-based and structure based virtual screening (VS) against *Leishmania donovani.* A diverse set selected from CHEMBL databank of 1397 structures, with tested antileishmanial activity against promastigote *L. donovani,* were classified according pIC_{50} values in order to generate and validate Random Forest model that show higher statistical indices values. The structure of six different *L. donovani* enzymes were downloaded from PDB databank and alkaloids structures were submitted to molecular docking. From the six azaphenanthrene alkaloids, sampangine, 3-methoxy, and 4-methoxy were indicated as potential actives by the RF model. Docking results gave similar values for all six azaphenanthrene alkaloids. *In vitro* tests with sampangine, imbiline 1, imbiline 4, and eupolaramine showed significant values of IC_{50} (> 5.26).

The effects of the Securinega alkaloid (+)-phyllanthidine on *Leishmania amazonensis* (Fig. 6.2C) and the first chemical investigation of *Margaritaria nobilis L.f.* (Phyllanthaceae) was realized by Moraes *et al.* (2015). Treating the parasites with this alkaloid caused a dose-dependent reduction in promastigote growth of 67.68% (IC_{50} 82.37 μg/ml or 353 μM) and in amastigote growth of 83.96% (IC_{50} 49.11 μg/ml or 210 μM), together with ultrastructural alterations in the promastigotes. No cytotoxic effect was detected in mammalian cells (CC_{50} 1727.48 μg/ml or CC_{50} 5268 μM). Classical chromatographic techniques and spectral methods led to the isolation and identification of betulinic acid, kaempferol, corilagin, gallic acid and its methyl ester, besides (+)-phyllanthidine from *M. nobilis* leaves and stems. *Margaritaria nobilis* is another source of the small group of Securinega alkaloids, together with other Phyllanthaceae (Euphorbiaceae s.l.) species. The low toxicity to macrophages and the effects against promastigotes and amastigotes are suggestive that (+)-phyllanthidine could be a promising antileishmanial agent for treating cutaneous leishmaniasis.

On the other hand, chemical investigation of the stem of *Thalictrum foliolosum* resulted in the isolation of two new bisbenzylisoquinoline alkaloids along with known protoberberine group of isoquinoline alkaloids thalifendine and berberine. The structures of the new compounds were established by detailed 2D NMR spectral analysis with their configurations determined from their optical rotation values and confirmed using circular dichroism. Inhibitory activities of these four compounds against DNA topoisomerase IB of *Leishmania donovani* were evaluated. One of bisbenzylisoquinoline alkaloid exhibited almost complete inhibition of the enzyme activity at 50 µM concentration and it was found to be effective in killing both wild type as well as SAG resistant promastigotes of the parasite (Kumar *et al.*, 2016).

The crude extracts of tubers of *Aconitum spicatum* (Bruhl) Stapf were investigated for *in vitro* antileishmanial activity against *Leishmania major*. The dichloromethane extract at pH 2.5 showed antileishmanial activity with IC_{50} value of 27.10 ± 0.0 µg/ml. Chromatographic purification of the dichloromethane extract led to isolation of three C-19 norditerpenoid alkaloids indaconitine, chasmaconitine and ludaconitine. The last two compounds showed antileishmanial activity with IC_{50} = 56.30 ± 2.1 and 36.10 ± 3.4 µg/ml, respectively. Indaconitine was less effective (IC_{50} > 100 µg/ml). The cytotoxicity of compounds studied against MCF7, HeLa and PC3 cancer cell lines and 3T3 normal fibroblast cell line did not show cytotoxicity at 30 µM (Shyaula *et al.*, 2015).

HPLC guided fractionation of the anti-parasitic extract obtained from the marine sponge *Monanchora arbuscula*, led to the isolation of a series of guanidine and pyrimidine alkaloids (Santos *et al.*, 2015). The pyrimidines monalidine A and batzelladines D, F, L (Fig. 6.2D) and norbatzelladine L were found to be active against *L. infantum* promastigotes, with IC_{50} values in the range 1.5–4.6 µg/ml. However, no activity was observed against intracellular *L. infantum* amastigotes for any of these compounds.

Naman *et al.*, 2015, showed *in vitro* antileishmanial activity by three bisbenzyltetrahydroisoquinoline alkaloids that were purified previously from *Thalictrum alpinum*. Of the compounds tested, northalrugosidine showed the most potent *in vitro* activity against *Leishmania donovani* promastigotes (0.28 µM) and the highest selectivity (29.3-fold) versus its general cytotoxicity against HT-29 human colon adenocarcinoma cells. Northalrugosidine was tested *in vivo* using a murine model of visceral leishmaniasis, resulting in the observation of a dose-dependent reduction of the parasitic burden in the liver and spleen without overt toxicity effects at 2.8, 5.6, and 11.1 mg/kg per animal when administered intravenously.

Quinones

Recently was showed the effects of epoxy-α-lapachone on *L. amazonensis*, this compound was able to cause measurable effects on promastigote and amastigote forms of the parasite, affecting plasma membrane organization and leading to death after 3 h of exposure (Souza-Silva *et al.*, 2015). Also, it had an effect in experimentally infected BALB/c mice, causing reductions in paw lesions 6 weeks after treatment with 0.44 mM epoxy-α-lapachone compared to untreated animals or animals treated with Glucantime. The effects of this compound on the serine proteinase activities of the parasite was evaluated. Serine proteinase-enriched fractions were extracted from both promastigotes and amastigotes form and they seem to act on specific serine proteinase substrates and they were sensitive to classic serine proteinase inhibitors (phenylmethylsulfonyl fluoride, aprotinin, and antipain). These fractions were also affected by epoxy-α-lapachone. Furthermore, in silico simulations indicated that epoxy-α-lapachone can bind to oligopeptidase B (OPB) of *L. amazonensis*, a serine proteinase, in a manner similar to that of antipain, interacting with an S1 binding site. This evidence suggests that OPB may be a potential target for epoxy-α-lapachone and, as such, may be related to the compound's effects on the parasite.

Mahmoudvand *et al.*, 2015 evaluated the anti-leishmanial effects of *Nigella sativa L.* (Ranunculaceae) against *Leishmania tropica* and *Leishmania infantum* with an *in vitro* model. Antileishmanial effects of essential oil and methanolic extract of *N. sativa* (0–200 µg/ml) and thymoquinone (0–25 µg/ml) on promastigotes of both species. As well as, cytotoxicity activities against murine macrophages were evaluated using the MTT assay and their leishmanicidal effects against amastigotes were investigated in a macrophage model. Results showed that essential oil (*L. tropica* IC_{50} 9.3 µg/ml and *L. infantum* IC_{50} 11.7 µg/ml) and methanolic extract (*L. tropica* IC_{50} 14.8 µg/ml and *L. infantum* IC_{50} 15.7 µg/ml) of *N. sativa*, particularly thymoquinone (*L. tropica* IC_{50} 1.16 µg/ml and *L. infantum* IC_{50} 1.47 µg/ml), had potent anti-leishmanial activity on promastigotes of both species after 72 h. In addition, essential oil (*L. tropica* IC_{50} 21.4 µg/ml and *L. infantum* IC_{50} 26.3 µg/ml), methanolic extract (*L. tropica* IC_{50} 30.8 µg/ml and *L. infantum* IC_{50} 34.6 µg/ml), and thymoquinone (*L. tropica* IC_{50} 2.1 µg/ml and *L. infantum* IC_{50} 2.6 µg/ml) mediated a significant decrease in the growth rate of amastigote forms of both species. Thymoquinone (CC50 38.8 µg/ml) exhibited higher cytotoxic effects against murine macrophages than the other extracts.

Iridoids

Amoa-Bosompem *et al.*, 2016, determined the activity of three compounds with an iridoid tetracyclic skeleton against *Leishmania* using field strain, 010, and the lab strain *Leishmania hertigi*. They found two of the three compounds, molucidin and ML-F52, to have anti-*Leishmania* activity against both strains (Fig. 6.2E). The fluor-microscope observation with DAPI stain revealed that both Molucidin and ML-F52 induced an abnormal parasite formation, with two sets of nuclei and kinetoplast in a cell, suggesting that compounds might inhibit cytokinesis in *Leishmania* parasites. Molucidin and ML-F52 might be good lead compounds for the development of new anti-*Leishmania* chemotherapy.

Terpenes

The antiprotozoal activity of the terpenes dehydroabietic acid, dehydroabietinol, oleanolic acid, and 34 synthetic derivatives were evaluated by Pertino *et al.* (2017) against epimastigote forms of *Trypanosoma cruzi* and promastigotes of *Leishmania braziliensis* and *Leishmania infantum*. The activity of the compounds was moderate and the antiparasitic effect was associated with the linker length between the diterpene and the triazole in dehydroabietinol derivatives. For the oleanolic acid derivatives, a free carboxylic acid function led to better antiparasitic activity.

In another study, Dutra *et al.* (2016) showed the effects of linalool and eugenol, components of essential oils, on *Leishmania chagasi*, one of the causative agents of visceral leishmaniasis (Fig. 6.2F). They compared the effects of those compounds to the effects of glucantime, as a positive control. In *L. chagasi* killing assays, the LD_{50} for eugenol was 220 µg/ml, and that for linalool was 550 µg/ml. Eugenol and linalool significantly decreased the number of parasites within the macrophages. Eugenol and linalool enhanced the activities of the *L. chagasi* protein kinases PKA and PKC and Linalool, also, decreased parasite oxygen consumption. In conclusion, both linalool and eugenol promoted a decrease in the proliferation and viability of *L. chagasi*. These effects were more pronounced during the interaction between the parasites and peritoneal mouse.

Semisynthetic terpenoid derivatives from the major latex components of *Euphorbia officinarum* have been evaluated against *Leishmania infantum* and *Trypanosoma cruzi* by Bailen *et al.* (2016). This study showed that 87% of the test compounds had antiparasitic effects on both *L. infantum* and *T. cruzi*, with some of them being selective parasite toxicants.

On the other hand, three new tetracyclic triterpenoids namely, integracides H, I and J, along with integracides B and F have been isolated from the endophytic fungus *Fusarium* sp. isolated from the roots of *Mentha longifolia* L. (Labiatae) (Ibrahim *et al.*, 2016). Integracides H and J (Fig. 6.2G) showed significant anti-leishmanial activity towards *Leishmania donovani*, with IC_{50} values of 4.75 and 3.29 µM, respectively, compared with pentamidine (IC_{50} 6.35 µM).

The herbaceous shrub *Tetradenia riparia* has been traditionally used to treat inflammatory and infectious diseases. Recently, a study showed that *T. riparia* essential oil (TrEO) has antileishmanial effects, although these results could be influenced by seasonal variation (Cardoso *et al.*, 2015). This study evaluated the activity of the TrEO obtained in different seasons against *Leishmania amazonensis, in vitro* and *in vivo*. The majority compounds in the TrEO were terpenoids and oxygenated sesquiterpenes (55.28%). The TrEO from all seasons showed IC_{50} for promastigotes of about 15 ng/ml; at 30 ng/ml and 3 ng/ml, the TrEO reduced intracellular amastigote infection, independently of season. The TrEO from plants harvested in summer had the highest CC_{50}, 1,476 ng/ml for J774.A1 macrophages, and in spring (90.94 ng/ml) for murine macrophages. NO production did not change in samples of the TrEO from different seasons. The antileishmanial effect *in vivo* consisted of a reduction of the parasite load in the spleen. These results suggest that the TrEO has potential effects on *L. amazonensis*, consonant with its traditional use to treat parasitic diseases.

Croton cajucara Benth. (Euphorbiaceae) is a plant found in the Amazonian region with a history of safe use in folk medicine. The effects of clerodane diterpenes, trans-dehydrocrotonin (DCTN), trans-crotonin (CTN) and acetylaleuritolic acid (AAA) obtained from powdered bark of *C. cajucara* against promastigotes, axenic and intracellular amastigotes of *Leishmania amazonensis* was investigated (Lima *et al.*, 2015). The results showed very high efficacy, especially of the diterpene DCTN, against promastigotes ($IC_{50} = 6.30 \pm 0.06$ µg/ml) and axenic amastigotes ($IC_{50} = 19.98 \pm 0.05$ µg/ml) of *L. amazonensis*. The cytotoxic effect of the best active natural product was evaluated on mouse peritoneal infected macrophages ($IC_{50} = 0.47 \pm 0.03$ µg/ml in 24 h of culture), and the treatment revealed that DCTN never reaches toxic concentrations while reducing the infection and, most importantly, with no toxicity (> 100 µg/ml with 0% of macrophage kill) when compared to pentamidine (37.5 µg/ml with 100% of macrophage kill). Furthermore, all of the natural products assayed on the trypanothione reductase enzyme inhibited the enzyme activity compared to the control.

In other study, four diterpenes with biological activity were isolated from *Salvia deserta* roots (Búfalo *et al.*, 2016). Taxodione was considered leishmanicidal with an IC_{50} value of 1.46 µM (0.46 mg/l) against *Leishmania donovani* and also exhibited antifungal and antimicrobial activities (Fig. 6.2H).

Based on previous finding that certain cembranoid diterpenes possess selective toxicity against protozoan pathogens of tropical diseases, Thao *et al.*, 2015, have subjected a series of 34 cembranes isolated from soft corals living in the Vietnamese sea to an *in vitro* screening for anti-protozoal activity. Lobocrasol A and lobocrasol C from *L. crassum* and *L. laevigatum* exhibited particularly significant inhibitory effects on *L. donovani* with IC_{50} values < 0.2 μM (Fig. 6.2I). The cytotoxicity of the active compounds on L6 rat skeletal myoblast cell was also assessed and found to be insignificant in all cases.

Demarchi *et al.*, 2015, evaluated the potential of the essential oil (TrEO) and 6,7-dehydroroyleanone (TrROY) (Fig. 6.2J), isolated from *T. riparia*, on *L. amazonensis* promastigote and amastigote forms, cytotoxicity on human erythrocytes and murine macrophages, nitric production and inducible nitric oxide synthase (iNOS) mRNA expression. The lethal dose of TrEO and TrROY were 0.8 μg/ml and 3 μg/ml, respectively. TrEO and TrROY were not cytotoxic to human erythrocytes, but TrROY was toxic to murine macrophages resulting in a low selectivity index. The transmission electronic microscopy showed that TrEO (0.03 μg/ml) was able to modify the promastigote ultrastructures suggesting autophagy as chromatin condensation, blebbing, membranous profiles and nuclear fragmentation. Infected-macrophages treated with TrEO (0.03 μg/ml) or TrROY (10 μg/ml) had an infection index decreased in 65 and 48%. TrEO did not induce iNOS mRNA expression or nitrite production in macrophages infected with *Leishmania*. TrROY and mainly TrEO promoted the *Leishmania* death, and TrROY showed loss toxicity to erythrocytes cells.

3beta,6beta,16beta-trihydroxylup-20(29)-ene is a lupane triterpene isolated from *Combretum leprosum* fruit (Fig. 6.2K). The lupane group has been extensively used in studies on anticancer effects; however, it's possible activity against protozoa parasites is yet poorly known. The activity of this product was evaluated against *Leishmania amazonensis* by determining the cytotoxicity of the compound on murine peritoneal macrophages, as well as its effects on parasite survival inside host cells (Teles *et al.*, 2015). Lupane showed significant activity against the intracellular amastigotes of *L. amazonensis*. The treatment with 109 μM for 96 h reduced in 80% the survival index of parasites in BALB/c peritoneal macrophages. At this concentration, the triterpene caused no cytotoxic effects against mouse peritoneal macrophages. Ultrastructural analyses of *L. amazonensis* intracellular amastigotes showed that lupane induced some morphological changes in parasites, such as cytosolic vacuolization, lipid body formation and mitochondrial swelling. Bioinformatic analyses through molecular docking suggest that this lupane has high-affinity binding with DNA topoisomerase.

The effect of astrakurkurone, a novel triterpene isolated from Indian mushroom *Astraeus hygrometricus*, has been investigated to elucidate the mechanisms involved in selective cell death of *Leishmania donovani* (Mallick *et al.*, 2015). The time dependent elevation of astrakurkurone-induced reactive oxygen species (ROS) was found intimately associated with apoptosis. The involvement of ROS in promastigote death was found confirmed as NAC and GSH could decrease the ROS level and restored the mitochondrial membrane potential ($\Delta\Psi(m)$). It also inhibited the intracellular amastigotes.

Phenoles and steroles

Lipid-rich microalgal species from the Red Sea belonging to three different genera (*Nannochloris*, *Picochlorum* and *Desmochloris*) were bioprospected for high-value, bioactive molecules. Nannochloris sp. SBL1 significantly reduced the viability of *Leishmania infantum* down to 62% (250 μg/ml) (Pereira *et al.*, 2015).

Tracanna *et al.* (2016) isolated a new phenolic derivative, 2,8-dihydroxy-7H-furo[2,3-f] chromen-7-one, together with isoquercitrin, from the aerial parts of *Tibouchina paratropica* (Fig. 6.2L). Both compounds showed antimicrobial activity towards a panel of bacterial and fungal pathogens, and the first compound displayed potent anti-parasitic activity against *Leishmania donovani* (IC_{50}=0.809 μg/ml).

Pentalinonsterol (PEN), a novel sterol isolated from a plant used by Mayan traditional healers, is effective in treatment of visceral leishmaniasis caused by *Leishmania donovani* (Gupta *et al.*, 2015) (Fig. 6.2M). PEN, which occurs naturally in the root of a Mexican medicinal plant, *Pentalinon andrieuxii*, is effective in the experimental BALB/c mouse model of VL induced by infection with *L. donovani*. Intravenous treatment with liposome-encapsulated PEN (2.5 mg/kg) led to a significant reduction in parasite burden in the liver and spleen. Furthermore, infected mice treated with liposomal PEN showed a strong host-protective TH1 immune response characterized by IFN-γ production and formation of matured hepatic granulomas. These results indicate that PEN could be developed as a novel drug against VL.

Sterols isolated from *Trametes versicolor* showed leishmanicidal activiy (Leliebre-Lara *et al.*, 2016). Two ergostanes, 5α,8α-epidioxy-22E-ergosta-6,22-dien-3β-ol and 5α-ergost-7,22-dien-3β-ol, and a lanostane, 3β-hydroxylanostan-8,24-diene-21-oic acid (trametenolic acid), were isolated from an *n*-hexane extract prepared from the fruiting body of *Trametes versicolor* (Bres. Rivarden) (Fig. 6.2N). The activity of the isolated sterols was evaluated against promastigotes and amastigotes of *Leishmania amazonensis* and, the lanostane, compound, showed the best inhibitory response (IC_{50} promastigotes 2.9 ± 0.1 μM and IC_{50} amastigotes 1.6 ± 0.1 μM). This effect was 25-fold higher compared with its cytotoxic effect on peritoneal macrophages from BALB/c mice.

References

Alvar, J., Aparicio, P., Aseffa, A., Den Boer, M., Cañavate, C., Dedet, J.P., Gradoni, L., Ter Horst, R., López-Vélez, R., and Moreno, J. (2008). The relationship between leishmaniasis and AIDS: the second 10 years. Clin. Microbiol. Rev. *21*, 334–359, table of contents. https://doi.org/10.1128/CMR.00061-07.

Alvar, J., Croft, S., and Olliaro, P. (2006). Chemotherapy in the treatment and control of leishmaniasis. Adv. Parasitol. *61*, 223–274.

Alvar, J., Vélez, I.D., Bern, C., Herrero, M., Desjeux, P., Cano, J., Jannin, J., and den Boer, M. (2012). Leishmaniasis worldwide and global estimates of its incidence. PLOS ONE *7*, e35671. https://doi.org/10.1371/journal.pone.0035671.

Alvar, J., Yactayo, S., and Bern, C. (2006). Leishmaniasis and poverty. Trends Parasitol. *22*, 552–557.

Alviano, D.S., Barreto, A.L., Dias, F.d.e.A., Rodrigues, I.d.e.A., Rosa, M.d.o.S., Alviano, C.S., and Soares, R.M. (2012). Conventional therapy and promising plant-derived compounds against trypanosomatid parasites. Front. Microbiol. *3*, 283. https://doi.org/10.3389/fmicb.2012.00283.

Amoa-Bosompem, M., Ohashi, M., Mosore, M.T., Agyapong, J., Tung, N.H., Kwofie, K.D., Ayertey, F., Owusu, K.B., Tuffour, I., Atchoglo, P., *et al.* (2016). *In vitro* anti-Leishmania activity of tetracyclic iridoids from *Morinda lucida*, benth. Trop. Med. Health *44*, 25. https://doi.org/10.1186/s41182-016-0026-5.

Bailen, M., Khamlichi, M.D., Benharref, A., Martinez-Diaz, R.A., and Gonzalez-Coloma, A. (2016). New bioactive semisynthetic derivatives of 31-norlanostenol and obtusifoliol from *Euphorbia officinarum*. Nat. Prod. Commun. *11*, 733–738.

Ben Salah, A., Ben Messaoud, N., Guedri, E., Zaatour, A., Ben Alaya, N., Bettaieb, J., Gharbi, A., Belhadj Hamida, N., Boukthir, A., Chlif, S., *et al.* (2013). Topical paromomycin with or without gentamicin for cutaneous leishmaniasis. N. Engl. J. Med. *368*, 524–532. https://doi.org/10.1056/NEJMoa1202657.

Blum, J., Desjeux, P., Schwartz, E., Beck, B., and Hatz, C. (2004). Treatment of cutaneous leishmaniasis among travellers. J. Antimicrob. Chemother. *53*, 158–166. https://doi.org/10.1093/jac/dkh058.

Brito, A.M., Dos Santos, D., Rodrigues, S.A., Brito, R.G., and Xavier-Filho, L. (2013). Plants with anti-*Leishmania* activity: Integrative review from 2000 to 2011. Pharmacogn. Rev. *7*, 34–41. https://doi.org/10.4103/0973-7847.112840.

Búfalo, J., Cantrell, C.L., Jacob, M.R., Schrader, K.K., Tekwani, B.L., Kustova, T.S., Ali, A., and Boaro, C.S. (2016). Antimicrobial and Antileishmanial activities of diterpenoids isolated from the roots of *Salvia deserta*. Planta Med. *82*, 131–137. https://doi.org/10.1055/s-0035-1557875.

Cardoso, B.M., de Mello, T.F., Lopes, S.N., Demarchi, I.G., Lera, D.S., Pedroso, R.B., Cortez, D.A., Gazim, Z.C., Aristides, S.M., Silveira, T.G., *et al.* (2015). Antileishmanial activity of the essential oil from *Tetradenia riparia* obtained in different seasons. Mem. Inst. Oswaldo Cruz *110*, 1024–1034. https://doi.org/10.1590/0074-02760150290.

Chawla, B., Jhingran, A., Panigrahi, A., Stuart, K.D., and Madhubala, R. (2011). Paromomycin affects translation and vesicle-mediated trafficking as revealed by proteomics of paromomycin -susceptible -resistant *Leishmania donovani*. PLOS ONE *6*, e26660. https://doi.org/10.1371/journal.pone.0026660.

Chouhan, G., Islamuddin, M., Sahal, D., and Afrin, F. (2014). Exploring the role of medicinal plant-based immunomodulators for effective therapy of leishmaniasis. Front. Immunol. *5*, 193. https://doi.org/10.3389/fimmu.2014.00193.

Demarchi, I.G., Thomazella, M.V., de Souza Terron, M., Lopes, L., Gazim, Z.C., Cortez, D.A., Donatti, L., Aristides, S.M., Silveira, T.G., and Lonardoni, M.V. (2015). Antileishmanial activity of essential oil and 6,7-dehydroroyleanone isolated from *Tetradenia riparia*. Exp. Parasitol. *157*, 128–137. https://doi.org/10.1016/j.exppara.2015.06.014.

Demicheli, C., Frézard, F., Lecouvey, M., and Garnier-Suillerot, A. (2002). Antimony(V) complex formation with adenine nucleosides in aqueous solution. Biochim. Biophys. Acta *1570*, 192–198.

Demicheli, C., Frézard, F., Mangrum, J.B., and Farrell, N.P. (2008). Interaction of trivalent antimony with a CCHC zinc finger domain: potential relevance to the mechanism of action of antimonial drugs. Chem. Commun. *39*, 4828–4830. https://doi.org/10.1039/b809186b.

Denton, H., McGregor, J.C., and Coombs, G.H. (2004). Reduction of anti-leishmanial pentavalent antimonial drugs by a parasite-specific thiol-dependent reductase, TDR1. Biochem. J. *381*, 405–412.

Dutra, F.L., Oliveira, M.M., Santos, R.S., Silva, W.S., Alviano, D.S., Vieira, D.P., and Lopes, A.H. (2016). Effects of linalool and eugenol on the survival of *Leishmania* (*L.*) *infantum chagasi* within macrophages. Acta Trop. *164*, 69–76.

European Centre for Diseases Control/WHO Regional Office for Europe (2013). HIV/AIDS Surveillance in Europe 2012. Available online: www.euro.who.int/__data/assets/pdf_file/0018/235440/e96953.pdf. Accessed 10 March 2017.

Feasey, N., Wansbrough-Jones, M., Mabey, D.C., and Solomon, A.W. (2010). Neglected tropical diseases. Br. Med. Bull. 93: 179–200.

Fernández, O.L., Diaz-Toro, Y., Ovalle, C., Valderrama, L., Muvdi, S., Rodríguez, I., Gomez, M.A., and Saravia, N.G. (2014). Miltefosine and antimonial drug susceptibility of *Leishmania Viannia* species and populations in regions of high transmission in Colombia. PLOS Negl. Trop. Dis. *8*, e2871. https://doi.org/10.1371/journal.pntd.0002871.

Ferreira, C.d.o.s.S., Martins, P.S., Demicheli, C., Brochu, C., Ouellette, M., and Frézard, F. (2003). Thiol-induced reduction of antimony(V) into antimony(III): a comparative study with trypanothione, cysteinyl-glycine, cysteine and glutathione. Biometals *16*, 441–446.

Ferreira, C., Passos, C.L., Soares, D.C., Costa, K.P., Rezende, M.J., Lobão, A.Q., Pinto, A.C., Hamerski, L., and Saraiva, E.M. (2017). Leishmanicidal activity of the alkaloid-rich fraction from *Guatteria latifolia*. Exp. Parasitol. *172*, 51–60.

Gupta, G., Peine, K.J., Abdelhamid, D., Snider, H., Shelton, A.B., Rao, L., Kotha, S.R., Huntsman, A.C., Varikuti, S., Oghumu, S., *et al.* (2015). A novel sterol isolated from a plant used by Mayan traditional healers is effective in treatment of visceral leishmaniasis caused by *Leishmania donovani*. ACS Infect. Dis. *1*, 497–506. https://doi.org/10.1021/acsinfecdis.5b00081.

Hussain, H., Al-Harrasi, A., Al-Rawahi, A., Green, I.R., and Gibbons, S. (2014). Fruitful decade for antileishmanial compounds from 2002 to late 2011. Chem. Rev. *114*, 10369–10428. https://doi.org/10.1021/cr400552x.

Ibrahim, S.R., Abdallah, H.M., Mohamed, G.A., and Ross, S.A. (2016). Integracides H-J: New tetracyclic triterpenoids from the endophytic fungus *Fusarium* sp. Fitoterapia *112*, 161–167. https://doi.org/10.1016/j.fitote.2016.06.002.

Jain, K., and Jain, N.K. (2013). Novel therapeutic strategies for treatment of visceral leishmaniasis. Drug Discov. Today *18*, 1272–1281. https://doi.org/10.1016/j.drudis.2013.08.005.

Kamiński, D.M. (2014). Recent progress in the study of the interactions of amphotericin B with cholesterol and ergosterol in lipid environments. Eur. Biophys. J. *43*, 453–467. https://doi.org/10.1007/s00249-014-0983-8.

Khademvatan, S., Gharavi, M.J., and Saki, J. (2011). Miltefosine induces metacaspase and PARP genes expression in *Leishmania infantum*. Braz. J. Infect. Dis. *15*, 442–448.

Koolen, H.H., Pral, E.M., Alfieri, S.C., Marinho, J.V., Serain, A.F., Hernández-Tasco, A.J., Andreazza, N.L., and Salvador, M.J. (2017). Antiprotozoal and antioxidant alkaloids from *Alternanthera littoralis*. Phytochemistry *134*, 106–113.

Krauth-Siegel, R.L., and Comini, M.A. (2008). Redox control in trypanosomatids, parasitic protozoa with trypanothione-based thiol metabolism. Biochim. Biophys. Acta *1780*, 1236–1248. https://doi.org/10.1016/j.bbagen.2008.03.006.

Kumar, A., Chowdhury, S.R., Sarkar, T., Chakrabarti, T., Majumder, H.K., Jha, T., and Mukhopadhyay, S. (2016). A new bisbenzylisoquinoline alkaloid isolated from *Thalictrum foliolosum*, as a potent inhibitor of DNA topoisomerase IB of *Leishmania donovani*. Fitoterapia *109*, 25–30. https://doi.org/10.1016/j.fitote.2015.11.021.

Leliebre-Lara, V., Monzote Fidalgo, L., Pferschy-Wenzig, E.M., Kunert, O., Nogueiras Lima, C., and Bauer, R. (2016). *In vitro* antileishmanial activity of sterols from *Trametes versicolor* (Bres. Rivarden). Molecules *21*, E1045. https://doi.org/10.3390/molecules21081045.

Lima, G.S., Castro-Pinto, D.B., Machado, G.C., Maciel, M.A., and Echevarria, A. (2015). Antileishmanial activity and trypanothione reductase effects of terpenes from the Amazonian species *Croton cajucara* Benth (Euphorbiaceae). Phytomedicine *22*, 1133–1137. https://doi.org/10.1016/j.phymed.2015.08.012.

Lindoso, J.A., Costa, J.M., Queiroz, I.T., and Goto, H. (2012). Review of the current treatments for leishmaniases. Res. Rep. Trop. Med. *3*, 69–77.

Lorenzo, V.P., Lúcio, A.S., Scotti, L., Tavares, J.F., Filho, J.M., Lima, T.K., Rocha, J.D., and Scotti, M.T. (2016). Structure- and ligand-based approaches to evaluate aporphynic alkaloids from annonaceae as multi-target agent against *Leishmania donovani*. Curr. Pharm. Des. *22*, 5196–5203.

Mahmoudvand, H., Tavakoli, R., Sharififar, F., Minaie, K., Ezatpour, B., Jahanbakhsh, S., and Sharifi, I. (2015). Leishmanicidal and cytotoxic activities of *Nigella sativa* and its active principle, thymoquinone. Pharm. Biol. *53*, 1052–1057. https://doi.org/10.3109/13880209.2014.957784.

Mallick, S., Dey, S., Mandal, S., Dutta, A., Mukherjee, D., Biswas, G., Chatterjee, S., Mallick, S., Lai, T.K., Acharya, K., *et al.* (2015). A novel triterpene from *Astraeus hygrometricus* induces reactive oxygen species leading to death in *Leishmania donovani*. Future Microbiol. *10*, 763–789. https://doi.org/10.2217/fmb.14.149.

Marinho, F.d.e.A., Gonçalves, K.C., Oliveira, S.S., Oliveira, A.C., Bellio, M., d'Avila-Levy, C.M., Santos, A.L., and Branquinha, M.H. (2011). Miltefosine induces programmed cell death in *Leishmania amazonensis* promastigotes. Mem. Inst. Oswaldo Cruz *106*, 507–509.

Martins, L.F., Mesquita, J.T., Pinto, E.G., Costa-Silva, T.A., Borborema, S.E., Galisteo Junior, A.J., Neves, B.J., Andrade, C.H., Shuhaib, Z.A., Bennett, E.L., *et al.* (2016). Analogues of marine guanidine alkaloids are *in vitro* effective against *Trypanosoma cruzi* and selectively eliminate *Leishmania* (*L.*) *infantum* intracellular amastigotes. J. Nat. Prod. *79*, 2202–2210. https://doi.org/10.1021/acs.jnatprod.6b00256.

Meheus, F., Balasegaram, M., Olliaro, P., Sundar, S., Rijal, S., Faiz, M.A., and Boelaert, M. (2010). Cost-effectiveness analysis of combination therapies for visceral leishmaniasis in the Indian subcontinent. PLOS Negl. Trop. Dis. *4*, e818. https://doi.org/10.1371/journal.pntd.0000818.

Monge-Maillo, B., Norman, F.F., Cruz, I., Alvar, J., and López-Vélez, R. (2014). Visceral leishmaniasis and HIV coinfection in the Mediterranean region. PLOS Negl. Trop. Dis. *8*, e3021. https://doi.org/10.1371/journal.pntd.0003021.

Moraes, L.S., Donza, M.R., Rodrigues, A.P., Silva, B.J., Brasil, D.S., Zoghbi, M.d., Andrade, E.H., Guilhon, G.M., and Silva, E.O. (2015). Leishmanicidal activity of (+)-phyllanthidine and the phytochemical profile of *Margaritaria nobilis* (Phyllanthaceae). Molecules *20*, 22157–22169. https://doi.org/10.3390/molecules201219829.

Mukherjee, A., Padmanabhan, P.K., Sahani, M.H., Barrett, M.P., and Madhubala, R. (2006). Roles for mitochondria in pentamidine susceptibility and resistance in *Leishmania donovani*. Mol. Biochem. Parasitol. *145*, 1–10.

Mukhopadhyay, R., Dey, S., Xu, N., Gage, D., Lightbody, J., Ouellette, M., and Rosen, B.P. (1996). Trypanothione overproduction and resistance to antimonials and arsenicals in *Leishmania*. Proc. Natl. Acad. Sci. U.S.A. *93*, 10383–10387.

Naman, C.B., Gupta, G., Varikuti, S., Chai, H., Doskotch, R.W., Satoskar, A.R., and Kinghorn, A.D. (2015). Northalrugosidine is a bisbenzyltetrahydroisoquinoline alkaloid from *Thalictrum alpinum* with *in vivo* antileishmanial activity. J. Nat. Prod. *78*, 552–556. https://doi.org/10.1021/np501028u.

Olliaro, P.L., Guerin, P.J., Gerstl, S., Haaskjold, A.A., Rottingen, J.A., and Sundar, S. (2005). Treatment options for visceral leishmaniasis: a systematic review of clinical studies done in India, 1980-2004. Lancet Infect. Dis. *5*, 763–774.

Paris, C., Loiseau, P.M., Bories, C., and Bréard, J. (2004). Miltefosine induces apoptosis-like death in *Leishmania donovani* promastigotes. Antimicrob. Agents Chemother. *48*, 852–859.

Patel, T.A., and Lockwood, D.N. (2009). Pentamidine as secondary prophylaxis for visceral leishmaniasis in the immunocompromised host: report of four cases. Trop. Med. Int. Health *14*, 1064–1070. https://doi.org/10.1111/j.1365-3156.2009.02329.x.

Pereira, H., Custódio, L., Rodrigues, M.J., de Sousa, C.B., Oliveira, M., Barreira, L., Neng, N.d.a.R., Nogueira, J.M., Alrokayan, S.A., Mouffouk, F., *et al.* (2015). Biological activities and chemical composition of methanolic extracts of selected autochthonous microalgae strains from the Red Sea. Mar. Drugs *13*, 3531–3549. https://doi.org/10.3390/md13063531.

Pertino, M.W., Vega, C., Rolón, M., Coronel, C., Rojas de Arias, A., and Schmeda-Hirschmann, G. (2017). Antiprotozoal activity of triazole derivatives of dehydroabietic acid and oleanolic acid. Molecules *22*, E369.

Polonio, T., and Efferth, T. (2008). Leishmaniasis: drug resistance and natural products (review). Int. J. Mol. Med. *22*, 277–286.

Porcheddu, A., Giacomelli, G., and De Luca, L. (2012). New pentamidine analogues in medicinal chemistry. Curr. Med. Chem. *19*, 5819–5836.

Rai, S., Bhaskar Goel, S.K., Nath Dwivedi, U., Sundar, S., and Goyal, N. (2013). Role of efflux pumps and intracellular thiols in natural antimony resistant isolates of *Leishmania donovani*. PLOS ONE *8*, e74862.

Reithinger, R., Dujardin, J.C., Louzir, H., Pirmez, C., Alexander, B., and Brooker, S. (2007). Cutaneous leishmaniasis. Lancet Infect. Dis. *7*, 581–596.

Santos, M.F., Harper, P.M., Williams, D.E., Mesquita, J.T., Pinto, É.G., da Costa-Silva, T.A., Hajdu, E., Ferreira, A.G., Santos, R.A., Murphy, P.J., *et al.* (2015). Anti-parasitic guanidine and pyrimidine alkaloids from the marine sponge *Monanchora arbuscula*. J. Nat. Prod. *78*, 1101–1112. https://doi.org/10.1021/acs.jnatprod.5b00070.

Seifert, K. (2011). Structures, targets and recent approaches in anti-leishmanial drug discovery and development. Open Med. Chem. J. *5*, 31–39. https://doi.org/10.2174/1874104501105010031.

Shaked-Mishan, P., Ulrich, N., Ephros, M., and Zilberstein, D. (2001). Novel intracellular SbV reducing activity correlates with antimony susceptibility in *Leishmania donovani*. J. Biol. Chem. *276*, 3971–3976. https://doi.org/10.1074/jbc.M005423200.

Shyaula, S.L., Tamang, T., Ghouri, N., Adhikari, A., Marasini, S., Bajracharya, G.B., Manandhar, M.D., and Choudhary, M.I. (2015). Antileishmanial diterpenoid alkaloids from *Aconitum spicatum* (Bruhl) Stapf. Nat. Prod. Res. *30*, 1–4 [Epub ahead of print]. https://doi.org/10.1080/14786419.2015.1114941.

Singh, N., Mishra, B.B., Bajpai, S., Singh, R.K., and Tiwari, V.K. (2014). Natural product based leads to fight against leishmaniasis. Bioorg. Med. Chem. *22*, 18–45. https://doi.org/10.1016/j.bmc.2013.11.048.

Sosa, N., Capitán, Z., Nieto, J., Nieto, M., Calzada, J., Paz, H., Spadafora, C., Kreishman-Deitrick, M., Kopydlowski, K., Ullman, D., *et al.* (2013). Randomized, double-blinded, phase 2 trial of WR 279,396 (paromomycin and gentamicin) for cutaneous leishmaniasis in Panama. Am. J. Trop. Med. Hyg. *89*, 557–563. https://doi.org/10.4269/ajtmh.12-0736.

Souza-Silva, F., Bourguignon, S.C., Pereira, B.A., Côrtes, L.M., de Oliveira, L.F., Henriques-Pons, A., Finkelstein, L.C., Ferreira, V.F., Carneiro, P.F., de Pinho, R.T., *et al.* (2015). Epoxy-α-lapachone has *in vitro* and *in vivo* anti-leishmania (*Leishmania*) *amazonensis* effects and inhibits serine proteinase activity *in* this parasite. Antimicrob. Agents Chemother. *59*, 1910–1918. https://doi.org/10.1128/AAC.04742-14.

Sundar, S., and Chakravarty, J. (2013). Leishmaniasis: an update of current pharmacotherapy. Expert Opin. Pharmacother. *14*, 53–63. https://doi.org/10.1517/14656566.2013.755515.

Sundar, S., Chakravarty, J., Agarwal, D., Rai, M., and Murray, H.W. (2010). Single-dose liposomal amphotericin B for visceral leishmaniasis in India. N. Engl. J. Med. *362*, 504–512. https://doi.org/10.1056/NEJMoa0903627.

Sundar, S., Jha, T.K., Thakur, C.P., Sinha, P.K., and Bhattacharya, S.K. (2007). Injectable paromomycin for visceral leishmaniasis in India. N. Engl. J. Med. *356*, 2571–2581.

Teles, C.B., Moreira-Dill, L.S., Silva, A.d.e.A., Facundo, V.A., de Azevedo, W.F., da Silva, L.H., Motta, M.C., Stábeli, R.G., and Silva-Jardim, I. (2015). A lupane-triterpene isolated from *Combretum leprosum* Mart. fruit extracts that interferes with the intracellular development of *Leishmania* (*L.*) *amazonensis in vitro*. BMC Complement. Altern. Med. *15*, 165. https://doi.org/10.1186/s12906-015-0681-9.

Thao, N.P., Luyen, B.T., Brun, R., Kaiser, M., Van Kiem, P., Van Minh, C., Schmidt, T.J., Kang, J.S., and Kim, Y.H. (2015). Anti-Protozoal activities of cembrane-type diterpenes from Vietnamese soft corals. Molecules *20*, 12459–12468. https://doi.org/10.3390/molecules200712459.

Tracanna, M.I., Fortuna, A.M., Cárdenas, A.V., Marr, A.K., McMaster, W.R., Gómez-Velasco, A., Sánchez-Arreola, E., Hernández, L.R., and Bach, H. (2015). Anti-leishmanial, anti-inflammatory and antimicrobial activities of phenolic derivatives from *Tibouchina paratropica*. Phytother. Res. *29*, 393–397. https://doi.org/10.1002/ptr.5263.

Ullah, N., Nadhman, A., Siddiq, S., Mehwish, S., Islam, A., Jafri, L., and Hamayun, M. (2016). Plants as Antileishmanial agents: current scenario. Phytother. Res. *30*, 1905–1925. https://doi.org/10.1002/ptr.5710.

Verma, N.K., and Dey, C.S. (2004). Possible mechanism of miltefosine-mediated death of *Leishmania donovani*. Antimicrob. Agents Chemother. *48*, 3010–3015. https://doi.org/10.1128/AAC.48.8.3010-3015.2004.

Walker, J., and Saravia, N.G. (2004). Inhibition of *Leishmania donovani* promastigote DNA topoisomerase I and human monocyte DNA topoisomerases I and II by antimonial drugs and classical antitopoisomerase agents. J. Parasitol. *90*, 1155–1162. https://doi.org/10.1645/GE-3347.

WHO. (2018). Leishmaniasis. Available online: www.who.int/mediacentre/factsheets/fs375/en/. Accessed 12 January 2018.

Zhou, Y., Messier, N., Ouellette, M., Rosen, B.P., and Mukhopadhyay, R. (2004). *Leishmania major* LmACR2 is a pentavalent antimony reductase that confers sensitivity to the drug pentostam. J. Biol. Chem. *279*, 37445–37451. https://doi.org/10.1074/jbc.M404383200.

Zuo, X., Djordjevic, J.T., Bijosono Oei, J., Desmarini, D., Schibeci, S.D., Jolliffe, K.A., and Sorrell, T.C. (2011). Miltefosine induces apoptosis-like cell death in yeast via Cox9p in cytochrome c oxidase. Mol. Pharmacol. *80*, 476–485. https://doi.org/10.1124/mol.111.072322

Naegleria fowleri: Biology and Pathogenesis

7

Ruqaiyyah Siddiqui and Naveed A Khan*

Department of Biological Sciences, Sunway University, Bandar Sunway, Malaysia.

*Correspondence: naveed5438@gmail.com

https://doi.org/10.21775/9781910190838.06

Abstract

Naegleria fowleri is a free-living protist pathogen, that can infect the central nervous system to produce fatal primary amoebic meningoencephaltis (PAM). The amoeba enters through the nasal cavity via contaminated water and reaches the brain via olfactory neuroepithelial route to produce infection within days and almost always results in death within days (Symmers, 1969). A complete knowledge of this parasite, how it lives in the environment, and produces diseases is crucial for the rational development of preventative and therapeutic strategies.

Cell biology and speciation

N. fowleri is named after Dr Malcom Fowler, who first described this parasite as the causative agent of PAM. *N. fowleri* has three stages in its life cycle, a trophozoite or amoeboid stage, a flagellate form and a dormant cyst stage (Schuster, 1963; Symmers, 1969; Fulton and Dingle, 1967, 1971). During the trophozoite stage, the amoeba feeds and reproduces asexually under favourable conditions. In the absence of food, but presence of water, *N. fowleri* trophozoites transform into the flagellate form to travel long distance in search of nutrients. The flagellate form is a transient stage that can neither encyst, nor reproduce. Under harsh conditions, trophozoites transform into the cyst form. The cyst is a dormant stage during which they remain inactive with little metabolic activity, but remain viable, for years. Both the flagellate and the cyst are non-feeding, and non-reproductive forms. Only the trophozoite form is able to feed, divide, and encyst. Cysts will excyst to yield trophozoites under favourable environmental conditions (Fig. 7.1).

The trophozoite stage has typical characteristics of eukaryotes consisting of nucleus, mitochondria, food vacuoles, endoplasmic reticulum, contractile vacuole, cytoplasmic organelles (Schuster, 1963; Symmers, 1969; Fulton and Dingle, 1967; 1971). The reproduction of *Naegleria* is via promitotic binary fission starting with elongation of the nucleolus and ending with two polar masses. In terms of morphology, the amoeboid stage undertakes a long and slender shape 8–15 μm in diameter and movement is facilitated by the formation of

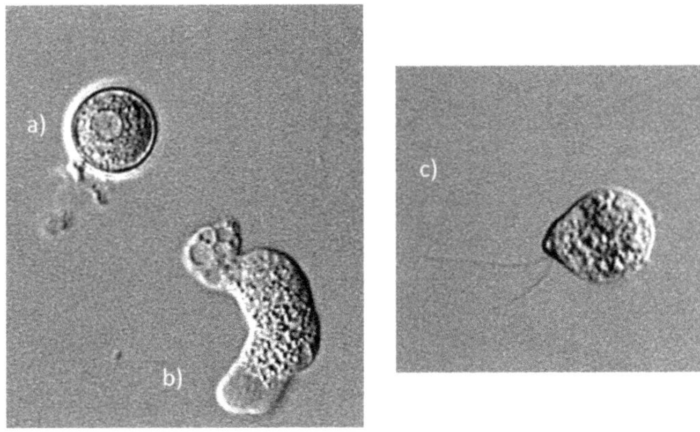

Figure 7.1 *Naegleria fowleri* in (a) cyst form, (b) trophozoite form, and (c) flagellate form, viewed under phase contrast microscopy, 1000×. (courtesy: G. S. Visvesvara, CDC, USA).

multiple lobopodia (Schuster, 1963; Forrester *et al.*, 1967). *Naegleria* cysts are spherical in structure and are usually seen in clusters. Mucoid-plugged pores are seen on the cyst surface that are a portal of emergence for trophozoites (John, 1982). Two flagella are seen in the flagellate form with a pear-shaped body. *De novo* enflagellation forms a flagellar apparatus comprising two basal bodies with corresponding flagella, a flagellar rootlet and cytoplasmic microtubules. Centrioles and tubulins are involved in the composition of flagella in *Naegleria.*

Compared to non-pathogenic species, *N. fowleri* is highly motile at 37°C that probably assists them in tissue invasion and migration and has been suggested as a marker for pathogenicity (Thong and Ferrante, 1986). For molecular basis of locomotion in *Naegleria*, chemotaxis, chemokinesis and food cup formation are particularly important in driving the amoeba towards the food source such as bacterial extracts and live cells via cytoplasmic streaming, pseudopod, cell polarity and focal contacts. Other factors that dictate adhesion and translocation of the amoeba include osmolarity of ions and surface tension with latter intimately related to conversion from swimming flagellate to amoeboid stage.

Biochemical properties of *N. fowleri* are dependent on the culture conditions and growth phase. Approximately, 80–85% of dry cell mass is made up of protein (120 pg per amoeba) in log phase whereas carbohydrate and RNA are 15 pg and 18 pg per amoeba respectively in both phases. Total DNA levels (0.2–0.3 pg per amoeba) fluctuates from log phase to stationary phase and is correlated with number of nuclei per amoeba (Weik and John, 1977, 1978; John, 1982).

On the contrary to human non-pathogenic *N. lovaniensis*, the ability of adhesion and invasion into extracellular matrix (ECM) is higher in *N. fowleri.* Higher integrin-like proteins are expressed and co-localized to a focal adhesion-like structure. Cathepsin B (*nscpb*) and cathepsin B-like cysteine proteases (*nfcpb*-L) and mucinolytic activity is reported in pathogenic *N. fowleri* that likely contribute to tissue colonization and invasion. Pyrophosphate dependent phosphofructokinase (PP_i-PFK) is reported in *N. fowleri.* It is sensitive to AMP. The PPi-PFK system in *Naegleria* is known to have two forms, i.e. tetramer and monomer forms respectively. AMP concentration is important for complete activation of the enzyme (Wessberg *et al.*, 1995). Using lectins, it is shown that the compositions of glycoconjugates on cell membranes differ among *Naegleria* species, but notably contain α-D-glucosyl, α-D-mannosyl, α-L-fucosyl, N-acetyl-α-D-galactosaminyl, and α-D-galactose. These glycoconjugates are attributed to adhesion capacity of pathogenic *N. fowleri.* Importantly, Nfa-1 protein on the cell surface membrane has been associated with pseudopodia movement and food cup formation. It has been implicated in contact-dependent killing mechanism of *N. fowleri*. A cytosolic heat shock protein 70 (Nf-cHSP70) has been demonstrated in *N. fowleri*. *N. fowleri* possesses CD45 which is a type of membrane glycoprotein that is implicated in cell cycle regulation and differentiation. Selenoproteins that protect other unicellular organisms from oxidative stress are also found in *Naegleria* species. Sterol synthesis in *Naegleria* species has been reported involving sterol precursors, cycloartenol, lanosterol, parkeol and their derivatives that are common in eukaryotes (Raederstorff *et al.*, 1987). Ergosterol is found to be the main 4,4-desmethylsterols gathered in cells via sterol biosynthesis (Raederstorff and Rohmer, 1987).

Naegleria fowleri belongs to the genus *Naegleria* in the family Vahlkampfiidae, belonging to the class Heterolobosea. *Naegleria fowleri* is the only known species capable of causing PAM in humans. *Naegleria* genus has a complex banding pattern with distinctive chromosomal pattern varying across species and subspecies (Byers, 1986). Nuclear and mitochondrial DNA content of *N. fowleri* fluctuates depending on culture condition though remain constant during log to post-log transition. *Naegleria* species possess plant like-pentatricopeptide repeat (PPR) that is employed for RNA editing. The rRNA genes consists of an internal transcribed spacer (ITS1), 5.8S rDNA, ITS2, and 28S rDNA in order. The evolution of ITS2 sequence enabled classifying about 40 species of *Naegleria* (De Jonckheere, 2004). Similarly, this genus has been categorized at the molecular level into eight different clusters, with *Naegleria fowleri* falling into cluster 8. It has been extrapolated that *Naegleria lovaniensis* belonging to the same cluster as *N. fowleri* evolves from *N. fowleri* judging from the genetic information available on both species. The length of the internal transcribed spacer 1 (ITS1) and a one bp transition in the 5.8S rDNA have been used to discern the differences in strains of *N. fowleri.* Despite the greater variation in sequences of mitochondrial small subunit rRNA (mtSSU rRNA), ITS1 gains an extra edge against it with length polymorphism, hence allowing advent of molecular techniques for genotyping.

Cellular differentiation in *N. fowleri*

Naegleria fowleri is an intriguing microorganism with three life forms that possess distinctive physiology. The amoeboid form is the feeding and reproductive units whereas flagellated form is the motile form for food hunting. Akin to many free-living protists, the cyst form is the dormant stage when harsh conditions set in to ensure survival of the organisms. Precisely due to these cellular differentiations, *Naegleria fowleri* is an excellent model to study mechanisms of phagocytosis, motility properties in other flagellates and cellular differentiation in term of cell dormancy (Fulton and Dingle, 1967). The movement of trophozoites of *N. fowleri* is facilitated by multiple lobose pseudopodias with higher negative charge at anterior end from other parts of the trophozites. Morphologically trophozoites of *N. fowleri* have the size of 8–15 μm with a slender shape with irregular surface (Forrester *et al.*, 1967; Antonios, 2010). They possess surface extensions believed to be associated with contact and cytolysis of cells. Food deprivation plays significant role in flagella differentiation and the numbers vary with majority having around two flagella (John *et al.*, 1991). The flagellation seems to be temporary with reversion to amoeba occurring within 1 hour and the amount of flagella is temperature dependent. Cyst forms have numerous flushed pores estimated between 1.2 and 7.2. Their surfaces can be smooth or rough with two-layered cyst walls which cannot be seen clearly under light microscope. Ideal isoelectric focused pH for existence of most polypeptides in amoeboflagellates is 6–7 with size between 20 and 60 kDa.

The process of encystation in *N. fowleri* involves concentration of cytoplasmic density from dehydration and organelle proximities towards each other including vacuoles with low electron density (Schuster, 1963b). The outer layer of the two-layered cyst walls is thinner (25 µm) whereas the inner layer is thicker (200–450 µm) with space consisting of a spongy network linked to cyst pores plugged by electron-transparent material. Alveolar transformation of the cyst allows for the amoeba to undergo excystation exiting the pores, following dissolution of the plug. Under lab conditions, the transformation calls for gain of WGA-specific saccharide and loss of UEA I-specific saccharides. Compaction of fibrillar material is observed in encystation of *N. fowleri* together with sealing of operculum structure at two to three ostioles during completion of encystation. Several factors are involved in excystation including CO_2 at molecular level, temperature, steroids and enolase. Ideal temperature and pCO_2 in the air for excystation are 30°C and 5% respectively. Concentration over 5% CO_2, actinomycin D and carbonic anhydrase antagonist downregulate excystation (Averner and Fulton, 1966). Progesterone and deoxycorticosterone activities on cell differentiation in enflagellation are concentration dependent with higher dosage hampering enflagellation. Lower doses of both hormones stimulate transformation individually while the effects of oestradiol and progesterone are addictive when applied simultaneously in inhibition of the flagellates (Pearson and Willmer, 1963). Immunological studies have revealed that enolase is part of the chemical makeup of the cell wall in the cysts of *N. fowleri* as well as cytoplasmic vesicles (Chávez-Munguía *et al.*, 2011).

The flagella of *Naegleria* are a part of a mostly single endoplasmic protrusion that extends into the filamentous structure. Flagella are absorbed alongside the disposal of the small part of the body when flagellates revert to trophozoites (Chang, 1958b). Culture age and temperature are pertinent to the process of flagellation with a higher age and temperature imposing inhibitory effects on flagellation. The phase of the amoeboid stage and amoeba population also affects flagellation with the stationary phase being favourable for flagellation. The process of flagellation initiates with the appearance of basal body followed by extension and elongation of axoneme filaments after evagination of cell membrane surface forming flagellum sheath. RNA and protein synthesis are necessary in the manufacture of the flagella which is hinged upon the transcription. This is substantiated by the antagonistic action of actinomycin D and cycloheximide, as RNA synthesis inhibitor and as protein synthesis inhibitor respectively, albeit once past certain concentrations and time intervals of these inhibitors, the antagonistic effects falter (Fulton and Walsh, 1980). Transcription activities correspond with the synthesis of flagellar tubulins as well as two calmodulin-like calcium-binding proteins. The balance of RNA moves to degradation past 30 min after initiation of flagellation, prompting loss of 6% in total RNA. Messenger-like RNA predominates in the synthesis of RNA as evident by EDTA and ribonuclease sensitivity in the polysomes in cytoplasm. Migration of RNA follows such paths from nucleoplasm to cytoplasm and equilibrates within 10 minutes. Nuclear DNA synthesis is primarily in decline during flagellation marked by the plunge in thymidine metabolism except for the phosphorylation of thymidine to dTMP which is maintained despite the ongoing mitochondrial DNA synthesis

(Corff and Yuyama, 1976). Flagellum formation occurs in mid-S phase with cessation of the cell cycle. A microtubule–nucleating complex is assembled consisting of gamma-tubulin, pericentrin and myosin II complex (GPM complex) before *de novo* basal bodies are formed (Kim *et al.*, 2005). The action is regulated by phosphorylation of the complex and with the interruption of dephosphorylation, the number of flagella increases with formation of basal bodies. The cytoskeletal microtubule network (CSMT) is yet another distinctive features of flagellate that is absent from the amoeboid form. The CSMT stems from the base of the flagella and leads a curving path the entire length of the cell, 61 minutes is spent for half the cell population in order to complete the synthesis of the protein required for CSMT prior to the beginning of differentiation (Walsh, 1984). Another structure that has been observed intensively is the flagellar rootlet which can only be visualized about 65 minutes after initiation of transformation with a length of 1.5–2 μm. It is exclusive to flagellates as it disappears in reversion to amoeboid form and reappears during flagellate differentiation. Its appearance coincides with basal bodies of flagella and its protein condensation is fluorescent in cytoplasm at the border of nucleus at time interval of 58 minutes (Larson and Dingle, 1981b). Centrin as a critical element of basal bodies is only expressed in *Naegleria* around the time of differentiation which is different from the majority of eukaryotes. As opposed to calmodulin and tubulin, centrin mRNA is not regulated synchronously throughout differentiation. The centrin focuses to the basal bodies when differentiation takes place as detected with anti-tubulin antibodies and is lowered in quantity with the transition of flagella to amoeba. Localization of centrin can be disrupted by oryzalin, an inhibitor of microtubule synthesis, nevertheless it does not halt the stockpile of centrin antigen. This finding verifies centrin as a flagella-exclusive protein in *Naegleria* (Levy *et al.*, 1998). Actin however is not actively expressed for differentiation of flagella as the translation of the protein declines with the initiation of differentiation. An actomyosin network is important in sustaining the shape of flagellate as blockage by myosin light chain kinase will result in prompt loss of flagellate shape and the contraction of the cells (Sussman *et al.*, 1984b). One of the alpha tubulins found in *Naegleria*, alpha6-tubulin is associated with spindle-fibre microtubules in mitosis but not with flagellar axonemes and cytoskeletal microtubules. This implies that *Naegleria* possess a variety of tubulins, flagellar tubulin synthesis in particular correlates intimately with mRNA events (Chung *et al.*, 2002). Beta-tubulin in *Naegleria* on the other hand has a conserved encoding region with the terminus of tyrosine with functional implications. The regional localization of two proteins N-gammaTRP (*Naegleria* gamma-tubulin-related protein) and N-PRP (*Naegleria* pericentrin-related protein) dictated by actin spot serve as foundation for microtubule nucleation in incorporation of flagella and its basal bodies (Suh *et al.*, 2002). As components of flagella, two types of calmodulin, calmodulin-1 (flagellar calmodulin) and calmodulin 2 (cell body calmodulin), have been found in different locations and each with different molecular weights 16 kDa and 15.3 kDa respectively. These calmodulins are revealed to only be displayed during flagellate differentiation as shown with translatable mRNA (Fulton *et al.*, 1986). The sequenced gene for calmodulin-1 (flagellar calmodulin) is shown to be single and devoid of intron encoding calmodulin with four calcium-binding domains and an amino-terminal extension of 10 amino acids in front of conserved calmodulin residue 4. Flagellar calmodulin expression coincides with flagellar tubulins in differentiation from amoebae to flagellates. Two other identifiable genes are discovered in association with flagellation, namely Clp, Class I on plasma membrane and Clb, Class I at basal bodies, each are expressed in CLB and CLP protein respectively. They share close similarities in amino acid sequences despite being wide apart in regional distribution. Oryzalin, the microtubule inhibitor, has independent actions on the two protein end-products with CLP production being truncated while CLB remaining at its site. For flagellar rootlet expression of protein, *Naegleria* has a homologue of UNC-119 protein which is usually found in neural tissues and ciliated single cell organisms. Another interesting fact uncovered in enzymatic activity involving flagellation is that in conjunction with decrease in nuclear DNA synthesis, thymidine kinase activity is lowered as well since the enzymatic activity is related to active division and flagella differentiation lacks the reproductivity (Bols *et al.*, 1977). The inactivation of the enzyme is independent of transcription or translation as it is not stopped by actinomycin D or cycloheximide. As flagellation and the flagella numbers are responsive to temperature, heat shock protein found in *Naegleria* can prompt multiple flagellation optimized past 38.2°C (Walsh, 1980). These proteins are found homologous in size ranging from 6 to 96 kDa to chick embryo fibroblast, mouse L, and baby hamster kidney cells. Other important factors in flagellation apart from aforementioned factors are ionic concentration, hydrostatic pressure, bacterial suspensions and β-mercaptoethanol. 80 mM NaCl, 90 mM KCl, 50 mM $CaCl_2$ and 60 mM $MgCl_2$ are the molarity of each salts for electrolyte balance in order to maintain amoeboid form from flagellation while non-electrolytes sucrose and glucose impose osmotic effect with 250 mM to achieve similar action (Jeffery and Hawkins, 1976). The presence of bacteria serves to deter transformation to flagellates regardless of the living or heat-inactivated state. B-mercaptoethanol has the capability of blocking the morphogenesis by disrupting the incorporation of precursor protein and polymerization of microtubules occurring after RNA synthesis (Jeffery and Hawkins, 1976). Interestingly enough, the reversion to amoeboid form takes place within short amount of time if the conditions are met such as the existence of 75% D_2O and chemical stimuli produced by *Naegleria* itself (Preston and O'Dell, 1974).

Growth and life cycle

The food source for *Naegleria* originates from yeast, algae and both Gram-negative and Gram-positive bacteria including cyanobacteria with filaments such as *Anabaena* and *Cylindrospermum* (Xinyao *et al.*, 2006). Nevertheless, food vacuoles are selective when it comes to morphology of some microorganisms such as the tight threads (*Oscilltoria*) and dense formations (*Aphanizomenon*). Cultivation of *Naegleria* is feasible with both bacterial and axenic cultures, for the former, it is grown on non-nutrient agar (1.5%) plated with living or inactivated *Enterobacter aerogenes* or *E. coli* (Anderson and Jamieson, 1972). In axenic culture, mediums for *Naegleria fowleri* such as cerva medium contains 2% Bacto-Casitone and 10% fresh horse serum in distilled water whereas Nelson medium has liver digest, glucose, and calf serum (Haight and John, 1980). The ideal pH for the agitated culture is 5.5 and heat-inactivated calf serum provides the best cell yields (John, 1982). Division requires 5.5 hours with amino acids serving as major carbon and energy sources. Protoporphyrin IX in the medium results in better mobility, virulence and resistance to host complements. Other elements of the chemically defined minimal medium are amino acids such as arginine, glycine, histidine, isoleucine, leucine, methionine, phenylalanine, proline, threonine, tryptophan, and valine. The medium also includes vitamins such as biotin, folic acid, haemin, pyridoxal, riboflavin, thiamine as well as guanosine, glucose and salts (Nerad *et al.*, 1983;

Bradley *et al.*, 1996). Physiologically, the temperature and nutrient level as defined by bacteria and amoeba ratio are critical in deciding the cell population of *Naegleria* within biofilms. A vigorous growth phase reaching 30–900 *N. fowleri* per cm^2 is only achievable when the temperature is set at 42°C, meanwhile 10^4 bacteria/amoeba in ratio is the cut-off point for minimum nutrient level for the detectable growth in a freshwater biosphere (Goudot *et al.*, 2012). Unlike *Naegleria gruberi*, *Naegleria fowleri* is not affected by high viscosity of surroundings while high concentration of chloride anions has negative impact on their growth. *Naegleria* species however has preferential pH of 6.5 (Cerva, 1978). Cryptomitosis has been described in division of nucleus in *Naegleria* at ultrastructure level, which illustrates persistence of nucleus membrane throughout mitosis. Centrioles are nowhere to be found and a spindle of microtubules extend to the length of the nucleus from pole to pole without meeting a focal point. Respiration wise, *N. fowleri* adopts aerobic metabolism and accomplishes so by possessing numerous mitochondria utilizing up to 30 ng/min/mg of cell protein of O_2 during log growth under agitated culture while non-pathogenic *N. gruberi* requires higher up to 80 ng signifying the evolution of *N. fowleri* to adapt to poorer environmental oxygen in less dissolved state in heated water condition (Ginger *et al.*, 2010; Opperdoes *et al.*, 2011; Tsaousis *et al.*, 2014). Lactate and aglycerophosphate are rate-limiting substrates for oxidization with the rates being ADP dependent. Phosphorylative efficiencies (ADP:O ratios) are about 1.4 for NAD-linked substrates and 1.0 for succinate. Pi, Mg^{2+} and serum albumin dictate the respiratory control ratios, which are usually 1.5–3 for 11 substrates. Inhibitory substances for mitochondrial oxidations such as cyanide, azide, malonate, amytal, rotenone, pentachlorophenol and DNP act upon phosphorylation, glutamate and succinate oxidation as well as their uncoupling respectively (Weik and John, 1979). Stability of mitochondria remains at pH 7 to 8. Oddly *N. gruberi* is found to be capable of anaerobic respiration by housing cytosolic [Fe–Fe] hydrogenase, an enzyme responsible for facultative anaerobic metabolism and HydE, a key component of hydrogenase maturases, despite lacking pyruvate-ferredoxin oxidoreductase for its mitochondria–hydrogenosome hybrid. *Naegleria* can be cryopreserved by the two-step cooling process in freezers and stored in liquid nitrogen. The cooling process entails −20°C for 60 min, followed by 60 min at −70°C and then stored in liquid nitrogen. 8% of cells recover with rapid thawing at 37°C (Kilvington and White, 1991). The viability and virulence of *Naegleria* can be maintained for a long period of time with 21% viability upon 10 years of cryopreservation, as described by John and John (2006).

Ecology

As a ubiquitous protist, it has a wide range of habitats, from soil and water to air. *Naegleria* species have been found throughout the globe with the exception of Antarctica for *N. fowleri* (De Jonckheere, 2011). *Naegleria* populations are influenced by season, with peak distribution during the late summer. In the freshwater environment, largest populations of *Naegleria* reside in the detrital layer with the depth of 3.0–3.4 m lasting though out the season (Kyle and Noblet, 1986). The preference is due to the anoxic condition of the hypolimnion and poor nutrition beneath the layer. The Littoral sediment is also favourable for *Naegleria*. Seropositivity among animals depends upon exposure to *Naegleria* when assuming ecological roles and behaviour in the environment. Younger wild mammalians are less likely to produce anti-*Naegleria* spp. antibodies and thus are rendered more susceptible to *Naegleria* infection. Several free-living amoebae have also been isolated including *N. fowleri* from reptilians and fishes, vindicating their ubiquity (Kollars and Wilhelm, 1996). *N. fowleri* can colonize nasopharynx and oral cavities of healthy individuals and has been isolated from hospital and dental treatment facilities. The titration of antibodies to *N. fowleri* can go up to 1:20 in saliva and sera in healthy population which are of classes IgA, IgG and IgM. They provide some resistance to *N. fowleri* colonization and attachment on the nasal mucosa. Molecular techniques with the immunoperoxidase test and Western blot analysis for hospital patients revealed parallel in serological results between *N. fowleri* and *N. lovaniensis* (Dubray *et al.*, 1987). The survivability of *Naegleria* is affected by pH and temperature, especially in the presence of thermal pollution (Cerva *et al.*, 1981). Thermally polluted wastewater can preserve the virulence of *N. fowleri* and assist in recolonization of the organism. *Naegleria fowleri* isolated from industrial waste was proven to be pathogenic to mice as described previously (Kasprzak *et al.*, 1982). Geothermal water also provides similar favourable conditions to *Naegleria fowleri*. *Naegleria* cysts can be recovered in inadequately treated swimming pools with a chloride concentration of less than 0.5 to 1.5 mg/l. 5.31 mg/ml is the recommended level of free chloride as a treatment for swimming pool which confers further prevention against soil contamination. A preventative measure of *Naegleria* infection is via identification of the organism in environmental water samples. The culture-based assays have disadvantage of differentiating between non-pathogenic and pathogenic amoebae, hence ELISA and flow cytometry has been employed for more specificity to *Naegleria fowleri* with the latter providing quantitative information on environmental samples. In regards to superiority, flow cytometry is statistically more precise with higher immunofluorescent titres and less subjective with the absence of fluorescent fading in comparison with conventional fluorescent microscope (Muldrow *et al.*, 1982). PCR has been developed in species specificity coupling with more efficient DNA extraction methods in order to identify *N. fowleri* in water samples with considerable sensitivity. Incubation on agar plate for DNA extraction gave a better yield in positivity. The advent of PCR techniques comes in the forms of multiplex PCR and duplex real-time PCR (qPCR) where differentiation between species and quantitative studies can be accomplished. The evaluation of performance for duplex real-time PCR in term of detection limit (DL) by fluorescent Taqman technique has revealed that qPCR has DL of 3 N with 100% specific to *N. fowleri*. However, qPCR is not without its drawbacks and culture method is still needed to be incorporated to offer a higher DL. Immunomagnetic separation (IMS) had been described by Mull in incorporation to qPCR yielding a recovery of 46% and DL of 14 amoebae/l of sample (Marciano-Cabral *et al.*, 2003).

Naegleria is known to maintain symbiosis relationships and harbours a wide spectrum of microorganisms, which allows microorganisms to thrive under harsh conditions such as in the presence of antibiotics and disinfectants and enables transmission to suitable hosts for propagation and multiplication. The advent in anti-amoebic regimens may potentially be beneficial not only in removing *Naegleria* as a threat on its own right but will also facilitate the elimination of a plethora of other pathogens harboured by the amoebae. Virus-like bodies had been found in the nucleus of *Naegleria* and they seemed to be transported between the nucleus and the cytoplasm. The existence of microtubule-like fibrils and degeneration of the nucleus and cytoplasm were found in connection to the virus-like particle. These particles were proven to have cytopathic tendencies. *Naegleria* also hosts a variety of bacteria from diphtheria to *Legionella*. Receptor-mediated endocytosis is involved in the process of uptake for bacteria, in order to do so, *Naegleria* is able to synthesize the required proteins for the interaction. Similar interactions with *Vibrio cholerae* and *Protochlamydia naegleriophila* had been discovered in *Naegleria gruberi* implicating a role in disease dissemination. However, *Naegleria* is still a predatory species to the bacteria; it is imperative for bacteria to come up with the strategy and mechanisms to evade *Naegleria*. *Pseudomonas fluorescens*, for example, produces the cyclic lipopeptide surfactants (CLPs), massetolide and viscosin that have proved to be harmful to *Naegleria*. Another study on *Bacillus licheniformis* has shown that several factors, such as pH, temperature, oxygenation of cultures and sporulation of bacteria, can affect the amoebicidal effect of bacteria. The amoeba counts and the media contents also play a role in killing of amoeba by bacteria. The amoebicins secreted are hydrophilic peptides comprising six heterologous amino acids which are resistant to heat and protease. Amoebolytic activity can be optimized by purification. The peptides have antimicrobial properties against other microorganisms (such as the yeasts *Saccharomyces heterogenicus* and *Cryptococcus neoformans*) and other bacteria (*Bacillus megaterium*, *Corynebacterium glutamicum* and *Sarcina* sp.). The antimicrobial mechanisms involved include inhibition of nucleotide and protein synthesis and the disruption of membrane potential. Hydrophobic peptides (amoebicins d13-A, d13-B, and d13-C) are also excreted by *Bacillus licheniformis*, which increases permeability of membrane via lytic activities. In conclusion, interactions between the microorganisms are complex yet fascinating and by further studies it will deepen the understanding of the impact of symbiotic relationships of microorganisms to the pathogenesis of human diseases.

Primary amoebic meningoencephalitis

Naegleria fowleri as an amphizoic free-living amoeba is able to thrive in human hosts and causes one of the deadliest central nervous system infections known as primary amoebic meningoencephalits (PAM). PAM is described as an acute haemorrhaging necrotizing meningoencephalitis. The susceptible demography is usually young adults and children with a history of engaging in water activities. Patients with PAM experience high fever of 38–41°C, headache, altered mental status and seizures, culminating in almost guaranteed death over a few days if treatment is not initiated. The incubation period is usually 1–2 weeks before onset of the disease; the inoculum size and virulence of *Naegleria* are among the factors that dictate the progression of the disease (Martínez, 1977). *Naegleria fowleri* is transmitted by inhalation of contaminated water. It enters through the nasal cavity and colonizes the olfactory neuroepithelium localized at the nasal mucosa. After colonization, it penetrates through the cribriform plate and travels along the olfactory nerve to reach the olfactory bulb. The olfactory bulb is seated in the subarachnoid space, which has an enriched blood and cerebrospinal supply; this allows dissemination of the organism throughout the central nervous system (Martínez, 1977). The first confirmed case of PAM was reported by Folwer and Carter in 1965 in Australia. Since then cases have been reported annually worldwide, predominantly in the United States, Europe and India. The trophozoites of *N. fowleri* can be found in the cerebrospinal fluid of PAM patients with the absence of the cyst form due to the rapid course of the disease. The onset of the disease is marked by severe headache and spiking fever reaching 41°C, nausea and vomiting. Upon examination, signs of meningism or meningeal irritation such as Kernig's sign, Brudzinski sign and neck stiffness can be seen in patients with PAM. The sensorium may be affected especially in taste and smell due to primary site of olfactory bulb being affected early in the course of the disease. Generalized seizures and altered consciousness are common events in cases of PAM leading up to a comatose state. The inflammation at the sites of the brain is extensive.

The pathology of the brain in patients with PAM revealed rapid cerebral oedema with the consequence of life-threatening cerebral herniation of uncal and cerebellar origins (Fig. 7.2). The inflamed meninges are opaque, congested and exudative with necrosis of the olfactory bulb and adjacent frontal cortices (John, 1982; Rojas-Hernández *et al.*, 2004b). Trophozoites are visualized in the Virchow–Robin space and are seen in abundance at the hypothalamus and midbrain. Computed topography (CT) reveals meningeal enhancement, cisternae effacement, enhanced basilar exudates (Singh *et al.*, 1998). Cerebral infarctions have been reported in several occasions and some PAM cases are complicated by hydrocephalus. Histologically inflammation is predominantly neutrophilic with distribution of amoebae at the perivascular space suggesting blood vessels as vehicles for dissemination of the amoebae (Visvesvara, 2010) (Fig. 7.3). Intracellular amoebae are seen in the macrophages upon examination of the exudates from the meninges and brain parenchyma, the uptake of cell debris and erythrocytes is also seen in the amoebae. Demyelination can occur in PAM at the white matter and is thought to be induced by enzymatic activities of the amoebae located at the surrounding grey matter.

Contact history is important to elicit when it comes to diagnosis of PAM as PAM is contracted via swimming in contaminated water, particularly in geothermal springs, freshwater lake, river and poorly managed swimming pools. Amoeba densities correlate intimately with climate, surprisingly given the affinity of *Naegleria fowleri* to high temperature, most cases were reported in the temperate regions. There is underreporting as public health is more accessible in temperate regions. Recreational water is most often not being treated properly. The limited potency of available drugs also contributes to overall challenges in overcoming the disease. It has been made known that ablution practices in Muslim community have also attributed to spreading of the disease when the water supplies are not clean of pathogens. Ablution without access to clean water directly introduces the pathogen to the nasal cavity which is the portal of entry for the amoeba; thus it is imperative to promote disinfection of water by boiling, filtering and chlorination. Poor maintenance of water tanks and other water storage units in public places, a lack in consistent water supply pose a risk in the prevalence of PAM. Religious practices other than ablution practices involving contaminated water

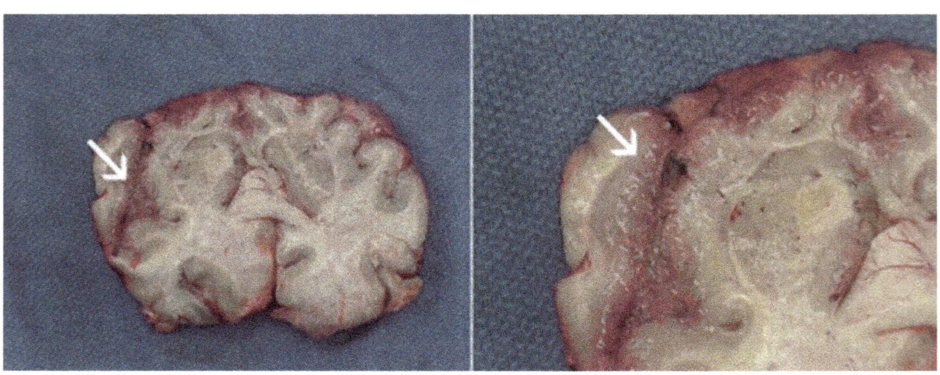

Figure 7.2 *Naegleria fowleri*-infected brain exhibiting extensive haemorrhage and necrosis in the frontal cortex (courtesy: G. S. Visvesvara, CDC, USA).

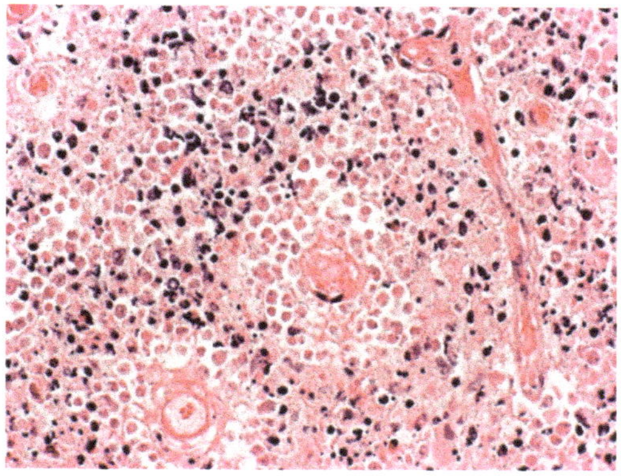

Figure 7.3 A section of the cerebral portion of the brain of *Naegleria fowleri*-infected patient, stained with haematoxylin and eosin, showing *N. fowleri* trophozoites, 400× (courtesy: G. S. Visvesvara, CDC, USA).

such as the Kumbh Mela provide a route of transmission to a large susceptible population. Nasal irrigation and cleansing for healthy purposes and Ayurveda practices with unsterilized water have been noted as risk factors for contracting *N. fowleri*. The importance of awareness of these risk factors cannot be stressed enough among health care professionals as well as the public. In terms of susceptibility, active young adult males with contact of contaminated water are most at risk. Other precipitating factors should also be explored in order to devise better preventative solutions such as genetic and anatomical predispositions.

Clinical and laboratory diagnosis

In light of the fast progress of primary amoebic meningoencephalitis, the survival of patients hinges upon a high index of suspicion and expeditious diagnosis followed by early treatment. Unfortunately, many PAM cases are diagnosed post-mortem due to the rapid course of the disease, lack of awareness and effective clinical regimens to eradicate the organisms. There are several diagnostic modalities for PAM and they should be combined in consideration of diagnosis. The usual imaging modalities for PAM are computed topography (CT) and magnetic resonance imaging (MRI), with the latter offering higher specificity to PAM. The findings in CT are often cisternae surrounding the midbrain and subarachnoid space obliterations with leptomeningeal enhancement after contrast administration. In some cases, exudates were presented as enhancement as well. The sulci and corresponding grey matter were described in CT images with enhancement in several cases of PAM. The CT also revealed sequelae of PAM such as cerebral oedema and acute hydrocephalus which are not specific to PAM alone. It is possible to have normal or false negative in imaging studies in the early stage of PAM, which comprised of 40% of cases as described previously, therefore imaging studies alone should be combined with the contact history and other lab diagnostic methods in consideration of PAM diagnosis. It is agreed upon that CT is to be used as exclusion diagnosis in illustrating complications of the disease (subdural empyema, infarctions and hydrocephalus) and to rule out intracranial mass effect rather than establishing diagnosis (Schumacher *et al.*, 1995). There are several techniques used in clinical settings to identify the amoeba in clinical specimens and each has their pros and cons. Collection of the nasal discharge with microscopy examination is simple and less costly. The cultivation method is timely and costly without guaranteed success. Immunofluorescence and PCR assays have been carried out for PAM for cerebrospinal fluid and brain tissues from biopsies to overcome problems with requirement in aseptic techniques and temperature. They have an edge against amoeba culture and microscopy in yielding genotype information. Nevertheless, they are not without their problems, particularly in sophistication and tendency of contaminations in samples. The serological method is limited in usage due to rapid progression of the disease without changes in serological response. On the contrary to microscopic detection, antigen detection is species specific and is not subjected to observers; however, its pitfall is the lower sensitivity as opposed to the PCR-based method. Cerebrospinal fluid (CSF) is demonstrated to resemble other central nervous system infection, on gross examination, the colour of CSF can be of greyish, yellowish or tinged red. The red blood cells in CSF which are usually rarely seen will be increased from 250 to 25,000 cells per mm^3 while the leucocytes soar in numbers with polymorphonuclear cells predominance. PAM is complicated by increased intracranial pressure, hence the elevation in opening pressure can be seen in the range of 300–600 mmH_2O. Usually central nervous system infection leads to a surge in protein, PAM is no exception with an increase in protein from 100 to 1000 mg per 100 ml. Another noticeable biochemical change is in glucose where the values vary in 10 mg per 100 ml or lower. Active trophozoites can be seen in freshly collected samples for wet mount preparation. The storage of CSF is at room temperature of 25°C if microscopic examination is not available. In the examination of motility, the samples are warmed in a 37°C incubator and staining of Giemsa or Wright helps in better visualization of the amoeba and distinguishing host cells. In facilities where brain biopsies are available, brain tissues can be obtained and examined microscopically. The exudates are usually comprised of polymorphonuclear leucocytes, granulocytes and lymphocytes in mainly cerebral hemispheres, brain stem and cerebellum. Upon gross examination, the brain is necrotic and oedematous. In the later stages of infection, granular substance between inflammatory cells and the amoebae, leucocyte lysis and erythrophagocytosis are uncovered in experimental PAM as described previously in ultrastructural analysis of inflammation in the brain of CD $38^{-/-}$ knockout mice (Cervantes-Sandoval *et al.*, 2008). In the same experiment, the amoebae were also found to be damaged with widespread granular contents, poorly defined organelles and several cytoplasmic dense bodies suggesting mutual destructions of hosts and the pathogen. The inflammation could be delayed due to immune privilege of the brain and the reduced virulence of the amoebae and is often accentuated by increased number of pathogens and concurrent augmentation of proinflammatory factors. Amoebae can be found in multiple sites including the Virchow–Robin spaces, the brain parenchyma and meninges and are identifiable by large nucleus and central large nucleolus. The amoebae can be cultured from brain tissue specimens and CSF by placing them on non-nutrient agar coated with bacterial lawn. They are also viable in axenic culture and tissue monolayer culture. Growth is monitored every day for a week and the trophozoites are observable within 48 hours, however the completion of culture requires several days and is of low diagnostic value. Page's amoeba saline is preferable over physiological or phosphate-buffered saline due to being the viable condition for amoebic growth. The storage of samples requires high maintenance of aseptic techniques and under room temperature. Culture is of limited value in diagnosis as opposed to its research value.

Enflagellation can be induced experimentally to identify *Naegleria fowleri* as it is a distinguishable stage from other pathogenic free-living amoebae, namely *Acanthamoeba* spp. and *Balamuthia mandrillaris.* Initially, CSF is mixed and shaken with 1 ml of distilled water in a sterile tube before it is transferred to coverslip, edge coated with petroleum jelly and placed in a humidified incubator for 1–2 hours until the transformation of trophozoites to biflagellate take place. Due to the fast progression of the disease and early mortality, serology bears little value in diagnosis as the antibody response usually is inadequate for detection. However, IgM had been discovered to be major class of antibody found in PAM survivors, the role of protection and indicator of past exposure of the antibody has yet to be determined. Immunofluorescence assays have the sensitivity and specificity required for detection of *N. fowleri.* The monoclonal antibodies (mAb) produced can be used in multiple techniques such as indirect immunofluorescence assay, dot immunobinding assay, and enzyme-linked immunotransfer blot technique. The mAb has the advantage of inter-regions reactivity of same species with interspecies cross-reactivity. It is further proven that staining bands at the 160, 104, 93 and 66 kDa regions are produced in enzyme-linked immunotransfer blot technique in *N. fowleri,* although a 160 kDa band is shared with *N. lovaniensis,* other *Naegleria* species do not have the reactivity. Flow cytometry corroborates the results with other strains of same species and multiple forms of *N. fowleri* coupled with enzyme-linked immunosorbent assay. Antigen detection tests can be used wide across the brain tissues as well as CSF and provide early diagnosis without reliance on maintenance of samples. Biochemical analysis, for example isoenzyme analysis and molecular techniques like PCR, nested and real time PCR provide an alternative to microscopy visualization in specificity of detection of the pathogen in clinical specimens. Repetitive mitochondrial DNA, genomic DNA and 18S small subunit ribosomal RNA (SSU rRNA) gene (McLaughlin *et al.*, 1991; Réveiller *et al.*, 2002; Qvarnstrom *et al.*, 2006) have been established for highly sensitive PCR and real time PCR assays. The SSU rRNA has copy numbers of approximately several hundreds and is sensitive to the point of a single amoeba. The wide range of genotypes in *N. fowleri* is detectable by real time PCR as well.

The patients who present with classic meningitis clinical manifestation without evidence of bacteria, viruses and fungi and a contact history with contaminated water should raise a high suspicion index, hence the cerebrospinal fluid should be tested by advanced molecular techniques for *N. fowleri* to improve overall survival. As *N. fowleri* portal of entry is via the nasal cavity, nasal discharge collection can be diagnostically valuable with the invention of the devices for collection. More advanced techniques such as lateral flow point-of-care (POC) test and expression of Single Chain Fragment Variable (scFv) protein molecules have been designed for more efficiency and reduction of cost. Loop-mediated isothermal amplification assays are designed by Mahittikorn *et al.* (2015) for the same objectives as readily detectable by hydroxyl naphthol blue within 90 minutes. On conclusion note, next generation sequencing is believed to be the future of PAM diagnosis.

Chemotherapeutic and disinfection strategies

The high mortality rate of primary amoebic meningoencephalitis despite the rarity of the disease, warrants a better management of the disease primarily in definitive treatment since the effectiveness of current chemotherapeutic regimens are limited in clinical settings. The basis of current studies is based on scarce survivor cases, *in vivo* animal infection models and *in vitro* cell culture are still in development. Therefore further research still needs to be done to devise plans to overcome the fatality of the disease and public health issues especially in preventative medicine of the disease. In the empirical treatment of primary amoebic meningoencephalitis, Amphotericin B has been highlighted as the first drug of choice for the infection, either by intravenous delivery or by combination with intrathecal administration, nevertheless most outcomes remained poor throughout the years. The combination of other drugs with Amphotericin B is often used, among them are the azole class such as fluconazole, miconazole and rifampicin. Other drugs like the macrolides, azithromycin and rokitamycin and the aminoglycoside hygromycin B (Goswick *et al.*, 2003; Kim *et al.*, 2008) also achieved some similar effectiveness in experimental settings. Miltefosine has been used to a varying degree of success. Randomized clinical trials and case control studies in large scales are yet to be carried out for optimization of therapy due to the scarcity of PAM and different therapeutic strategies employed among physicians. As originally an antifungal agent, the mechanism of action for Amphotericin B is via ergosterol binding forming pores in cell membranes of protists and killing by cell efflux and influx, this is in tune with the fact that *Naegleria fowleri* also possesses sterol synthesis pathway. Ultrastructure observations of *N. fowleri* in contact with Amphotericin B revealed pseudopodia malformation, distortion of nucleus morphology, mitochondria degeneration and the presence of autophagic vacuoles. (Schuster *et al.*, 1975). It was shown that 0.78 µg/ml amphotericin B is needed to achieve an MIC of 100% amoebicidal effects in the previous study involving azithromycin. Treatment in mouse models of PAM described by Goswick indicated that intravenous amphotericin B of 7.5 mg/kg of body weight/day improved survival in the infected mice. It is recommended that therapy should last for 10 days with dosage of 0.25–1.5 mg/kg/day while a narrower therapeutic window for children in 0.5–0.7 mg/kg/day is recommended by the Centers for Disease Control and Prevention. As there are availabilities of different preparations of amphotericin B, liposomal amphotericin B is thought to have lower systemic toxicities and better tissue penetration across the blood–brain barrier which is often served as an obstacle for drug delivery to the central nervous system due to the fat solubility. However, deoxycholated or conventional Amphotericin B is not without its merits over liposomal preparations as it offers a better dose–response in mouse models and shows superiority in MIC 10-fold higher than the liposomal form. Amphotericin B is known to have multiple adverse effects, the most severe form of which are renal and hepatic toxicities. The factors contributing to the problems of amphotericin B include aqueous insolubility affecting compartmental concentration and clearance (Grace *et al.*, 2015), hence lesser crossing of blood–brain barrier resulting in effective dosage in tissues needed to eradicate the pathogens. Miltefosine had been used in several survivor cases but it is not known whether the efficacy is due to early diagnosis and treatment or the drugs per se. It belongs in the alkylphosphocholine class where it is amphiphilic and acts as protein kinase B (PKB) suppressor. It has been established in animal models that the oral bioavailability of miltefosine is > 80% involving active and passive transport in absorption of the drug. Milfosine is easily bound to plasma proteins and distributes in high concentration in the target tissues. Drug–drug interactions in miltefosine is believed to be low, i.e. 40 µg of miltefosine is required to accomplish amoebistatic actions while the amoebicidal effect is reached when the concentration is above 55 µg as minimal amoebicidal concentration (Schuster *et al.*, 2006). Azole antifungal agents are usually used as an adjuvant therapy for amphotericin B, most commonly used azoles are fluconazole and miconazole. Fluconazole synergises well with amphotericin B in recruiting neutrophils to infection sites. Voriconazole > 1 μg/ml is proved effective against *N. fowleri in vitro* (Schuster *et al.*, 2006). One of the macrolides, azithromycin, has been used *in vitro* and in experimental animal models and it requires MIC of 13.4 μM (10 μg/ml) *in vitro* and allowed for 100% survival in mice at 75 mg/kg/day where amphotericin B as high as 7.5 mg/kg/day succeed in protecting only 50% of mice (Goswick *et al.*, 2003). While azithromycin has lower MIC activities, it has relatively less threatening adverse effects than amphotericin B with superior central nervous system penetration. The study with mouse models showed that the combination of amphotericin B of 2.5 mg/kg and azithromycin of 25 mg/kg offered absolute protection for mouse overall survival, denoting the synergistic actions against *N. fowleri* (Soltow and Brenner, 2007). An *in vitro* study with rifampin exhibited paradoxical effect against different strains derived from different geographical locations. Moreover, rifampin is a liver enzyme inducer with complex drug–drug interactions especially with the 14α-demethylase inhibitors, an azole fungistatics. The changes in pharmacokinetics in azoles when combined with rifampin are counterproductive in therapeutic effects (Lee *et al.*, 1979). Animal sera have been looked into for amoebicidal activities throughout the years for alternatives to chemotherapeutics. Among the sera, alligators, bullfrogs, muskrats and raccoons exhibited excellent activities as opposed to sera from toads, box turtles and sparrows. The animal sera show time-dependent and concentration dependent eradications of amoebae, with the alligator serum amoebicidal effect peaking at 20 minutes. The sera are also temperature sensitive and are subjected to degradation of proteases suggestive of complement activity. General measures are often undertaken for complications of the disease such as reduction of brain oedema and seizures. Dexamethasone had been used clinically to counteract inflammatory response and ultimately decreased intracranial pressure. Mechanical methods with external ventricular drainage and osmolar diuretics such as mannitol had been performed in the cases with PAM. Other measures such as hyperventilation and medically induced hyperthermia facilitate lowering of intracranial pressure. Preventative measures can be taken to reduce exposure to *Naegleria fowleri* such as avoidance of freshwater lake, river during summer and hot spring in temperate zones. It is advisable to refrain from jumping and submerging into water and the usage of nose clips to render nasal passage inaccessible to the pathogens as portal of entry. Proper treatment of water with chlorination and saltwater offer protection against *N. fowleri* as the conditions are hostile to the pathogens. Nasal rinsing for sinus should be done with meticulous care by using distilled, boiled or filtered water.

Pathogenesis

Naegleria fowleri has an interesting chain of infection as it gains access via nasal cavity and attach to olfactory neuroepithelium of nasal mucosa before they propagate to the olfactory bulb and the brain. A better understanding of neuropathogenesis and factors of virulence are needed in order to devise plans to manage the disease accordingly. Culture conditions play a significant role in pathogenicity of *N. fowleri* as axenic culture reduces virulence in virulent strains, particularly in plummets within first 2 years of passage. Virulence is usually reinstated when it is passaged serially in live hosts such as mice apart from cell cultures for low virulent strains. Non-virulent strains from environmental isolates will not be virulent by the means of cell cultures and live host passages. Cyst form storage however will not diminish virulence. The growth in high temperature is a good determinant of virulence as described by Griffin in 1972. The compositions of axenic culture media also affect behaviours and pathogenicity of *Naegleria* strains. An axenic media with contents of 10 g of casein, 1.325 g of Na_2HPO_4, 2.5 g of glucose, 0.8 g of KH_2PO_4, 5 g of yeast extract, 900 ml of distilled water, and 100 ml of fetal calf serum manage to differentiate between weakly and non-pathogenic strains with the former cultivated successfully in the media. Nelson's medium showed similar pattern as opposed to low pathogenic potential provided by PYNFH medium; however, supplementation with liver hydrolysates in both media seemed to enhance the capability of proliferation and pathogenicity (Burri *et al.*, 2012). Late log phase and early stationary phase are phases where pathogenicity of virulent *Naegleria* peaks. To establish infection models of PAM, eastern grey squirrel, sheep, hispid cotton rat, muskrat and house mouse had been shown to be susceptible to the infection while opossum, raccoon, and eastern cottontail rabbit failed to develop infection even at 10^6 inoculum size. The olfactory bulb pathology was significant with haemorrhagic necrosis and microscopic trophozoite distribution. Meningitis and lymphocytic infiltration around the blood vessels were apparent in the cerebrum, cerebellum, pons and medulla in study involving sheep. The study with the rabbits suggested an infection kinetics in intracisternal inoculation for further understanding of the amoeba–host response in the cerebrospinal fluid and the optimal route of administration of chemotherapeutic agents. The mice models were used for many studies involving immunological and pathogenic aspects of the pathogens. The weight and age were proven to be important variables for selection of samples in animal study of PAM (Holbrook and Parker, 1979). New *in vitro* models such as organotypic slice cultures derived from rat brain have been developed for PAM to look at morphology and tissue damages corresponding to *in vivo* models (Gianinazzi *et al.*, 2005). There are certain distinctive features that separate axenically cultured *Naegleria* to the pathogens found in animal models. Morphologically axenically cultured amoebae have an irregular shape with oval, cylindrical mitochondria whereas amoebae found in tissues are round in appearance with dumbbell shaped mitochondria aside from standard shapes of *Naegleria* mitochondria. Other organelles exhibit difference in shapes as well. Rough endoplasmic reticulum is strictly tubular in CNS tissues but it can be of vesicular shape in culture. Food vacuoles are found in amoebae grown in brain tissues as vacuoles reflect active phagocytic activities. Polymorphonuclear cell infiltrates and amoebae are observed around blood vessels and destroyed brain tissues. The nature of inflammation in brains of PAM is mainly neutrophilic and is usually accompanied by degenerated neurones and glial processes. Clustering intranuclear bodies can be found in the size of 30 nm along with crystalloids in the amoebae residing brain tissues. Cell shrinkage and nuclear pycnosis are found in cells infected with *N. fowleri*. Intraoral, intraconjunctival and intradermal routes had been explored in the past but none of the routes developed the disease suggesting that intranasal route is exclusive to entry of the amoeba. Central nervous system infection by *Naegleria folweri* produced among animals involves a similar process. The amoebae attach to the mucous layers where olfactory neuroepithelium is seated at 8 hours post inoculation. Mucous layer shedding removes the pathogens but some will reach the epithelium and eventually paving way to submucosal nerve plexus, olfactory nerves, cribriform plate and olfactory bulb within 24 hours post infection. The inflammatory response peaks at 96 hours of infection. There are two mechanisms of pathogenicity for *N. fowleri*, contact dependent and contact independent (Fig. 7.4). For contact dependent, adhesion refers to the ability of the amoeba to attach the nasal mucosa, increased locomotive function and neurotaxis of the organism with mediation of adhesions. Glycoproteins linking to α-D-mannose and

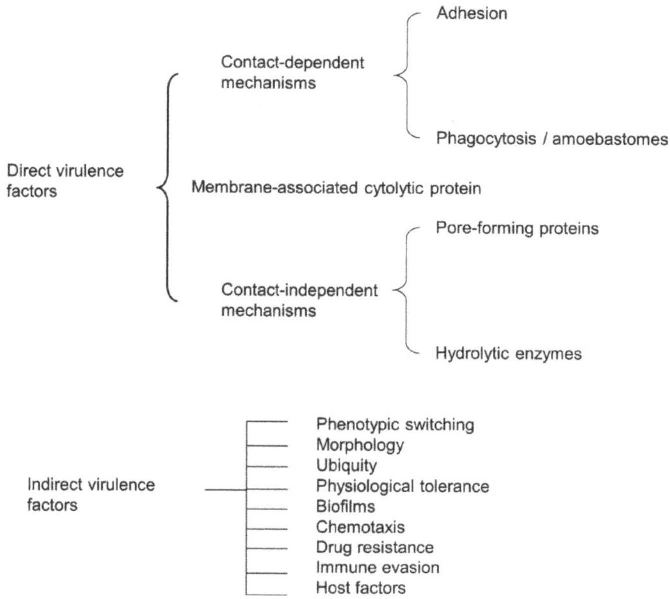

Figure 7.4 Factors associated with the pathogenesis of *Naegleria fowleri* infection.

α-D-glucose were suggested to contribute to adherence by Carrasco-Yepez *et al.* (2013). Targets of adhesion are usually found in extracellular matrix components, laminin-1, fibronectin and collagen I. Integrin-like proteins anchor to focal adhesion-like structure in extracellular matrix components and the process can be disrupted by anti-integrin antibodies. Fibronectin binding protein involvement in the association with cells are essential in localization of amoebae to host cells. Other elements such as protein kinase C and reactive oxygen species are found interconnected with contact dependent cytotoxicity. Actin dependent phagocytosis and amoebastomes by expression of Nf-actin gene are crucial to host cell damages as well. Nfa-protein expressions on pseudopodias and food cups are required for cytopathic ability for *Naegleria* species as Jeong *et al.* (2005) uncovered. *Naegleria fowleri* cytolytic activity lies with some fractions of surface membrane proteins and it is proven to be heat resistant up to 75°C lasting 30 minutes. It is different from lysosomal cytolytic pathway, though the membrane-associated cytolysin is activated by lysosomal fractions. Mp2CL5 of size 17 kDa expressed during log phase in particular is attributable to pathogenicity as it is found in pathogenic *N. fowleri* but not in non-pathogenic *Naegleria* species.

In the contact-independent mechanism, N-PFP as cytolytic pore-forming proteins lyse erythrocytes as well as other cell lines by depolarizing the membrane potential. Two pore-forming polypeptides isolated previously are isoforms named as *Naegleriapores* A and B and share traits with granulysin from human natural killer cells and amoebapores from *Entamoeba histolytica*. Their molecular structures are of amphipathic alpha-helices combined with framework of cysteine residues for forming disulfide bonds. The precursors of the proteins contain *N-glycosylated* links and are variably processed at the C-terminus. These *Naegleria* pores are also active against bacteria. As pathogenic species, *Naegleria fowleri* releases phospholipase A, phospholipase C, sphingomyelinase, neuroaminidase, elastase and other proteolytic enzymes. Their cytototoxicity activities are negatively influenced by the presence of high pH, high temperature (50°C for 15 min), ethylenediamine tetraacetate (EDTA) and phospholipase A inhibitor such as Rosenthal's reagent, but not by multiple freeze–thawing and inhibitors of proteolytic enzymes. The presence of 5% ethanol accentuates the cytotoxicity. The direct result of the aforementioned cytotoxic enzymes is phospho-lipolysis, leading to the diffuse demyelination of the brain and the disruption of membrane integrity. Proteases such as cysteine proteases and serine proteases have been implicated in tissue invasion in *N. fowleri* infection. Acid proteinase, N-acetylglucosaminidase, acid phosphatase, 5′-nucleotidase, aspartate aminotransferase, alpha-D-glucosidase and aminopeptidase are scattered in different organelles of the cells and are often found to be liable for the pathogenicity of *N. fowleri*. Other established factors such as nitric oxide can cause harm to host cells themselves despite the initial amoebicidal effect, in addition *N. fowleri* also releases nitric oxide with the reaction to NOS2 antibody *in vivo* and *in vitro*, suggesting links of nitric oxide to the pathogenicity of PAM. Several other novel findings of pathogenicity factors in recent years are believed to be Nf-cHSP70, cyclophilin, apoptosis-linked gene-2-interacting protein X1 (AIP1), Golgi-localized transmembrane protein HID-1, Ras-related protein Rab-1 and villin-1 protein among others.

The host-damage response to *N. fowleri*

In PAM, host damage is not just merely a result of virulence from *N. fowleri*, the exaggerated immune response against *Naegleria* by host cells also contributes to the whole clinical picture of PAM. Going in depth in host responses against the infection will provide an outlook to the improvement and management of this life-threatening condition from immunological perspective. Humoral and cellular immunity each play their part as the defence system against *N. fowleri*. Animal studies suggested that non-pathogenic *Naegleria gruberi* can offer immunogenicity and cross protection to *N. fowleri* (John, 1982). The first line of defence is usually the mucous secretion which contains mucins that disrupt the adhesions and progression of the pathogens. Nevertheless, there are possibilities where *N. fowleri* is able to successfully invade the tissue and the inflammation which is predominantly neutrophilic ensue. Infection with $CD38^{-/-}$ knockout mice have indicated that mortality delay in immune-deficient mice stem from attenuated host response and polymorphonuclear cell lysis. *Naegleria fowleri* is able to provoke the release of pro-inflammatory cytokines tumour necrosis factor-alpha, interleukin 6, and interleukin 1β without the production of chemokine c-c motif ligand 11. As a result of the release, microglial cells undergo cythopathic changes. Astrocytes are known for their ability to induce inflammatory response by secreting interleukin 8 and interleukin 6. The regulation of cytokine level is achieved by the expression of DNA binding activity of activator protein-1 (AP-1) which requires the activation of extracellular signal regulated kinase (ERK), c-Jun N-terminal kinase (JNK), p38 mitogen-activated protein kinase (MAPKs). Cell-mediated immunity is less understood in countering *Naegleria* infection. Neutrophils however are stimulated by specific antibodies and complement for their amoebicidal functions (Ferrante and Mocatta, 1984). Phagocytosis mediated in such a manner is incomplete but it destroys the integrity of the amoeba. The adherence of the neutrophils is facilitated by tumour necrosis factor which is believed to be achieved via increase in oxygen radical production. Lysosomal enzyme release in the neurophils responding to *Naegleria* infection is also on the rise with the presence of tumour necrosis factor-alpha. Chloride ions are involved in neutrophilic amoebicidal activity as arginine, the hypochlorite scavenger deters the activity. Two primary mechanisms are involved in macrophage killing of *N. fowleri*, the nitric oxide production from arginine-dependent cytolysis and the release of non-oxidative cytokines such as TNF-α and interleukin 1. Both contact dependent process and secretion of cytolytic molecules are required for the eradication of *N. fowleri* by the immunomodulatory-activated macrophages. In term of lymphocytes, the preparation of whole-killed *Naegleria fowleri* is mitogenic to T lymphocytes, hinting role of T lymphocytes in defence against PAM (Ferrante and Symth, 1984). As to the subsets of T-cell populations, L3T4+ T cells and Ly2+ T cells saw the surge in numbers during *Naegleria* infection in contrast to the control group in experimental infection models, denoting a connection of the T cell subsets to immunity against the infection. A study by Rivera-Aguilar (2000) on human saliva and serum suggested that mucosal sIgA recognizes Nf proteins and confers protection by preventing the adherence of the amoeba to nasal mucosa with cross-reaction and specific immunization. Antibodies produced for *Naegleria* species show antigenic surface specificity across species (Marciano-Cabral *et al.*, 1987). Ferrante and Thong (1979) however stated that the antibody protection contributes little to the whole picture of the immune system against PAM as passive immunization with anti-*Naegleria* IgG or *Naegleria* monoclonal antibody results in no cure but delay in mortality. The antibodies can only immobilize the amoeba, assist in complement-mediated lysis and neutralize cytotoxins produced by the amoeba. In regards to complement pathways, the complement-mediated lysis seem to be effective in weakly pathogenic *Naegleria fowleri*. Inactivation of C3 or C9 complement components by heat and various other methods seem to diminish serum amoebicidal activity (Whiteman and Marciano-Cabral, 1987), which, in contrast, can be enhanced following the exposure of specific agglutinating antibody to *N. fowleri*. *Naegleria fowleri* evolves to counteract the immunity of the host. The amoeba develops mucinolytic ability to prevent the hindrance to adherence of the nasal mucosa via action of a 37 kDa protein. The mucinolytic protein is inhibited by cysteine protease inhibitors. To overcome potent immune system, highly pathogenic *N. fowleri* is able to resist complement-mediated lysis as well as suppression of cell-mediated immunity. The lack of haemin in Nelson media cultivated strains may prompt the strains to lose the resistance. Similarly, calcium ions affect complement resistance within certain range of threshold (Chu *et al.*, 2002). Two major pathways have been identified for complement resistance, one being the synthesis of complement-regulatory proteins and the other being vesicular shedding of the membrane attack complex (C5b–C9) (Toney and Marciano-Cabral, 1992, 1994a,b). A protein expressed on the membrane of the amoeba is implicated to be reactive with antibodies to human CD59 and is immunoprecipitate to complement component C9, thus preventing the amoeba from complement lysis (Fritzinger *et al.*, 2006). Protein kinases and their phosphorylations play similar protective role for complement resistance of *N. fowleri*. Vesiculation, on the other hand, can also remove the complement proteins depositing membrane of the amoeba, it is substantiated by blocking actin-dependent exocytic processes with cytochalasins hence rendering the amoeba susceptible to complement attack. The 89, 60, 44, and 28 kDa glycoproteins are suggested to participate in this pathway (Toney and Marciano-Cabral, 1992). *N. fowleri* has other interesting means of preventing host destruction by secreting the lymphokine macrophage inhibition factor and internalizing surface-bound antibody. Extensive study of immunological aspects of *N. fowleri* will allow for the advent in immunizations. One protein of interest, Cry1Ac protoxin promotes antibody responses to proteins or polysaccharides as an adjunctive to mucosa and systemic defence (Rojas-Hernández *et al.*, 2004).

The proportion of activated lymphocytes with highly expressed CD25 and CD69 elevates in nasal mucosa when Cry1Ac is administered. CD4 and C8T cells are also on the rise with the addition of Cry1Ac. STAT-6 Th2 immune response is thus essential for the survival against *N. fowleri* infection as shown in mouse model studies (Carrasco-Yepez *et al.*, 2010). Cholera toxin is another potential immunization candidate as it confers 100% protection to the mice with experimental PAM (Carrasco-Yepez *et al.*, 2014). Interleukin 6 (IL-6), interleukin 10 and interferon-γ expressions were upregulated while tumour necrosis factor-α level declined when cholera toxin was administered intranasally. Cytokines produced provided favourable conditions for subsequent IgA, IgG, pIgR and activation of polymorphonuclear leucocytes in nasal mucosa. *Naegleria fowleri* culture supernatant gives rise to certain degree of protection scaled to level of anti-*Naegleria* antibodies. The secondary serum and the IgG fraction from the serum offer mucosal protection as well (Thong *et al.*, 1983; Ferrante *et al.*, 1988). An attempt at intracisternal passive immune therapy with anti-*Naegleria* immune serum, fractionated immunoglobulin G, and a monoclonal antibody to *N. fowleri* had been carried out in rabbits and the results of multiple doses was satisfactory with prolonged survival, the passive immune therapy might be a prospective vaccination strategy for primary amoebic meningoencephalitis. Other potential vaccine candidates such as rNfa1 protein had been explored in different routes. Both intranasal and intraperitoneal rNfa1 protein introduced into virus vector stimulated Th1 cytokines production, interferon-γ and regulatory inerleukin 10. The immunization also conferred protection by the release of IgG, IgA and IgE antibodies (Lee *et al.*, 2011; Kim *et al.*, 2013). The rNfa1 protein could be investigated further in hope of vaccination development.

Conclusion and future studies

As current diagnosis and management of *N. fowleri* infection are still faced with many challenges, more extensive studies are required for prevention, pathogenesis, and therapy. Furthermore, the awareness of this deadly organism needs to be raised for the concerns of medical professionals as well as general population by means of educational channels. Research into the molecular basis of *N. fowleri* neuropathogenesis and immunity are yet to be illustrated to devise strategies to counter the fatal infection. One of the many problems with *Naegleria fowleri* is the delay in diagnosis as available diagnostic methods are hindered by contraindications. Primary amoebic meningoencephalitis is often complicated by increased intracranial pressure and its sequela brain herniation, in this instance lumbar puncture which is usually used for central nervous system infection is contraindicated as it worsens the herniation. The dilemma prompts the search for alternatives and using nasal secretion by saline flush could potentially be invaluable as non-invasive tool bypassing lumbar puncture to diagnose PAM. This is primarily based on the entry route of the amoeba where the colonization occurs at olfactory neuroepithelium in nasal cavity. In addition to diagnostic advantages, the intranasal route or transcribial route as a novelty in PAM can also be used in alternative drug delivery system. As current available therapeutic agents are mostly administered via the intravenous route, they have many pitfalls of systemic drug delivery such as failure to achieve minimal inhibitory concentration in the brain due to the biological blood–brain barrier. The blood brain barrier is highly selective to therapeutic molecules and most current antiamoebic agents have poor central nervous system penetration. The intravenous route is also frowned upon due to the systemic adverse effects of the therapeutic agents especially for hepatotoxicity and nephrotoxicity. The intranasal route could overcome many a problems encountered with intravenous modality. Lack of new biomarkers for detection of *Naegleria* is a challenge due to poor understanding of the pathogenesis and pathophysiology. As PAM is a rare infection and the resources are scarce for diagnosis, it had been overlooked frequently with major public health implications. Other major strategies such as drug repurposing and drug target identification are currently under investigations for their outlooks in treatment of PAM. In drug repurposing, development of automated approach for matching databases of drug targets against proteomes of pathogens holds a promising prospect. Another cost effective in-silico method is a proposed alternative to proteome-based screening which has its own merits for constructing the library that allow molecular screening, structural modelling and docking programing (Sateriale *et al.*, 2014). Enzymatic pathways had been contemplated recently as drug targets including hydrolytic enzymes, glycolytic enzymes, thiol synthesis and enzymes of redox metabolism.

References

Anderson, K., and Jamieson, A. (1972). Agglutination test for the investigation of the genus *Naegleria*. Pathology *4*, 273–278.

Antonios, S.N. (2010). Scanning electron microscopic study of trophozoite and cyst stages of *Naegleria fowleri*. J Egypt Soc Parasitol. *40*, 271–276.

Averner, M., and Fulton, C. (1966). Carbon dioxide: signal for excystment of *Naegleria* gruberi. J. Gen. Microbiol. *42*, 245–255. https://doi.org/10.1099/00221287-42-2-245.

Bols, N.C., Corff, S., and Yuyama, S. (1977). Rapid decline thymidine kinase activity during differentiation of *Naegleria*. J. Cell. Physiol. *90*, 271–280.

Bradley, S.G., Toney, D.M., Zhang, Y., and Marciano-Cabral, F. (1996). Dependence of growth, metabolic expression, and pathogenicity of *Naegleria fowleri* on exogenous porphyrins. J. Parasitol. *82*, 763–768.

Burri, D.C., Gottstein, B., Zumkehr, B., Hemphill, A., Schürch, N., Wittwer, M., and Müller, N. (2012). Development of a high-versus low-pathogenicity model of the free-living amoeba *Naegleria fowleri*. Microbiology *158*, 2652–2660.

Byers, T.J. (1986). Molecular biology of DNA in *Acanthamoeba*, Amoeba, Entamoeba, and *Naegleria*. Int. Rev. Cytol. *99*, 311–341.

Carrasco-Yepez, M., Rojas-Hernandez, S., Rodriguez-Monroy, M.A., Terrazas, L.I., and Moreno-Fierros, L. (2010). Protection against *Naegleria fowleri* infection in mice immunized with Cry1Ac plus amoebic lysates is dependent on the STAT6 Th2 response. Parasite Immunol. *32*, 664–670. https://doi.org/10.1111/j.1365-3024.2010.01222.x.

Carrasco-Yepez, M., Campos-Rodriguez, R., Godinez-Victoria, M., Rodriguez-Monroy, M.A., Jarillo-Luna, A., Bonilla-Lemus, P., De Oca, A.C., and Rojas-Hernandez, S. (2013). *Naegleria fowleri* glycoconjugates with residues of α-D-mannose are involved in adherence of trophozoites to mouse nasal mucosa. Parasitol. Res. *112*, 3615–3625. https://doi.org/10.1007/s00436-013-3549-2.

Carrasco-Yepez, M., Campos-Rodriguez, R., Lopez-Reyes, I., Bonilla-Lemus, P., Rodriguez-Cortes, A.Y., Contis-Montes de Oca, A., Jarillo-Luna, A., Miliar-Garcia, A., and Rojas-Hernandez, S. (2014). Intranasal coadministration of Cholera toxin with amoeba lysates modulates the secretion of IgA and IgG antibodies, production of cytokines and expression of pIgR in the nasal cavity of mice in the model of *Naegleria fowleri* meningoencephalitis. Exp. Parasitol. *145*, S84–92. https://doi.org/10.1016/j.exppara.2014.04.002.

Cerva, L. (1978). Some further characteristics of the growth of *Naegleria fowleri* and N. gruberi in axenic culture. Folia Parasitol. *25*, 1–8.

Cerva, L., Kasprzak, W., and Mazur, T. (1981). *Naegleria fowleri* in cooling waters of power plants. J. Hyg. Epidemiol. Microbiol. Immunol. *26*, 152–161.

Cervantes-Sandoval, I., Serrano-Luna, J.d.e.J., García-Latorre, E., Tsutsumi, V., and Shibayama, M. (2008). Characterization of brain inflammation during primary amoebic meningoencephalitis. Parasitol. Int. *57*, 307–313. https://doi.org/10.1016/j.parint.2008.01.006.

Chang, S.L. (1958). Cultural, cytological and ecological observations on the amoeba stage of *Naegleria* gruberi. J. Gen. Microbiol. *18*, 565–578. https://doi.org/10.1099/00221287-18-3-565.

Chávez-Munguía, B., Segovia-Gamboa, N., Salazar-Villatoro, L., Omaña-Molina, M., Espinosa-Cantellano, M., and Martínez-Palomo, A. (2011). *Naegleria fowleri*: enolase is expressed during cyst differentiation. J. Eukaryot. Microbiol. *58*, 463–468. https://doi.org/10.1111/j.1550-7408.2011.00574.x.

Chu, D.M., Woodward, J., Fritzinger, A., and Marciano-Cabral, F. (2002). Calcium-dependent protection from complement lysis in *Naegleria fowleri* amebae. Cell Calcium *31*, 105–114. https://doi.org/10.1054/ceca.2001.0256.

Chung, S., Cho, J., Cheon, H., Paik, S., and Lee, J. (2002). Cloning and characterization of a divergent alpha-tubulin that is expressed specifically in dividing amebae of *Naegleria* gruberi. Gene *293*, 77–86.

Corff, S., and Yuyama, S. (1976). Cessation of nuclear DNA synthesis in differentiating *Naegleria*. J. Protozool. *23*, 587–593.

De Jonckheere, J.F. (2004). Molecular definition and the ubiquity of species in the genus *Naegleria*. Protist *155*, 89–103.

De Jonckheere, J.F. (2011). Origin and evolution of the worldwide distributed pathogenic amoeboflagellate *Naegleria fowleri*. Infect. Genet. Evol. *11*, 1520–1528. https://doi.org/10.1016/j.meegid.2011.07.023.

Dubray, B.L., Wilhelm, W.E., and Jennings, B.R. (1987). Serology of *Naegleria fowleri* and *Naegleria lovaniensis* in a hospital survey. J. Protozool. *34*, 322–327.

Ferrante, A., and Mocatta, T.J. (1984). Human neutrophils require activation by mononuclear leucocyte conditioned medium to kill the pathogenic free-living amoeba, *Naegleria fowleri*. Clin. Exp. Immunol. *56*, 559–566.

Ferrante, A., and Smyth, C. (1984). Mitogenicity of *Naegleria fowleri* extract for murine T lymphocytes. Immunology *51*, 461–468.

Ferrante, A., and Thong, Y.H. (1979). Antibody induced capping and endocytosis of surface antigens in *Naegleria fowleri*. Int. J. Parasitol. *9*, 599–601.

Ferrante, A., Carter, R.F., Lopez, A.F., Rowan-Kelly, B., Hill, N.L., and Vadas, M.A. (1988). Depression of immunity to *Naegleria fowleri* in mice by selective depletion of neutrophils with a monoclonal antibody. Infect. Immun. *56*, 2286–2291.

Forrester, J.A., Gingell, D., and Korohoda, W. (1967). Electrophoretic polarity exhibited by amoeboid cells of *Naegleria* gruberi. Nature *215*, 1409–1410.

Fritzinger, A.E., Toney, D.M., MacLean, R.C., and Marciano-Cabral, F. (2006). Identification of a *Naegleria fowleri* membrane protein reactive with anti-human CD59 antibody. Infect. Immun. *74*, 1189–1195.

Fulton, C., and Dingle, A.D. (1967). Appearance of the flagellate phenotype in populations of *Naegleria* amebae. Dev. Biol. *15*, 165–191.

Fulton, C., and Dingle, A.D. (1971). Basal bodies, but not centrioles, in *Naegleria*. J. Cell Biol. *51*, 826.

Fulton, C., and Walsh, C. (1980). Cell differentiation and flagellar elongation in *Naegleria* gruberi. Dependence on transcription and translation. J. Cell Biol. *85*, 346–360.

Fulton, C., Cheng, K.L., and Lai, E.Y. (1986). Two calmodulins in *Naegleria* flagellates: characterization, intracellular segregation, and programmed regulation of mRNA abundance during differentiation. J. Cell Biol. *102*, 1671–1678.

Gianinazzi, C., Schild, M., Müller, N., Leib, S.L., Simon, F., Nunez, S., Joss, P., and Gottstein, B. (2005). Organotypic slice cultures from rat brain tissue: a new approach for *Naegleria fowleri* CNS infection in vitro. Parasitology *132*, 797–804.

Ginger, M.L., Fritz-Laylin, L.K., Fulton, C., Cande, W.Z., and Dawson, S.C. (2010). Intermediary metabolism in protists: a sequence-based view of facultative anaerobic metabolism in evolutionarily diverse eukaryotes. Protist *161*, 642–671. https://doi.org/10.1016/j.protis.2010.09.001.

Goswick, S.M., and Brenner, G.M. (2003). Activities of azithromycin and amphotericin B against *Naegleria fowleri* in vitro and in a mouse model of primary amebic meningoencephalitis. Antimicrob. Agents Chemother. *47*, 524–528.

Goudot, S., Herbelin, P., Mathieu, L., Soreau, S., Banas, S., and Jorand, F. (2012). Growth dynamic of *Naegleria fowleri* in a microbial freshwater biofilm. Water Res. *46*, 3958–3966. https://doi.org/10.1016/j.watres.2012.05.030.

Grace, E., Asbill, S., and Virga, K. (2015). *Naegleria fowleri*: pathogenesis, diagnosis, and treatment options. Antimicrob. Agents Chemother. *59*, 6677–6681. https://doi.org/10.1128/AAC.01293-15.

Griffin, J.L. (1972). Temperature tolerance of pathogenic and nonpathogenic free-living amoebas. Science *178*, 869–870.

Haight, J.B., and John, D.T. (1980). Growth of *Naegleria fowleri* in several axenic media. Folia Parasitol. *27*, 207–212.

Holbrook, T.W., and Parker, B.W. (1979). *Naegleria fowleri* in Chick Embryos. Am. J. Trop. Med. Hyg. *28*, 984–987.

Jeffery, S., and Hawkins, S.E. (1976). Studies on transformation in *Naegleria* gruberi: effects of ions and bacterial suspensions. Microbios *15*, 27–36.
Jeong, S.R., Lee, S.C., Song, K.J., Park, S., Kim, K., Kwon, M.H., Im, K.I., and Shin, H.J. (2005). Expression of the nfa1 gene cloned from pathogenic *Naegleria fowleri* in nonpathogenic *N. gruberi* enhances cytotoxicity against CHO target cells in vitro. Infect. Immun. *73*, 4098–4105.
John, D.T. (1982). Primary amebic meningoencephalitis and the biology of *Naegleria fowleri*. Annu. Rev. Microbiol. *36*, 101–123. https://doi.org/10.1146/annurev.mi.36.100182.000533.
John, D.T., Cole, T.B., and John, R.A. (1991). Flagella number among *Naegleria* flagellates. Folia Parasitol. *38*, 289–295.
John, D.T., and John, R.A. (2006). Viability of pathogenic *Naegleria* and *Acanthamoeba* isolates during 10 years of cryopreservation. Folia Parasitol. *53*, 311–312.
Kasprzak, W., Mazur, T., and Cerva, L. (1982). *Naegleria fowleri* in thermally polluted waters). Folia Parasitol. *29*, 211–218.
Kilvington, S., and White, D. (1991). A simple method for the cryopreservation of free-living amoebae belonging to the genera *Naegleria* and *Acanthamoeba*. Eur. J. Protistol. *27*, 115–118. https://doi.org/10.1016/S0932-4739(11)80332-9.
Kim, H.K., Kang, J.G., Yumura, S., Walsh, C.J., Cho, J.W., and Lee, J. (2005). De novo formation of basal bodies in *Naegleria gruberi*: regulation by phosphorylation. J. Cell Biol. *169*, 719–724.
Kim, J.H., Jung, S.Y., Lee, Y.J., Song, K.J., Kwon, D., Kim, K., Park, S., Im, K.I., and Shin, H.J. (2008). Effect of therapeutic chemical agents in vitro and on experimental meningoencephalitis due to *Naegleria fowleri*. Antimicrob. Agents Chemother. *52*, 4010–4016. https://doi.org/10.1128/AAC.00197-08.
Kim, J.H., Sohn, H.J., Lee, J., Yang, H.J., Chwae, Y.J., Kim, K., Park, S., and Shin, H.J. (2013). Vaccination with lentiviral vector expressing the nfa1 gene confers a protective immune response to mice infected with *Naegleria fowleri*. Clin. Vaccine Immunol. *20*, 1055–1060. https://doi.org/10.1128/CVI.00210-13.
Kollars, T.M., and Wilhelm, W.E. (1996). The occurrence of antibodies to *Naegleria* species in wild mammals. J. Parasitol. *82*, 73–77.
Kyle, D.E., and Noblet, G.P. (1986). Seasonal distribution of thermotolerant free-living amoebae. I. Willard's Pond. J. Protozool. *33*, 422–434.
Larson, D.E., and Dingle, A.D. (1981). Isolation, ultrastructure, and protein composition of the flagellar rootlet of *Naegleria* gruberi. J. Cell Biol. *89*, 424–432.
Lee, K.K., Karr, S.L., Wong, M.M., and Hoeprich, P.D. (1979). In vitro susceptibilities of *Naegleria fowleri* strain HB-1 to selected antimicrobial agents, singly and in combination. Antimicrob. Agents Chemother. *16*, 217–220.
Lee, Y.J., Park, C.E., Kim, J.H., Sohn, H.J., Lee, J., Jung, S.Y., and Shin, H.J. (2011). *Naegleria fowleri* lysate induces strong cytopathic effects and pro-inflammatory cytokine release in rat microglial cells. Korean J. Parasitol. *49*, 285–290. https://doi.org/10.3347/kjp.2011.49.3.285.
Levy, Y.Y., Lai, E.Y., Remillard, S.P., and Fulton, C. (1998). Centrin is synthesized and assembled into basal bodies during *Naegleria* differentiation. Cell Motil. Cytoskeleton *40*, 249–260.
Mahittikorn, A., Mori, H., Popruk, S., Roobthaisong, A., Sutthikornchai, C., Koompapong, K., Siri, S., Sukthana, Y., and Nacapunchai, D. (2015). Development of a rapid, simple method for detecting *Naegleria fowleri* visually in water samples by loop-mediated isothermal amplification (LAMP). PLOS ONE *10*, e0120997. https://doi.org/10.1371/journal.pone.0120997.
Marciano-Cabral, F., Cline, M.L., and Bradley, S.G. (1987). Specificity of antibodies from human sera for *Naegleria* species. J. Clin. Microbiol. *25*, 692–697.
Marciano-Cabral, F., MacLean, R., Mensah, A., and LaPat-Polasko, L. (2003). Identification of *Naegleria fowleri* in domestic water sources by nested PCR. Appl. Environ. Microbiol. *69*, 5864–5869.
Martínez, A.J. (1977). Free-living amebic meningoencephalitides: comparative study. Neurol. Neurocir. Psiquiatr. 18 (Suppl. 2-3), 391–401.
McLaughlin, G.L., Vodkin, M.H., and Huizinga, H.W. (1991). Amplification of repetitive DNA for the specific detection of *Naegleria fowleri*. J. Clin. Microbiol. *29*, 227–230.
Muldrow, L.L., Tyndall, R.L., and Fliermans, C.B. (1982). Application of flow cytometry to studies of pathogenic free-living amoebae. Appl. Environ. Microbiol. *44*, 1258–1269.
Nerad, T.A., Visvesvara, G., and Daggett, P.M. (1983). Chemically defined media for the cultivation of *Naegleria*: pathogenic and high temperature tolerant species. J. Protozool. *30*, 383–387.
Opperdoes, F.R., De Jonckheere, J.F., and Tielens, A.G. (2011). *Naegleria gruberi* metabolism. Int. J. Parasitol. *41*, 915–924. https://doi.org/10.1016/j.ijpara.2011.04.004.

Pearson, J.L., and Willmer, E.N. (1963). Some observations on the actions of steroids on the metaplasia of the amoeba, *Naegleria gruberi*. J. Exp. Biol. *40*, 493–515.
Preston, T.M., and O'Dell, D.S. (1974). Studies on the amoeboid and flagellate stages of *Naegleria*. Ann. Soc. Belg. Med. Trop. *54*, 279–286.
Qvarnstrom, Y., Visvesvara, G.S., Sriram, R., and da Silva, A.J. (2006). Multiplex real-time PCR assay for simultaneous detection of *Acanthamoeba* spp., *Balamuthia* mandrillaris, and *Naegleria fowleri*. J. Clin. Microbiol. *44*, 3589–3595.
Raederstorff, D., and Rohmer, M. (1987). Sterol biosynthesis via cycloartenol and other biochemical features related to photosynthetic phyla in the amoebae *Naegleria lovaniensis* and *Naegleria gruberi*. Eur. J. Biochem. *164*, 427–434.
Réveiller, F.L., Cabanes, P.A., and Marciano-Cabral, F. (2002). Development of a nested PCR assay to detect the pathogenic free-living amoeba *Naegleria fowleri*. Parasitol. Res. *88*, 443–450.
Rivera-Aguilar, V., Hernández-Martínez, D., Rojas-Hernández, S., Oliver-Aguillón, G., Tsutsumi, V., Herrera-González, N., and Campos-Rodríguez, R. (2000). Immunoblot analysis of IgA antibodies to *Naegleria fowleri* in human saliva and serum. Parasitol. Res. *86*, 775–780.
Rojas-Hernández, S., Rodríguez-Monroy, M.A., López-Revilla, R., Reséndiz-Albor, A.A., and Moreno-Fierros, L. (2004). Intranasal coadministration of the Cry1Ac protoxin with amoebal lysates increases protection against *Naegleria fowleri* meningoencephalitis. Infect. Immun. *72*, 4368–4375. https://doi.org/10.1128/IAI.72.8.4368-4375.2004.
Rojas-Hernández, S., Jarillo-Luna, A., Rodríguez-Monroy, M., Moreno-Fierros, L., and Campos-Rodríguez, R. (2004). Immunohistochemical characterization of the initial stages of *Naegleria fowleri* meningoencephalitis in mice. Parasitol. Res. *94*, 31–36. https://doi.org/10.1007/s00436-004-1177-6.
Schumacher, D.J., Tien, R.D., and Lane, K. (1995). Neuroimaging findings in rare amebic infections of the central nervous system. AJNR Am. J. Neuroradiol. 16, 930–935.
Sateriale, A., Bessoff, K., Sarkar, I.N., and Huston, C.D. (2014). Drug repurposing: mining protozoan proteomes for targets of known bioactive compounds. J. Am. Med. Inform Assoc. *21*, 238–244. https://doi.org/10.1136/amiajnl-2013-001700.
Schuster, F. (1963). An electron microscope study of the amoebo-flagellate, *Naegleria gruberi* (Schardinger). I. The amoeboid and flagellate stages. J. Protozool. *10*, 297–313.
Schuster, F. (1963). An Electron Microscope Study of the Amoebo-flagellate, *Naegleria* gruberi (Schardinger). II. The Cyst Stage. J. Protozool. *10*, 313–320.
Schuster, F.L., and Rechthand, E. (1975). In vitro effects of amphotericin B on growth and ultrastructure of the amoeboflagellates *Naegleria gruberi* and *Naegleria fowleri*. Antimicrob. Agents Chemother. *8*, 591–605.
Schuster, F.L., Guglielmo, B.J., and Visvesvara, G.S. (2006). In-vitro activity of miltefosine and voriconazole on clinical isolates of free-living amebas: *Balamuthia* mandrillaris, *Acanthamoeba* spp., and *Naegleria fowleri*. J. Eukaryot. Microbiol. 53, 121–126.
Singh, S.N., Patwari, A.K., Dutta, R., Taneja, N., and Anand, V.K. (1998). *Naegleria* meningitis. Indian pediatr. 35, 1012–1015.
Soltow, S.M., and Brenner, G.M. (2007). Synergistic activities of azithromycin and amphotericin B against *Naegleria fowleri* in vitro and in a mouse model of primary amebic meningoencephalitis. Antimicrob. Agents Chemother. *51*, 23–27.
Suh, M.R., Han, J.W., No, Y.R., and Lee, J. (2002). Transient concentration of a gamma-tubulin-related protein with a pericentrin-related protein in the formation of basal bodies and flagella during the differentiation of *Naegleria* gruberi. Cell Motil. Cytoskeleton *52*, 66–81. https://doi.org/10.1002/cm.10033.
Sussman, D.J., Lai, E.Y., and Fulton, C. (1984). Rapid disappearance of translatable actin mRNA during cell differentiation in *Naegleria*. J. Biol. Chem. *259*, 7355–7360.
Symmers, W.C. (1969). Primary amoebic meningoencephalitis in Britain. Br. Med. J. *4*, 449–454.
Thong, Y.H., Carter, R.F., Ferrante, A., and Rowan-Kelly, B. (1983). Site of expression of immunity to *Naegleria fowleri* in immunized mice. Parasite Immunol. *5*, 67–76.
Thong, Y.H., and Ferrante, A. (1986). Migration patterns of pathogenic and nonpathogenic *Naegleria* spp. Infect. Immun. *51*, 177–180.
Toney, D.M., and Marciano-Cabral, F. (1992). Alterations in protein expression and complement resistance of pathogenic *Naegleria* amoebae. Infect. Immun. *60*, 2784–2790.
Toney, D.M., and Marciano-Cabral, F. (1994a). Membrane vesiculation of *Naegleria fowleri* amoebae as a mechanism for resisting complement damage. J. Immunol. *152*, 2952–2959.

Toney, D.M., and Marciano-Cabral, F. (1994b). Modulation of complement resistance and virulence of *Naegleria fowleri* amoebae by alterations in growth media. J. Eukaryot. Microbiol. *41*, 337–343.

Tsaousis, A.D., Nyvltová, E., Sutak, R., Hrdy, I., and Tachezy, J. (2014). A nonmitochondrial hydrogen production in *Naegleria* gruberi. Genome Biol. Evol. *6*, 792–799. https://doi.org/10.1093/gbe/evu065.

Visvesvara, G.S. (2010). Amebic meningoencephalitides and keratitis: challenges in diagnosis and treatment. Curr. Opin. Infect. Dis. *23*, 590–594. https://doi.org/10.1097/QCO.0b013e32833ed78b.

Walsh, C. (1980). Appearance of heat shock proteins during the induction of multiple flagella in *Naegleria* gruberi. J. Biol. Chem. *255*, 2629–2632.

Walsh, C. (1984). Synthesis and assembly of the cytoskeleton of *Naegleria* gruberi flagellates. J. Cell Biol. *98*, 449–456.

Weik, R.R., and John, D.T. (1977). Cell size, macromolecular composition, and O2 consumption during agitated cultivation of *Naegleria* gruberi. J. Protozool. *24*, 196–200.

Weik, R.R., and John, D.T. (1978). Macromolecular composition and nuclear number during growth of *Naegleria fowleri*. J. Parasitol. *64*, 746–747.

Weik, R.R., and John, D.T. (1979). Cell and mitochondria respiration of *Naegleria fowleri*. J. Parasitol. *65*, 700–708.

Wessberg, K.L., Skolnick, S., Xu, J., Marciano-Cabral, F., and Kemp, R.G. (1995). Cloning, sequencing and expression of the pyrophosphate-dependent phosphofructo-1-kinase from *Naegleria fowleri*. Biochem. J. *307*, 143–149.

Whiteman, L.Y., and Marciano-Cabral, F. (1987). Susceptibility of pathogenic and nonpathogenic *Naegleria* spp. to complement-mediated lysis. Infect. Immun. *55*, 2442–2447.

Xinyao, L., Miao, S., Yonghong, L., Yin, G., Zhongkai, Z., Donghui, W., Weizhong, W., and Chencai, A. (2006). Feeding characteristics of an amoeba (Lobosea: *Naegleria*) grazing upon cyanobacteria: food selection, ingestion and digestion progress. Microb. Ecol. *51*, 315–325. https://doi.org/10.1007/s00248-006-9031-2

Entamoeba histolytica: from High-throughput Technology to New Drugs

8

Anjan Debnath[1]* and Sharon L Reed[2]

[1]Center for Discovery and Innovation in Parasitic Diseases, University of California San Diego, La Jolla, CA, USA.
[2]Department of Pathology and Department of Medicine, University of California San Diego, La Jolla, CA, USA.

*Correspondence: adebnath@ucsd.edu

https://doi.org/10.21775/9781910190838.07

Abstract

Amoebiasis, caused by the protozoan parasite *Entamoeba histolytica*, is a major contributor to morbidity and mortality worldwide. Current therapy mainly depends on metronidazole, but it is not active against *E. histolytica* cysts that transmit amoebiasis. Therefore, an additional luminal drug paromomycin is required for the treatment. Treatment with two drugs for 20 days leads to incomplete compliance. Moreover, in the absence of a back-up drug and a looming threat of drug resistance, the identification of new drugs for the treatment of amoebiasis is a top priority. With the advent of high-throughput screening technology, we focused on developing an assay that could utilize robotics for rapid interrogation of compounds against *E. histolytica*. We developed a high-throughput screen for *E. histolytica* and used the validated assay to screen a smaller FDA-approved drug library and a larger kinase-targeted library. The screen of the kinase-targeted library identified six amoebicidals and the FDA-approved drug library screen identified auranofin as amoebicidal. This opened the avenue for repurposing the anti-arthritic drug, auranofin, for the treatment of amoebiasis. Currently, auranofin has completed a phase I clinical trial for amoebiasis and is undergoing phase II clinical trial evaluation for the treatment of amoebiasis and giardiasis.

Introduction

Invasive amoebiasis, caused by the intestinal protozoan parasite *Entamoeba histolytica*, is a major health and social problem in the developing world. It is also a problem in the developed world in travellers, immigrants, and homosexual men (Haque *et al.*, 2006; Chen *et al.*, 2009). Approximately 500 million people worldwide are infected annually with *E. histolytica*, resulting in 50 million cases of invasive amoebiasis (WHO, 1997) and about 100,000 annual deaths (Haque *et al.*, 2003). The mortality from fulminating amoebic colitis is almost 70% and that from amoebic liver abscess is up to 10% (WHO, 1987). A recent study found that amoebiasis results in the loss of approximately 2.2 million disability-adjusted life years (DALYs) (Hotez *et al.*, 2014).

In spite of the substantial impact on human health, the treatment for amoebiasis is essentially dependent on one drug, metronidazole, for more than 50 years. Standard treatment with metronidazole requires at least 10 days at 750 mg three times a day to eradicate intestinal amebae. Unfortunately, metronidazole has many adverse effects such as nausea, vomiting, diarrhoea or constipation (Sweetman, 2002), change in the taste sensation, CNS toxicity including ataxia and encephalopathy, dryness of mouth, hypersensitivity, leukopenia, pancreatitis, peripheral neuropathy, thrombocytopenia, thrombophlebitis and urinary tract effects. In addition, metronidazole is not active on infective cysts, which transmit amoebiasis through faeces. Therefore, an additional luminal drug paromomycin is used for 5 to 10 days at 25 to 35 mg/kg daily in three doses (Knight, 1980) to eliminate the risks of disease transmission by shedding of cysts. Treatment with two drugs makes compliance difficult. In addition, treatment failure (Koutsaimanis *et al.*, 1979) and recurrence of amoebic liver abscess (Dooley and O'Morain, 1988) have been reported after treatment with metronidazole. Potential resistance to metronidazole may also compound the difficulties of treatment in the future. Metronidazole resistant *E. histolytica* lines were generated *in vitro* (Samarawickrema *et al.*, 1997; Wassmann *et al.*, 1999), and a recent study found that *E. histolytica* resistant to a lower concentration of metronidazole was also less sensitive to tinidazole (Penuliar *et al.*, 2015). This low level of metronidazole resistance *in vitro* resulted in reduced cell growth, adhesion, phagocytosis, virulence and glucose uptake, which were linked to specific transcriptional changes (Penuliar *et al.*, 2015). All these studies support the importance of finding an alternative drug for the treatment of amoebiasis.

Assay development for drug discovery

The major impediment to drug discovery research for *E. histolytica* was the absence of an assay that could interrogate the activity of several compounds in a relatively short period of time. Earlier studies relied on radioisotopes and staining coupled with microscopy (Singh *et al.*, 2005) but these methods were laborious, involved multiple steps and investigated the effect of only a handful of compounds. The low-throughput of these methods necessitated the development of an assay that can be compatible with workstation-based automation and examine the effect of larger compound libraries against *E. histolytica* trophozoites. Growing the trophozoites in a microtitre plate was the first requirement for developing a high-throughput screen (HTS) for *E. histolytica*, but the wells in the microtitre plate provided an oxygenated environment that threatened the survival of *E. histolytica* trophozoites. We overcame this obstacle by growing trophozoites under anaerobic conditions by incubating the microtitre plates at 37°C for 48 h using the GasPak EZ Anaerobe Gas Generating Pouch Systems (VWR). Maintaining an oxygen-free environment was not a prerequisite during robotic transfer of compounds or addition of trophozoites to the microtitier plates. The assay development was performed in both 96- and 384-well plates with logarithmically grown trophozoites of 50,000/ml and 30,000/ml, respectively. Since DMSO is the widely used solvent for compounds, we also optimized the tolerability of the percentage of DMSO for *E. histolytica* and found that they could withstand up to 0.5% DMSO with no significant reduction of growth rate. Our assay read-out was based on ATP-bioluminescence and measured by the CellTiter-Glo Luminescent Cell Viability Assay (Promega) technology.

Once the assay was developed, we calculated EC_{50} for metronidazole, the current drug of choice for amoebiasis, to ensure that the assay could be standardized with a known amoebicidal agent. This was performed in a 96-well microtitre plate with exponentially growing *E. histolytica* trophozoites. In our system, EC_{50} value for metronidazole (Sigma-Aldrich), defined as that concentration of compound necessary to reduce the culture density to 50% of that of a DMSO-treated culture, was 5 μM in the 96-well format. Standard assay metric such as Z′ was calculated for the plate during determination of EC_{50} for metronidazole. A Z′ of 0.9 was achieved with a signal to background ratio of 322.33. Per cent inhibition relative to maximum and minimum reference signal controls was calculated using the formula:

$$\%\,\text{Inhibition} = [(\text{mean of Maximum Signal Reference Control} - \text{Experimental Value}) / (\text{mean of Maximum Signal Reference Control} - \text{mean of Minimum Signal Reference Control})] \times 100$$

Compound screening in 96-well plates using an FDA-approved compound library

The HTS protocol for compound screening in 96-well plates has been outlined in Fig. 8.1. The protocol introduced a robotic liquid handling system (Biomek FX[P] Laboratory Automation Workstation, Beckman Coulter and Matrix WellMate bulk dispenser, Thermo Fisher Scientific) for compound dilution, seeding trophozoites, and adding luciferase reagent to the microtitre plates. We validated the utility of this HTS by screening 910 compound collections containing either pharmacologically active molecules or drugs already approved for human use. The use of compounds already in clinical use allowed the potential to rapidly and cost-effectively repurpose these drugs for the treatment of amoebiasis. From a technical

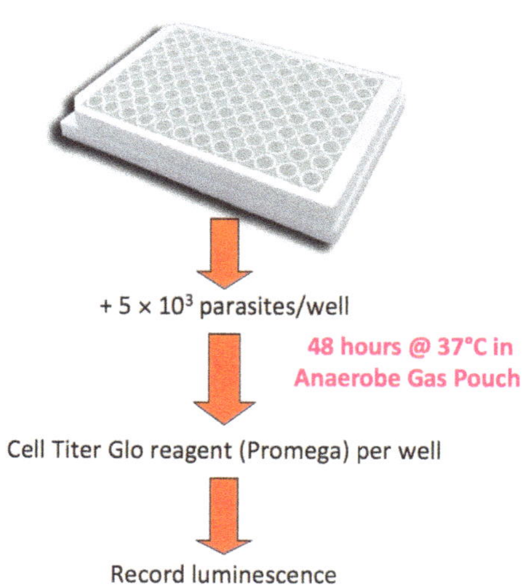

Figure 8.1 High-throughput compound screening protocol for *E. histolytica* trophozoites in 96-well microtitre plate.

and conceptual standpoint, drug repurposing is well received at the US National Institutes of Health (NIH) and is considered a 'pre-clinical innovation'. Repurposing is more relevant for amoebiasis because this is a disease primarily of poor people in developing countries, and there is little economic incentive to the pharmaceutical industry to develop new drugs for this disease.

For primary screening of trophozoites in 96-well plates, 4 µl of 125 µM compound in 12.5% DMSO was transferred to sterile screen plates followed by addition of 46 µl of TYI-S-33 complete medium. Fifty µl of exponentially growing *E. histolytica* trophozoites (5000 parasites) in TYI-S-33 complete medium were seeded into each well of the 96-well microtitre plates under sterile conditions so that the final concentrations of the test compound and DMSO per well were 5 µM and 0.5%, respectively. Negative controls in the screen plates contained 0.5% DMSO and positive controls contained 30 µM metronidazole. Assay plates were incubated for 48 h at 37°C in the GasPak^{TM} EZ Anaerobe Gas Generating Pouch Systems to maintain anaerobic condition throughout the incubation period. At the end of incubation, 50 µl of CellTiter-Glo Luminescent Cell Viability Assay were added in each well of the 96-well plates. The resulting ATP-bioluminescence of the trophozoites was measured at room temp. using an Analyst HT plate reader from Molecular Devices (Debnath *et al.*, 2012). Our screening strategy in 96-well microtitre plates employed a relatively high initial compound concentration of 5 µM to maximize the yield of active compounds. Since in our system, the EC_{50} value for the current drug metronidazole is 5 µM, we concluded that any compound showing more than 50% inhibition at primary screening at 5 µM should be investigated in secondary screening to validate activity.

The primary screening of the chemical library containing 746 compounds of FDA-approved drugs and 164 bioactive compounds identified auranofin as the most potent compound against *E. histolytica*. The amoebicidal activity of auranofin was reconfirmed in a secondary screen where auranofin was tested at different concentrations ranging from 0.01 µM to 25 µM. Testing the compound at this 8-point concentration range determined the EC_{50} of auranofin as 0.5 µM, ten-fold less than metronidazole. Based on its potency and oral availability, we next tested auranofin in animal models of amoebic colitis and amoebic liver abscess. An oral dose of 1 mg/kg q.d. for 7 days significantly reduced the parasite burden and damaging host inflammatory response in a mouse model of amoebic colitis (Debnath *et al.*, 2012). Auranofin also showed efficacy in a hamster model of amoebic liver abscess and an oral dose of 3 mg/kg q.d. for 7 days significantly reduced the hepatic damage caused by *E. histolytica* (Debnath *et al.*, 2012). A similar dose of metronidazole did not have a protective effect against colitis or liver abscess.

Auranofin: an alternative to 'one bug-one drug' approach

Since the discovery of auranofin as a new drug lead for *E. histolytica*, several investigators looked at its efficacy for related organisms and showed efficacy of auranofin against other protozoans including the diarrhoeagenic parasites *Giardia lamblia* (Tejman-Yarden *et al.*, 2013) and *Cryptosporidium parvum* (Debnath *et al.*, 2013), the sexually transmitted parasite *Trichomonas vaginalis* (Hopper *et al.*, 2016), and intracellular parasites *Leishmania donovani* (Ilari *et al.*, 2012; Sharlow *et al.*, 2014), *Toxoplasma gondii* (Andrade *et al.*, 2014) and *Trypanosoma cruzi* (da Silva *et al.*, 2016). Auranofin was equally effective against multicellular worms such as *Schistosoma mansoni* (Angelucci *et al.*, 2009), *Onchocerca*, *Brugia* (Bulman *et al.*, 2015) and *Taenia* (Martínez-González *et al.*, 2010). Together, these parasites infect more than 1 billion people worldwide, and auranofin with its broad-spectrum antiparasitic activity has the potential to replace the one bug for one drug paradigm.

Auranofin: indications for approval in 1985

Auranofin is an FDA-approved drug and has been in clinical use to treat rheumatoid arthritis since 1985. It is sold as a 3 mg oral tablet or capsule. Chemically, it is a monomeric gold species with a triethyl phosphine group and the chemical name is 2,3,4,6-tetra-O-acetyl-1-thio-β-D-glucopyranosato-*S*-(triethyl-phosphine) gold (Kean *et al.*, 1997). Because of the presence of gold, there may be a concern of higher price of this drug but on a weight basis, the content of gold in the compound is about 29% (Kean *et al.*, 1997). The dose of 1 mg/kg at which auranofin showed efficacy in a mouse model of amoebic colitis would correspond to a daily human equivalent dose for adults of about 5 mg. Based on gold content alone and the current gold price, a course of treatment with auranofin for 7 days would be cost-effective, making it an inexpensive option for the treatment of amoebiasis.

The early pharmacokinetic studies of auranofin after oral administration detected about 25% of the administered dose in plasma and the peak concentrations reached within 1–2 h (Kean *et al.*, 1997). Auranofin has long half-life, which ranges from 15–25 days and the total body elimination rate is 55–80 days (Blocka, 1983; Kean *et al.*, 1997).

Many adverse effects are listed on the package insert of long-term (> 1 year) auranofin therapy, including dermatological: rash (26%); gastrointestinal: loose stools (42%), abdominal pain (14%) and nausea (10%); haematological: anaemia, leucopenia and thrombocytopenia in up to 3%; hepatic: elevated liver enzymes (2%); mucous membranes: stomatitis (13%); renal: proteinuria (1%); and eye: conjunctivitis (10%). These early studies were all performed in small numbers of patients with rheumatoid arthritis given auranofin for months to years, therefore are unlikely to reflect the risk in short-term therapy for infectious diseases.

Auranofin is a Category C drug whose use should be avoided during pregnancy or nursing since gold has been found in the blood of newborns (Cohen *et al.*, 1981; Richards, 1977), and studies in pregnant rabbits and rats found decreased maternal and fetal weight and litter size with 4–50× the human dose (Medical Economics Co., 1996). Paediatric use of auranofin is not currently approved but it was not due to safety issues, rather because of ineffectiveness in the treatment of juvenile rheumatoid arthritis (Giannini *et al.*, 1990).

Recent clinical studies of auranofin

Amoebiasis has been listed as a rare disease by the NIH Office of Rare Diseases Research, having a prevalence of fewer than 200,000 affected individuals in the United States. The rare disease status of amoebiasis makes a drug developed against it to be eligible to receive orphan drug status under the US Orphan Drug Act. Based on the *in vitro* and *in vivo* activities of auranofin against *E. histolytica*, auranofin has received an Orphan Drug status from the FDA for the treatment of amoebiasis (Debnath *et al.*, 2012). The orphan drug status of auranofin may bring several economic incentives to stimulate the drug development such as availability of research and development grants, 50% tax credit for costs involved in clinical trial and 7-year market exclusivity provision (Seoane-Vazquez *et al.*, 2008). An

NIH-sponsored research and development grant recently conducted a phase I clinical trial of auranofin (NCT02089048) to characterize the pharmacokinetics of gold in a short-term therapy that is relevant to the treatment of amoebiasis.

Phase I clinical trial

As auranofin received FDA approval in 1985 for the treatment of rheumatoid arthritis, the majority of the pharmacokinetics data of auranofin originated from long-term therapy in rheumatoid arthritis patients with less sensitive methods such as radiolabelled gold tracer studies (Gottlieb *et al.*, 1974) or atomic absorption spectroscopy (Finkelstein *et al.*, 1976). Therefore, it was important to evaluate the safety and pharmacokinetics of auranofin in otherwise healthy patients for short-course therapy, using modern techniques to accurately measure gold levels in plasma and faeces. With these aims, a phase I, single-centre, open-label study was conducted in 15 healthy adults (Capparelli *et al.*, 2017). Each of them received 6 mg (p.o.) of auranofin daily for 7 days and was followed for 126 days. The following treatment-emergent adverse events were reported but all were resolved without treatment: gastrointestinal: abdominal discomfort, constipation, diarrhoea, flatulence, nausea (20%); headache (13.3%), and thrombocytopen (13.3%). These data suggested that short-term treatment with auranofin was safe and well tolerated.

Plasma and faecal concentrations of gold were measured by inductively coupled plasma-mass spectrometry (ICP-MS) and following 7 days of once-daily dosing, plasma gold maximum concentration was 0.312 μg/ml and the mean terminal half-life was approximately 35 days (Capparelli *et al.*, 2017). Once-daily dosing for 7 days led to a mean concentration of 8.80 μg/g of gold in faeces. This concentration was more than 25× the EC_{50} concentration for *E. histolytica* and 4× the EC_{50} concentration for *G. lamblia* (Capparelli *et al.*, 2017). Thus, the results of the phase I clinical trial justified the initiation of a phase II clinical trial of auranofin for amoebiasis and giardiasis.

Phase II clinical trial

A phase IIa, randomized, placebo-controlled, single-blinded superiority treatment study to compare placebo to once daily doses of 6 mg auranofin for adults with asymptomatic amoebiasis or giardiasis is under way and has been registered at ClinicalTrials.gov with identifier NCT02736968. The estimated duration of the study is 3.5 years and participants of 18–65 years old will receive 7 days' treatment for amoebiasis and 5 days' treatment for giardiasis with follow-up until day 28.

Compound screening in 384-well plates using kinase-targeted compound collections

The discovery of amoebicidal auranofin from the screening of 96-well plates encouraged us to interrogate the anti-*E. histolytica* activity of larger compound collections available in 384-well format. Kinase-targeted compound collections with 10,000 compounds were purchased from ChemDiv, Inc. We selected the kinase-targeted collections because *E. histolytica* possesses a substantial complement of protein kinases accounting for about 3% of the total proteome size, which is slightly more than the size of kinome of most other eukaryotes. The genome of *E. histolytica* encodes 307 putative protein kinases which is more than the number of kinases encoded in the genome of other parasites such as *Plasmodium falciparum*,

Leishmania major, Trypanosoma brucei, Trypanosoma cruzi. Moreover, the *E. histolytica* kinome differs in numerous ways from kinases in the mammalian host (Anamika *et al.*, 2008). So, we reasoned that any *E. histolytica* kinase-targeted hit would be more selective for *E. histolytica*, and not affect the human host.

Compound dilution scheme for primary screening

For primary screening of *E. histolytica* trophozoites in 384-well plates, 5 µl of 1 mM compound in 100% DMSO was transferred to 384-well polypropylene dilution plates (E&K Scientific) using a Biomek FX^{P} Laboratory Automation Workstation. 320 compounds from 384-well ChemDiv library stock plates were transferred to each 384-well dilution plate leaving columns 1, 2, 23, and 24 empty for addition of negative and positive controls. The Matrix WellMate bulk dispenser was then used to add 35 µl of sterile water to the dilution plates to yield 125 µM compound in 12.5% DMSO. FX^{P} transferred 2 µl of diluted compound to the 384-well screen plates followed by addition of 23 µl of TYI-S-33 complete medium by the WellMate.

Automated primary HTS using cell viability assay

The *E. histolytica* compound susceptibility assay was performed in a final volume of 50 µl using 384-well microtitre plate format. For automated HTS procedures, 25 µl of trophozoites (1500 parasites) in complete medium were seeded into each well using the WellMate to achieve final concentrations of test compound and DMSO per well 5 µM and 0.5%, respectively. Negative controls in the screen plates contained 0.5% DMSO and positive controls contained 30 µM metronidazole. Assay plates were incubated for 48 h at 37°C in the GasPak™ EZ Anaerobe Gas Generating Pouch Systems and at the end of incubation ATP-bioluminescence of the trophozoites was measured using CellTiter-Glo Luminescent Cell Viability Assay.

Secondary screen for potency determination

Hits from the primary screen were confirmed in a 96-well screen plates with a three-fold serial dilution of compounds covering a concentration range of 0.01–25 µM. The assays were performed in triplicate using CellTiter-Glo Luminescent Cell Viability Assay.

Cytotoxicity assay

Mammalian cell line toxicity assay was performed with Jurkat T cells, an established model for *in vitro* compound toxicity (Buenz, 2007). Cytotoxicity assays were performed with the compound dilution plates prepared earlier for secondary screening with a concentration range of 0.25–625 µM. From this dilution plate, 4 µl were transferred into the 96-well screen plates followed by addition of 96 µl of Jurkat (10,000 cells) to yield a final 8-point concentration range spanning 0.01–25 µM in final 0.5% DMSO. Negative control contained final 0.5% DMSO and positive control contained final concentration of 40 µM staurosporine (BioVision). The assays were performed in triplicate using CellTiter-Glo Luminescent Cell Viability Assay.

Identification of hits

After interrogating 10,000 compounds of the ChemDiv library at 5 µM in thirty-two 384-well plates, we identified 10 primary hits with both at least 50% inhibition in the single readout and 3 standard deviations above the mean of the population of compounds tested. The screening was prosecuted over three days and conformed to HTS standard with a Z′ close to 1. Nine primary hits, meeting our criteria, were identified after screening twenty-two plates, performed in two days (Fig. 8.2). The third day screening with ten plates yielded one primary hit, but this one (compound 51269) was found toxic to mammalian cell at 25 μM and was not considered for secondary screening. Six of the primary hits were confirmed in secondary screening. Four of these confirmed hits had similar or better EC_{50} than metronidazole. The structures and the EC_{50}s of the confirmed hits are shown in Table 8.1.

Jurkat cells are particularly responsive to the induction of cell death and they are easily maintained in cell culture. They were used to determine the toxicity of the confirmed hits. The mammalian cell line well tolerated six ChemDiv hits (compounds 146122, 147781, 147819, 147828, 147839, 147840) at concentration 25 μM for 48 h (Fig. 8.3).

Similarity ensemble approach

In order to identify the target for the ChemDiv library hits, we used a statistics-based chemoinformatics approach called similarity ensemble approach (SEA). SEA compares targets by the similarity of the ligands that bind to them, expressed as expectation values, adapting the BLAST algorithms (Keiser *et al.*, 2007; Keiser *et al.*, 2009). Using an online SEA search tool (http://sea.bkslab.org/search/) to rapidly search large compound databases and to build cross-target similarity maps, we predicted MAP kinase as the target for compounds 147828 and 147839.

Perspectives and challenges

We optimized a novel HTS assay for *E. histolytica* and identified new amoebicidal lead compounds. The HTS assay is an important advance in drug discovery efforts for amoebiasis. The assay is compatible with robotic liquid handling platforms. By contrast, other assays use long timescales utilizing staining methods. An analysis of first-in-class small-molecule drugs approved by the FDA between 1999 and 2008 showed that the contribution of phenotypic screening to the discovery pipeline exceeded that of target-based approaches, demonstrating the strength of our strategy (Swinney and Anthony, 2011). Our approach of investigating FDA-approved compound library successfully identified auranofin as a new amoebicidal compound. Auranofin is currently undergoing the first clinical trial for two parasitic infections.

Molecular targeting of protein kinases is a new paradigm in the treatment of human diseases. Protein kinase inhibitors represent an important and emerging class of targeted therapeutic agents. In spite of screening large number of kinase-targeted small molecules, only six compounds were identified as confirmed hits. This small number of hits with no toxicity against human cell further confirms the specificity of those small molecules for *E. histolytica*. Utilizing chemoinformatics, a new way to view molecular structures using the power of computer databases, we predicted *E. histolytica* MAP kinase might be the target for two of these small molecules. This is an encouraging finding from the point of view of designing inhibitors of *E. histolytica* protein kinases. Target validation and *in vivo* efficacy of kinase-targeted small molecules will be a prerequisite for future drug development.

In spite of some success, drug discovery for amoebiasis faces several challenges. Because *E. histolytica* mainly affects poor people and this infection is considered rare in the United

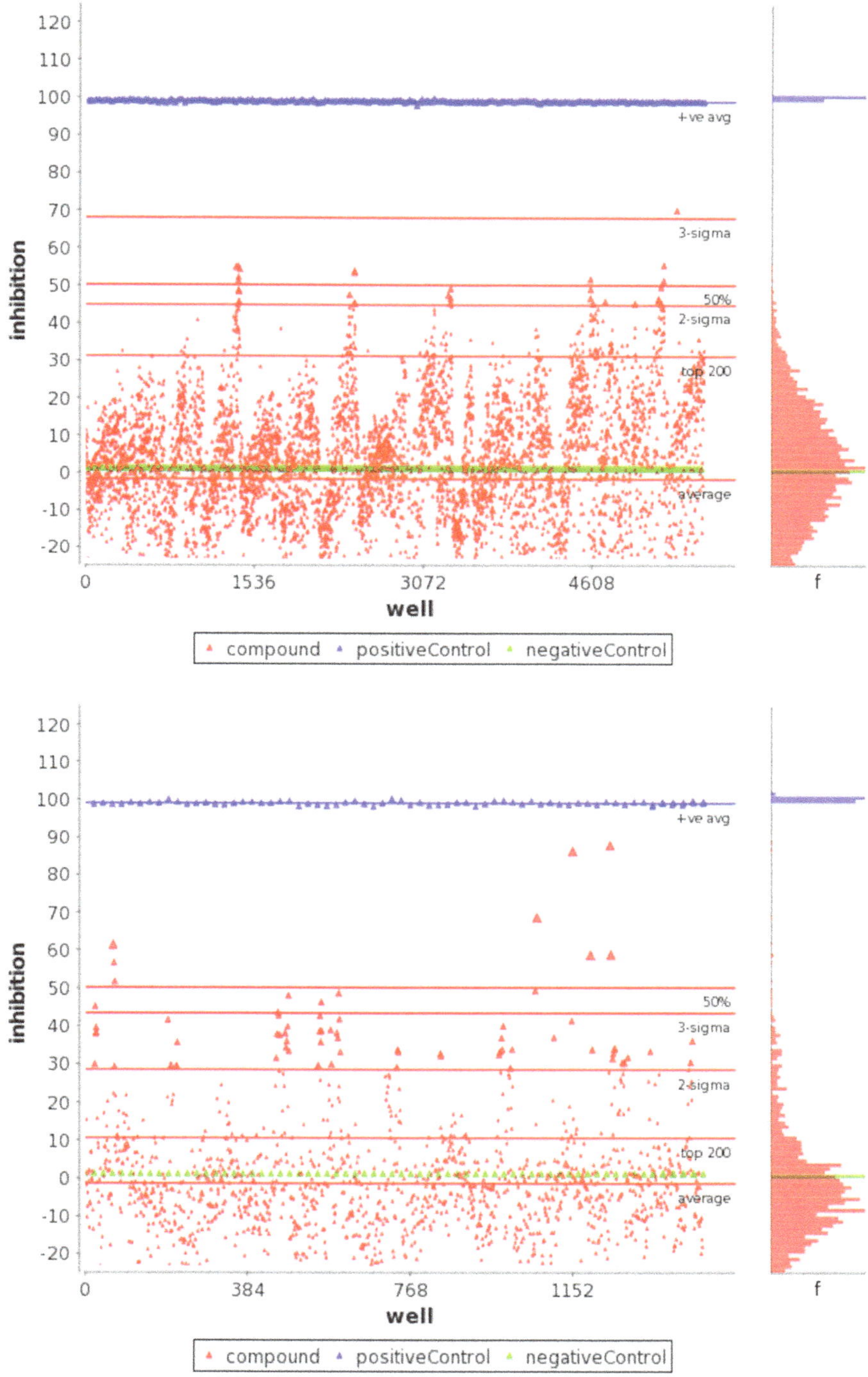

Figure 8.2 Scatterplot of percentage inhibition of each well from twenty-two 384-well plates of the ChemDiv library. Screening of 16 plates on day 1 (A) and six plates on day 2 (B) at 5 μM yielded total nine hits with both 50% inhibition and 3 standard deviations above the mean of the population of compounds tested. Positive controls included metronidazole-treated *E. histolytica* and negative controls included 0.5% DMSO-treated *E. histolytica*.

Table 8.1 Structures and EC_{50}s of the ChemDiv library hits

Compound	Structure	EC_{50} (μM)
146122	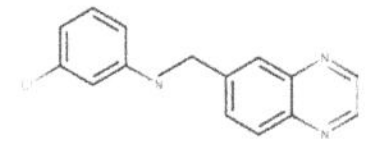	3.8
147781	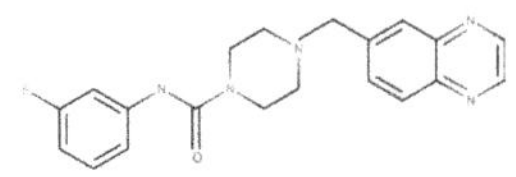	5.0
147819	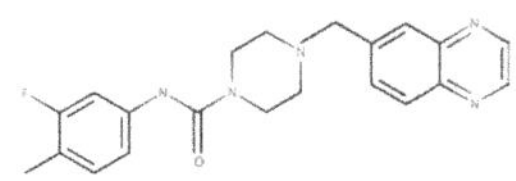	2.5
147828	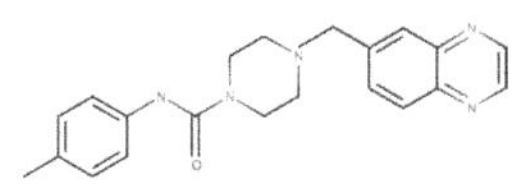	9.4
147839	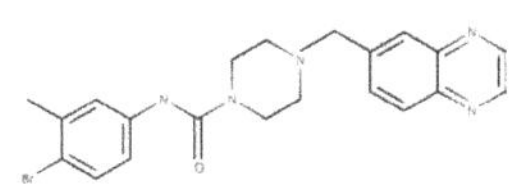	2.7
147840	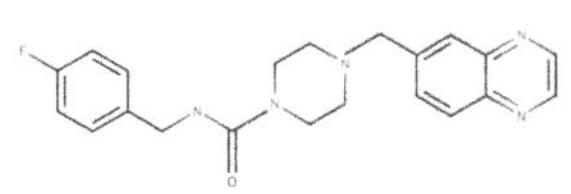	10.4

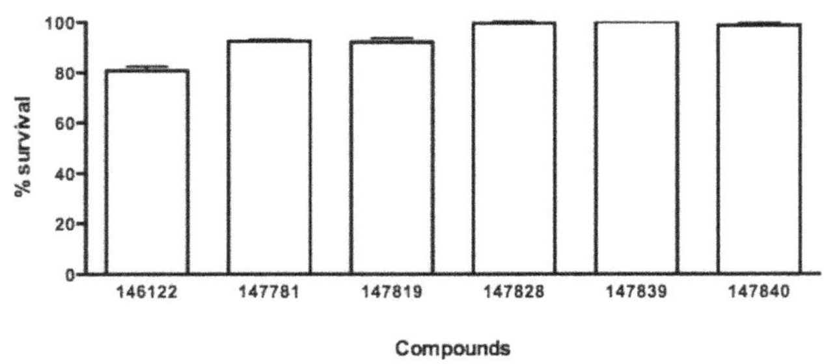

Figure 8.3 Jurkat cytotoxicity of ChemDiv hits. Jurkat cells well-tolerated ChemDiv library hits at 25 μM. Percentage survival was calculated from the means and standard deviations of triplicate wells.

States, there are no economic incentives to invest in development of amoebiasis drugs by the pharmaceutical industry. In the absence of economic advantages, drug discovery for amoebiasis largely relies on academic research centres with limited resources. Another challenge is the identification of a compound that can treat both cysts and trophozoites, thereby reducing the course of treatment. Though current drug discovery efforts primarily depend on phenotypic screen, developing an HTS targeting a biologically important protein and identifying target-based inhibitors for the treatment of amoebiasis should also be explored. Last but not least, no target product profile (TPP) is available for amoebiasis. In the absence of an established TPP for amoebiasis, we have outlined in Table 8.2 our current goals for a therapeutic, as compared to current standard of care.

Table 8.2 Target product profile for amoebiasis

Target	Metronidazole	Minimal requirement	Ideal requirement
Population	All	All	All
Adults/children	Adult/children	Adult	Adult/children
Clinical efficacy	Approximately 90% of patients with mild-to-moderate amoebic dysentery	At least equal to metronidazole	
Safety	Anorexia, nausea, vomiting, diarrhoea, abdominal discomfort; disulfuram-like intolerance reaction with alcohol; rarely, neurotoxicity, including seizures, peripheral neuropathy, dizziness, confusion, irritability	Less adverse effects than metronidazole, no diarrhoea or loose stools	Minimal adverse effects
Contraindications	Patients with a prior history of hypersensitivity to metronidazole or other nitroimidazole derivatives	Patients with a prior history of hypersensitivity	None
Precautions	Suspected mutagen/teratogen	No genotoxicity; no prolongation of QTc interval	No genotoxicity; no teratogenicity; no prolongation of QTc interval

Acknowledgements

This research was supported by the National Institutes of Health Study DMID 12–0101 under Phase 1 Unit Contract Number HHSN272200800024C to S. L. Reed, U01A1077822 to S. L. Reed, U01AI110435 to S. L. Reed. A. Debnath is supported by the National Institutes of Health, Grant #1KL2TR001444. Compound libraries were kindly provided by Michelle R. Arkin and Steven Chen of the University of California, San Francisco Small Molecule Discovery Center.

References

Anamika, K., Bhattacharya, A., and Srinivasan, N. (2008). Analysis of the protein kinome of *Entamoeba histolytica*. Proteins *71*, 995–1006. https://doi.org/10.1002/prot.21790.

Andrade, R.M., Chaparro, J.D., Capparelli, E., and Reed, S.L. (2014). Auranofin is highly efficacious against *Toxoplasma gondii* in vitro and in an in vivo experimental model of acute toxoplasmosis. PLOS Negl. Trop. Dis. *8*, e2973. https://doi.org/10.1371/journal.pntd.0002973.

Angelucci, F., Sayed, A.A., Williams, D.L., Boumis, G., Brunori, M., Dimastrogiovanni, D., Miele, A.E., Pauly, F., and Bellelli, A. (2009). Inhibition of *Schistosoma mansoni* thioredoxin-glutathione reductase by auranofin: structural and kinetic aspects. J. Biol. Chem. *284*, 28977–28985. https://doi.org/10.1074/jbc.M109.020701.

Blocka, K. (1983). Auranofin versus injectable gold. Comparison of pharmacokinetic properties. Am. J. Med. 75, 114–122.

Buenz, E.J. (2007). A high-throughput cell-based toxicity analysis of drug metabolites using flow cytometry. Cell Biol. Toxicol. 23, 361–365. https://doi.org/10.1007/s10565-007-0226-1.

Bulman, C.A., Bidlow, C.M., Lustigman, S., Cho-Ngwa, F., Williams, D., Rascón, A.A., Tricoche, N., Samje, M., Bell, A., Suzuki, B., *et al.* (2015). Repurposing auranofin as a lead candidate for treatment of lymphatic filariasis and onchocerciasis. PLOS Negl. Trop. Dis. *9*, e0003534. https://doi.org/10.1371/journal.pntd.0003534.

Capparelli, E.V., Bricker-Ford, R., Rogers, M.J., McKerrow, J.H., and Reed, S.L. (2017). Phase I clinical trial results of auranofin, a novel antiparasitic agent. Antimicrob. Agents Chemother. *61*, e01947–16.

Chen, L.H., Wilson, M.E., Davis, X., Loutan, L., Schwartz, E., Keystone, J., Hale, D., Lim, P.L., McCarthy, A., Gkrania-Klotsas, E., *et al.* (2009). Illness in long-term travelers visiting GeoSentinel clinics. Emerging Infect. Dis. *15*, 1773–1782. https://doi.org/10.3201/eid1511.090945.

Cohen, D.L., Orzel, J., and Taylor, A. (1981). Infants of mothers receiving gold therapy. Arthritis Rheum. *24*, 104–105.

da Silva, M.T., Silva-Jardim, I., Portapilla, G.B., de Lima, G.M., Costa, F.C., Anibal, F.d.e.F., and Thiemann, O.H. (2016). In vivo and in vitro auranofin activity against *Trypanosoma cruzi*: Possible new uses for an old drug. Exp. Parasitol. *166*, 189–193. https://doi.org/10.1016/j.exppara.2015.05.012.

Debnath, A., Ndao, M., and Reed, S.L. (2013). Reprofiled drug targets ancient protozoans: drug discovery for parasitic diarrheal diseases. Gut Microbes *4*, 66–71. https://doi.org/10.4161/gmic.22596.

Debnath, A., Parsonage, D., Andrade, R.M., He, C., Cobo, E.R., Hirata, K., Chen, S., García-Rivera, G., Orozco, E., Martínez, M.B., *et al.* (2012). A high-throughput drug screen for *Entamoeba histolytica* identifies a new lead and target. Nat. Med. *18*, 956–960. https://doi.org/10.1038/nm.2758.

Dooley, C.P., and O'Morain, C.A. (1988). Recurrence of hepatic amebiasis after successful treatment with metronidazole. J. Clin. Gastroenterol. *10*, 339–342.

Finkelstein, A.E., Walz, D.T., Batista, V., Mizraji, M., Roisman, F., and Misher, A. (1976). Auranofin. New oral gold compound for treatment of rheumatoid arthritis. Ann. Rheum. Dis. *35*, 251–257.

Giannini, E.H., Brewer, E.J., Kuzmina, N., Shaikov, A., and Wallin, B. (1990). Auranofin in the treatment of juvenile rheumatoid arthritis. Results of the USA-USSR double-blind, placebo-controlled trial. The USA Pediatric Rheumatology Collaborative Study Group. The USSR Cooperative Children's Study Group. Arthritis Rheum. *33*, 466–476.

Gottlieb, N.L., Smith, P.M., and Smith, E.M. (1974). Pharmacodynamics of ^{197}Au and ^{195}Au labeled aurothiomalate in blood. Correlation with course of rheumatoid arthritis, gold toxicity and gold excretion. Arthritis Rheum. *17*, 171–183.

Haque, R., Huston, C.D., Hughes, M., Houpt, E., and Petri, W.A. (2003). Amebiasis. N. Engl. J. Med. *348*, 1565–1573. https://doi.org/10.1056/NEJMra022710.

Haque, R., Mondal, D., Duggal, P., Kabir, M., Roy, S., Farr, B.M., Sack, R.B., and Petri, W.A. (2006). *Entamoeba histolytica* infection in children and protection from subsequent amebiasis. Infect. Immun. *74*, 904–909.

Hopper, M., Yun, J.F., Zhou, B., Le, C., Kehoe, K., Le, R., Hill, R., Jongeward, G., Debnath, A., Zhang, L., *et al.* (2016). Auranofin inactivates *Trichomonas vaginalis* thioredoxin reductase and is effective against trichomonads in vitro and in vivo. Int. J. Antimicrob. Agents *48*, 690–694.

Hotez, P.J., Alvarado, M., Basáñez, M.G., Bolliger, I., Bourne, R., Boussinesq, M., Brooker, S.J., Brown, A.S., Buckle, G., Budke, C.M., *et al.* (2014). The global burden of disease study 2010: interpretation and implications for the neglected tropical diseases. PLOS Negl. Trop. Dis. *8*, e2865. https://doi.org/10.1371/journal.pntd.0002865.

Ilari, A., Baiocco, P., Messori, L., Fiorillo, A., Boffi, A., Gramiccia, M., Di Muccio, T., and Colotti, G. (2012). A gold-containing drug against parasitic polyamine metabolism: the X-ray structure of trypanothione reductase from *Leishmania infantum* in complex with auranofin reveals a dual mechanism of enzyme inhibition. Amino Acids *42*, 803–811. https://doi.org/10.1007/s00726-011-0997-9.

Kean, W.F., Hart, L., and Buchanan, W.W. (1997). Auranofin. Br. J. Rheumatol. *36*, 560–572.

Keiser, M.J., Roth, B.L., Armbruster, B.N., Ernsberger, P., Irwin, J.J., and Shoichet, B.K. (2007). Relating protein pharmacology by ligand chemistry. Nat. Biotechnol. *25*, 197–206.

Keiser, M.J., Setola, V., Irwin, J.J., Laggner, C., Abbas, A.I., Hufeisen, S.J., Jensen, N.H., Kuijer, M.B., Matos, R.C., Tran, T.B., *et al.* (2009). Predicting new molecular targets for known drugs. Nature *462*, 175–181. https://doi.org/10.1038/nature08506.

Knight, R. (1980). The chemotherapy of amoebiasis. J. Antimicrob. Chemother. *6*, 577–593.

Koutsaimanis, K.G., Timms, P.W., and Rée, G.H. (1979). Failure of metronidazole in a patient with hepatic amebic abscess. Am. J. Trop. Med. Hyg. *28*, 768–769.

Martínez-González, J.J., Guevara-Flores, A., Alvarez, G., Rendón-Gómez, J.L., and Del Arenal, I.P. (2010). In vitro killing action of auranofin on *Taenia crassiceps* metacestode (cysticerci) and inactivation of thioredoxin-glutathione reductase (TGR). Parasitol. Res. *107*, 227–231. https://doi.org/10.1007/s00436-010-1867-1.

Medical Economics Co. (1996). Non-human toxicity excerpts, p. 2513. In Physicians' Desk Reference, 50th ed. Available online: https://pubchem.ncbi.nlm.nih.gov/compound/6333901#section=Non-Human-Toxicity-Excerpts. Accessed 30 March 2016.

Penuliar, G.M., Nakada-Tsukui, K., and Nozaki, T. (2015). Phenotypic and transcriptional profiling in *Entamoeba histolytica* reveal costs to fitness and adaptive responses associated with metronidazole resistance. Front. Microbiol. *6*, 354. https://doi.org/10.3389/fmicb.2015.00354.

Richards, A.J. (1977). Transfer of gold from mother to fetus. Lancet *1*, 99.

Samarawickrema, N.A., Brown, D.M., Upcroft, J.A., Thammapalerd, N., and Upcroft, P. (1997). Involvement of superoxide dismutase and pyruvate:ferredoxin oxidoreductase in mechanisms of metronidazole resistance in *Entamoeba histolytica*. J. Antimicrob. Chemother. *40*, 833–840.

Seoane-Vazquez, E., Rodriguez-Monguio, R., Szeinbach, S.L., and Visaria, J. (2008). Incentives for orphan drug research and development in the United States. Orphanet J. Rare Dis. *3*, 33. https://doi.org/10.1186/1750-1172-3-33.

Sharlow, E.R., Leimgruber, S., Murray, S., Lira, A., Sciotti, R.J., Hickman, M., Hudson, T., Leed, S., Caridha, D., Barrios, A.M., *et al.* (2014). Auranofin is an apoptosis-simulating agent with in vitro and in vivo anti-leishmanial activity. ACS Chem. Biol. *9*, 663–672. https://doi.org/10.1021/cb400800q.

Singh, S., Athar, F., and Azam, A. (2005). Synthesis, spectral studies and in vitro assessment for antiamoebic activity of new cyclooctadiene ruthenium(II) complexes with 5-nitrothiophene-2-carboxaldehyde thiosemicarbazones. Bioorg. Med. Chem. Lett. *15*, 5424–5428.

Sweetman, S.C. (2002). Martindale: The Complete Drug Reference, 33rd ed. (Pharmaceutical Press, London).

Swinney, D.C., and Anthony, J. (2011). How were new medicines discovered? Nat. Rev. Drug Discov. *10*, 507–519. https://doi.org/10.1038/nrd3480.

Tejman-Yarden, N., Miyamoto, Y., Leitsch, D., Santini, J., Debnath, A., Gut, J., McKerrow, J.H., Reed, S.L., and Eckmann, L. (2013). A reprofiled drug, auranofin, is effective against metronidazole-resistant *Giardia lamblia*. Antimicrob. Agents Chemother. *57*, 2029–2035. https://doi.org/10.1128/AAC.01675-12.

Wassmann, C., Hellberg, A., Tannich, E., and Bruchhaus, I. (1999). Metronidazole resistance in the protozoan parasite *Entamoeba histolytica* is associated with increased expression of iron-containing superoxide dismutase and peroxiredoxin and decreased expression of ferredoxin 1 and flavin reductase. J. Biol. Chem. *274*, 26051–26056.

WHO (1987). Prevention and control of intestinal parasitic infections. Report of a WHO Expert Committee. World Health Organ Tech Rep. Ser. *749*, 1–86.

WHO (1997). WHO/PAHO/UNESCO report. A consultation with experts on amoebiasis. Mexico City, Mexico 28-29 January, 1997. Epidemiol. Bull. *18*, 13-14.

Gene Silencing and Therapeutic Targets Against *Acanthamoeba* Infections

9

Jacob Lorenzo-Morales[1*], María Reyes-Batlle[1], Ines Sifaoui[1,2], Atteneri López-Arencibia[1], Olfa Chiboub[1,2], Aitor Rizo-Liendo[1], Carlos J Bethencourt-Estrella[1], Francisco Arnalich-Montiel[3], Pedro Rocha-Cabrera[1,4] and José E Piñero[1]

[1]Instituto Universitario de Enfermedades Tropicales y Salud Pública de Canarias, Universidad de La Laguna, Tenerife, Canary Islands, Spain.
[2]Preparatory Institute for Scientific and Technical Studies (IPEST), University of Carthage, La Marsa, Tunisia.
[3]Cornea Unit, Hospital Ramón y Cajal, Madrid, Spain.
[4]Ophthalmology Unit, Hospital Universitario de Canarias (HUC), Tenerife, Canary Islands, Spain.

*Correspondence: jmlorenz@ull.edu.es

https://doi.org/10.21775/9781910190838.08

Abstract

Pathogenic strains of *Acanthamoeba* genus are the causative agents of a sight-threating infection of the cornea known as *Acanthamoeba* keratitis (AK). The incidence of AK has risen worldwide in the recent years, which is highly related to an increase in the number of contact lens wearers, the main group of individuals at risk. Moreover, due to the ability of these amoebae to form a highly resistant cyst stage, most of the currently available treatments and contact lens maintenance solutions are not fully effective.

The development of *Acanthamoeba* genome project, releasing the sequences of the whole genome of this pathogen as well as the availability of novel therapeutic approaches such as gene silencing using RNAi technology, has allowed the application of this approach for the elucidation of novel drug targets in *Acanthamoeba*. In this chapter, the advances in this field as well as current and future therapeutic approaches against AK are described.

Introduction

Acanthamoeba genus is the causal agent of an even blinding infection of the cornea known as *Acanthamoeba* keratitis (AK) (Fig. 9.1). Furthermore, the current number of diagnosed AK cases is worryingly increasing worldwide (Lorenzo-Morales *et al.*, 2015; Cope *et al.*, 2016), even though AK is still considered a rare disease and it is included in the Orphanet database (ORPHA67043) and with an estimated prevalence of 1–9/100,000.

Acanthamoeba keratitis (AK) mainly affects contact lens wearers however pathogenic strains of these amoebae have also been reported to cause infections non-contact lens wearers (Khan, 2006, 2009; Lorenzo-Morales *et al.*, 2013, 2015). Moreover, *Acanthamoeba* is able to form a resistant dormant stage in its life cycle named cyst which is highly resistant to many physical and chemical agents (Fig. 9.2). Among the main risk factors besides contact lens use, AK is favoured by local immunosuppression due to corneal injury, lack and/or poor hygiene in the handling of contact lenses and their cases is the main risk factor for this infection (Khan, 2006, 2009; Lorenzo-Morales *et al.*, 2013, 2015). Patients suffering AK have been reported to experience photophobia, 20% of cases present ring-like stromal infiltrates, epithelial defects, lid oedema and if not treated on time loss of vision (Khan, 2009; Sun *et al.*, 2013; Lorenzo-Morales *et al.*, 2015).

Even though awareness on the devastating effects of AK is increasing worldwide, diagnosis of AK remains a challenge. Moreover, the available treatment options are lengthy and not fully effective against all strains (Dart *et al.*, 2009; Arnalich-Montiel *et al.*, 2012; Lorenzo-Morales *et al.*, 2015) and the pathogenic mechanisms of this emerging infectious agent is still under study.

Current status of therapy against *Acanthamoeba*

At present, the established standard treatment of AK in most countries consists on a combination of biguanides (chlorhexidine or PHMB) and a diamidine (propamidine or hexamidine) administered each hour initially, then reduced to 2 hourly by day for 3–4 weeks and then adjusted to each patient individually (Khan, 2006; Roberts and Henriquez, 2010; Lorenzo-Morales *et al.*, 2015). However, most of these drugs are associated with toxic side-effects to the cornea, and thus most of the patients, even following this treatment, need to undertake a keratoplasty for therapeutic purposes when the infection reaches severe levels (Roberts and Henriquez, 2010). Recently, voriconazole has been successfully used (Arnalich-Montiel *et al.*, 2012). Finally, and unfortunately, propamidine and hexamidine are not available in all countries.

Therefore, there is an urgent need to develop novel active compounds against AK and thus the need to elucidate therapeutic targets against these infections.

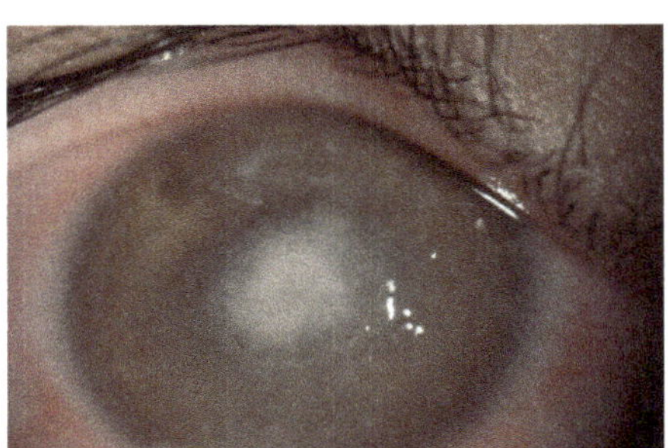

Figure 9.1 *Acanthamoeba* keratitis in a Spanish patient (Dr F. Arnalich collection).

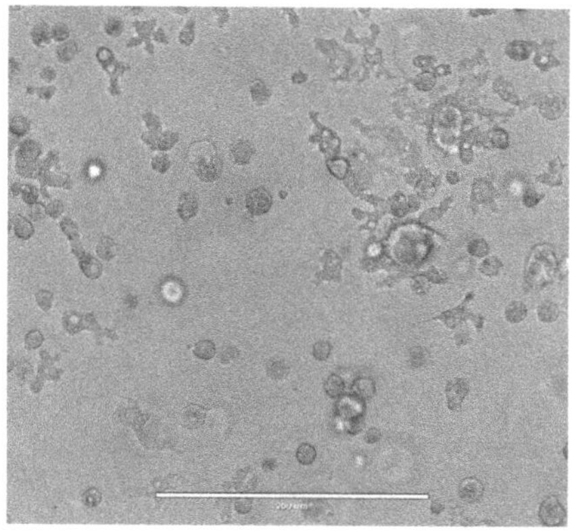

Figure 9.2 *Acanthamoeba* trophozoites and cysts in axenic culture observed under an inverted microscope. Magnification of 40×.

Drug and therapeutic targets discovery in *Acanthamoeba*

The major impediment to drug discovery research for *Acanthamoeba* was the lack of gene sequence data and also deep knowledge about the pathogenesis of this pathogen. Moreover, novel compounds or targets should be active in both stages of *Acanthamoeba,* trophozoite and cyst as well as low toxic to mammalian cells.

Drug sensitivity and activity screening

A key factor in the search of novel anti-*Acanthamoeba* compounds and also to check the efficacy of current AK treatments, is the establishment of a reliable and repetitive sensitivity and activity assay. Until recently, tests focused on this field were based on lengthy protocols involving dated methods such as manual counting and/or haemocytometers or the most probable number following agar plate spreading (Beattie *et al.*, 2003; Lorenzo-Morales *et al.*, 2016) After that, in 2005 a 96-well plate colorimetric assay based on Alamar Blue® was developed (McBride *et al.*, 2005) which is being standardized in most laboratories working on AK therapy worldwide (McBride *et al.*, 2005; Martín-Navarro *et al.*, 2008; Lorenzo-Morales *et al.*, 2016).

This assay is useful as it provides comparable and repetitive results in order to assess the number of viable cells in a culture based on innate cellular metabolic activity, reducing the Alamar Blue® dye and changing its colour (McBride *et al.* 2005). Moreover, this assay would provide reliable and repetitive results in order to establish novel and effective *in vitro* testing of novel compounds against *Acanthamoeba.*

Natural products against *Acanthamoeba* keratitis

Animal-origin compounds

From all the reported studies focused on the application of animal-origin compounds. The one which has shown to be more promising applied magainins. These molecules are defence peptides with antimicrobial activity secreted by the African clawed frog (*Xenopus laevi*).

In the case of *Acanthamoeba*, some magainins have been reported to eliminate amoebae (trophozoite and cysts) at concentrations ranging from 20 to 40 µg/ml (Feldman *et al.*, 1991; Schuster and Jacob, 1992).

Plant-origin compounds

Recently, a wide number of publications have been reporting the potential use of medicinal plants and herbs as a source of novel anti-*Acanthamoeba* agents. Most of the studies have focused on the screening of the activity of different plant extracts against various strains of *Acanthamoeba* (Degerli *et al.*, 2012; Vunda *et al.*, 2012; Sifaoui *et al.*, 2013, 2014). Further studies should be aimed in this field in order to elucidate and isolate the main molecules responsible for the reported anti-*Acanthamoeba* activity in these plant sources. In the later years 2016 and 2017, the number of publications reporting anti-*Acanthamoeba* activity of molecules obtained from plant extracts has dramatically risen worldwide. Nevertheless, as mentioned above, the main targets of these molecules as well as their mechanisms of action need to be further elucidated (Santos *et al.*, 2016; Hajaji *et al.*, 2017; Saoudi *et al.*, 2017; Sifaoui *et al.*, 2017).

Gene silencing (RNAi) and novel drug target validation

Drug target validation in most pathogens requires the elucidation of a target which are specific to the pathogenic agent itself in order to reduce possible side toxic effects in the human or animal host. After the development of *Acanthamoeba* genome project (Clarke *et al.*, 2013) and the rise of gene silencing approaches, a new field recently opened in *Acanthamoeba* and other free-living amoebae (FLA) (Lorenzo-Morales *et al.*, 2016).

Focusing on *Acanthamoeba*, the aim to elucidate such targets is of high priority and may be of different nature such as gene products, biological pathways of the amoeba, cell membranes or other physical attributes among others (Lorenzo-Morales *et al.* 2015, 2016). When the RNAi mechanisms were elucidated and also when the possibility to design and synthesize specific small interfering RNAs (siRNAs) was available in the market, a novel field was opened in the drug discovery against AK as mentioned above. In *Acanthamoeba*, the Free-Living Amoebae Laboratory at the University of La Laguna in Tenerife, Spain, was the first one to develop and apply this technology (siRNAs) in *Acanthamoeba* genus (Lorenzo-Morales *et al.*, 2005, 2016; Martín-Navarro *et al.*, 2013, 2014). Until now, a few therapeutic targets with potential use against AK have been validated and applied *in vitro* even against clinical strains of *Acanthamoeba* (Martín-Navarro *et al.*, 2014; Lorenzo-Morales *et al.*, 2016). Recently, a key enzyme in the synthesis of ergosterol (main lipid in the cell membrane of *Acanthamoeba*) was established as a good therapeutic target using specific siRNAs which also allowed to identify that statins are inhibitors of the same enzyme (3-hydroxy-3-methylglutaryl-coenzyme A reductase, HMG-CoA reductase) (Martín-Navarro *et al.*, 2013, 2014). In the same study, the *in vitro* activity of several statins was tested against clinical strains of *Acanthamoeba* showing high activities of these compounds against both the trophozoite and the cyst stage of *Acanthamoeba* (Martín-Navarro *et al.*, 2013, 2014). Currently, further studies are being developed in order to establish statins as a novel treatment option against AK. At present developments in order to spread the application of these compounds is being carried out.

Future perspectives

There is worryingly increase in the number of AK cases worldwide each year, mainly due to due to a higher numbers of contact lens wearers in the world and also a general lack of hygiene in the handling of contact lenses and their cases. Increased awareness combined with an early diagnostic of the disease is currently a good pathway towards a better outcome of AK and even social networks include now AK awareness and support groups which helping in the support and higher spread of AK knowledge worldwide. Nevertheless, AK is still considered a rare disease and lack of funding is a general issue among researchers working in this field worldwide as well as lack of interest from health authorities towards the control on the incidence increase of this infection.

Currently, developed efforts by researchers working on this field worldwide have highlighted that not only biguanides should be used a first line treatment and thus recently reported activities of compounds such as voriconazole, miltefosine and statins should be considered at least in the near future for a more effective treatment of AK. Moreover, novel techniques such as siRNA mediated gene silencing or even CRISPR/Cas9 state of the art technology are undoubtedly powerful tools to establish novel drug targets and assist in the development of more effective therapeutic measures against this emerging pathogen.

Acknowledgements

This work was supported by the grants Red de Investigation Colaborativa en Enfermedades Tropicales RICET (project no. RD16/0027/0001 of the programme of Redes Temáticas de Investigación Cooperativa, FIS), Spanish Ministry of Health, Madrid, Spain, and PI13/00490 'Protozoosis Emergentes por Amebas de Vida Libre: Aislamiento, Caracterización, Nuevas Aproximaciones Terapéuticas y Traslación Clínica de los Resultados' from the Instituto de Salud Carlos III. IS and ALA were supported by the Agustín de Betancourt Programme. JLM was supported by Ayudas Plan Propio de la Universidad de La Laguna 2017 (proyectos puente I+D+i).

References

Arnalich-Montiel, F., Martín-Navarro, C.M., Alió, J.L., López-Vélez, R., Martínez-Carretero, E., Valladares, B., Piñero, J.E., and Lorenzo-Morales, J. (2012). Successful monitoring and treatment of intraocular dissemination of acanthamoeba. Arch. Ophthalmol. *130*, 1474–1475. https://doi.org/10.1001/archophthalmol.2012.2376.

Beattie, T.K., Seal, D.V., Tomlinson, A., McFadyen, A.K., and Grimason, A.M. (2003). Determination of amoebicidal activities of multipurpose contact lens solutions by using a most probable number enumeration technique. J. Clin. Microbiol. *41*, 2992–3000.

Clarke, M., Lohan, A.J., Liu, B., Lagkouvardos, I., Roy, S., Zafar, N., Bertelli, C., Schilde, C., Kianianmomeni, A., Bürglin, T.R., *et al.* (2013). Genome of *Acanthamoeba castellanii* highlights extensive lateral gene transfer and early evolution of tyrosine kinase signalling. Genome Biol *14*, R11.

Cope, J.R., Collier, S.A., Schein, O.D., Brown, A.C., Verani, J.R., Gallen, R., Beach, M.J., and Yoder, J.S. (2016). *Acanthamoeba* Keratitis among Rigid Gas Permeable Contact Lens Wearers in the United States, 2005 through 2011. Ophthalmology *123*, 1435–1441. https://doi.org/10.1016/j.ophtha.2016.03.039.

Dart, J.K., Saw, V.P., and Kilvington, S. (2009). *Acanthamoeba* keratitis: diagnosis and treatment update 2009. Am. J. Ophthalmol. *148*, 487–499.e2. https://doi.org/10.1016/j.ajo.2009.06.009.

Degerli, S., Tepe, B., Celiksoz, A., Berk, S., and Malatyali, E. (2012). *In vitro* amoebicidal activity of *Origanum syriacum* and *Origanum laevigatum* on *Acanthamoeba castellanii* cysts and trophozoites. Exp. Parasitol. *131*, 20–24. https://doi.org/10.1016/j.exppara.2012.02.020.

Feldman, S.T., Speaker, M., and Cleveland, P. (1991). Effect of magainins on *Acanthamoeba castellanii*. Rev. Infect. Dis. 13 (Suppl. 5), S439.

Hajaji, S., Sifaoui, I., López-Arencibia, A., Reyes-Batlle, M., Valladares, B., Pinero, J.E., Lorenzo-Morales, J., and Akkari, H. (2017). Amoebicidal activity of α-bisabolol, the main sesquiterpene in chamomile (*Matricaria recutita* L.) essential oil against the trophozoite stage of *Acanthamoeba castellani* Neff. Acta Parasitol. *62*, 290–295. https://doi.org/10.1515/ap-2017-0036.

Khan, N.A. (2009). *Acanthamoeba* – Biology and Pathogenesis (Caister Academic Press, Norfolk).

Khan, N.A. (2006). *Acanthamoeba*: biology and increasing importance in human health. FEMS Microbiol. Rev. *30*, 564–595.

Kuźma, Ł., Derda, M., Hadaś, E., and Wysokińska, H. (2015). Abietane diterpenoids from *Salvia sclarea* transformed roots as growth inhibitors of pathogenic *Acanthamoeba* spp. Parasitol. Res. *114*, 323–327. https://doi.org/10.1007/s00436-014-4211-3.

Lim, N., Goh, D., Bunce, C., Xing, W., Fraenkel, G., Poole, T.R., and Ficker, L. (2008). Comparison of polyhexamethylene biguanide and chlorhexidine as monotherapy agents in the treatment of *Acanthamoeba* keratitis. Am. J. Ophthalmol. *145*, 130–135.

Lorenzo-Morales, J., Ortega-Rivas, A., Foronda, P., Abreu-Acosta, N., Ballart, D., Martínez, E., and Valladares, B. (2005). RNA interference (RNAi) for the silencing of extracellular serine proteases genes in *Acanthamoeba*: molecular analysis and effect on pathogenicity. Mol. Biochem. Parasitol. *144*, 10–15.

Lorenzo-Morales, J., Kliescikova, J., Martinez-Carretero, E., De Pablos, L.M., Profotova, B., Nohynkova, E., Osuna, A., and Valladares, B. (2008). Glycogen phosphorylase in *Acanthamoeba* spp.: determining the role of the enzyme during the encystment process using RNA interference. Eukaryotic Cell 7, 509–517. https://doi.org/10.1128/EC.00316-07.

Lorenzo-Morales, J., Martín-Navarro, C.M., López-Arencibia, A., Arnalich-Montiel, F., Piñero, J.E., and Valladares, B. (2013). *Acanthamoeba* keratitis: an emerging disease gathering importance worldwide? Trends Parasitol. *29*, 181–187. https://doi.org/10.1016/j.pt.2013.01.006.

Lorenzo-Morales, J., Khan, N.A., and Walochnik, J. (2015). An update on *Acanthamoeba* keratitis: diagnosis, pathogenesis and treatment. Parasite *22*, 10. https://doi.org/10.1051/parasite/2015010.

Lorenzo-Morales, J., Reyes-Batlle, M., Sifaoui, I., Arnalich-Montiel, F., López-Arencibia, A., Wagner, C., Rocha-Cabrera, P., del Castillo-Remiro, A., Martínez-Carretero, E., Piñero, J.E., and Valladares, B. (2016). Therapeutic targets and investigated treatment strategies in *Acanthamoeba* keratitis. Exp. Op. Orph Drugs *4*, 1069–1073.

Martin-Navarro C.M., Lorenzo-Morales J., Cabrera-Serra M.G., Rancel, F., Coronado-Alvarez N.M., Piñero, J.E., Valladares, B.*et al* (2008). The potential pathogenicity of chlorhexidine-sensitive *Acanthamoeba* strains isolated from contact lens cases from asymptomatic individuals in Tenerife, Canary Islands, Spain. J. Med. Microbiol. *57*, 1399–1404.

Martín-Navarro, C.M., Lorenzo-Morales, J., López-Arencibia, A., Reyes-Batlle, M., Piñero, J.E., Valladares, B., and Maciver, S.K. (2014). Evaluation of *Acanthamoeba* myosin-IC as a potential therapeutic target. Antimicrob. Agents Chemother. *58*, 2150–2155. https://doi.org/10.1128/AAC.01199-13.

Martín-Navarro, C.M., Lorenzo-Morales, J., Machin, R.P., López-Arencibia, A., García-Castellano, J.M., de Fuentes, I., Loftus, B., Maciver, S.K., Valladares, B., and Piñero, J.E. (2013). Inhibition of 3-hydroxy-3-methylglutaryl-coenzyme A reductase and application of statins as a novel effective therapeutic approach against *Acanthamoeba* infections. Antimicrob. Agents Chemother. *57*, 375–381. https://doi.org/10.1128/AAC.01426-12.

McBride, J., Ingram, P.R., Henriquez, F.L., and Roberts, C.W. (2005). Development of colorimetric microtiter plate assay for assessment of antimicrobials against *Acanthamoeba*. J. Clin. Microbiol. *43*, 629–634.

Polat, Z.A., Walochnik, J., Obwaller, A., Vural, A., Dursun, A., and Arici, M.K. (2014). Miltefosine and polyhexamethylene biguanide: a new drug combination for the treatment of *Acanthamoeba* keratitis. Clin. Exp. Ophthalmol. *42*, 151–158. https://doi.org/10.1111/ceo.12120.

Roberts, C.W., and Henriquez, F.L. (2010). Drug target identification, validation, characterisation and exploitation for treatment of *Acanthamoeba* (species) infections. Exp. Parasitol. *126*, 91–96. https://doi.org/10.1016/j.exppara.2009.11.016.

Santos, I.G., Scher, R., Rott, M.B., Menezes, L.R., Costa, E.V., Cavalcanti, S.C., Blank, A.F., Aguiar, J.d.o.s.S., da Silva, T.G., and Dolabella, S.S. (2016). Amebicidal activity of the essential oils of *Lippia* spp. (Verbenaceae) against *Acanthamoeba polyphaga* trophozoites. Parasitol. Res. *115*, 535–540. https://doi.org/10.1007/s00436-015-4769-4.

Saoudi, S., Sifaoui, I., Chammem, N., Reyes-Batlle, M., López-Arencibia, A., Pacheco-Fernández, I., Pino, V., Hamdi, M., Jiménez, I.A., Bazzocchi, I.L., *et al.* (2017). Anti-*Acanthamoeba* activity of Tunisian *Thymus capitatus* essential oil and organic extracts. Exp. Parasitol. *183*, 231–235.

Schuster, F.L., and Jacob, L.S. (1992). Effects of magainins on ameba and cyst stages of *Acanthamoeba polyphaga*. Antimicrob. Agents Chemother. *36*, 1263–1271.

Sifaoui, I., López-Arencibia, A., Martín-Navarro, C.M., Chammem, N., Mejri, M., Lorenzo-Morales, J., Abderabba, M., and Piñero, J.E. (2013). Activity assessment of Tunisian olive leaf extracts against the trophozoite stage of *Acanthamoeba*. Parasitol. Res. *112*, 2825–2829. https://doi.org/10.1007/s00436-013-3453-9.

Sifaoui, I., López-Arencibia, A., Ticona, J.C., Martín-Navarro, C.M., Reyes-Batlle, M., Mejri, M., Lorenzo-Morales, J., Jiménez, A.I., Valladares, B., Lopez-Bazzocchi, I., *et al.* (2014). Bioassay guided isolation and identification of anti-*Acanthamoeba* compounds from Tunisian olive leaf extracts. Exp. Parasitol. *145*, S111–4. https://doi.org/10.1016/j.exppara.2014.02.018.

Sifaoui, I., López-Arencibia, A., Martín-Navarro, C.M., Reyes-Batlle, M., Wagner, C., Chiboub, O., Mejri, M., Valladares, B., Abderrabba, M., Piñero, J.E., *et al.* (2017). Programmed cell death in *Acanthamoeba castellanii* Neff induced by several molecules present in olive leaf extracts. PLOS ONE *12*, e0183795. https://doi.org/10.1371/journal.pone.0183795.

Sun, Y., Hong, J., Zhang, P., Peng, R., and Xiao, G. (2013). Pathological characteristics of the different stages of *Acanthamoeba* keratitis. Histopathology *63*, 862–868. https://doi.org/10.1111/his.12237.

Timko, L., Fischer-Fodor, E., Garajová, M., Mrva, M., Chereches, G., Ondriska, F., Bukovský, M., Lukáč, M., Karlovská, J., Kubincová, J., *et al.* (2015). Synthesis of structural analogues of hexadecylphosphocholine and their antineoplastic, antimicrobial and amoebicidal activity. Eur J Med Chem *93*, 263–273. https://doi.org/10.1016/j.ejmech.2015.02.014.

Vunda, S.L., Sauter, I.P., Cibulski, S.P., Roehe, P.M., Bordignon, S.A., Rott, M.B., Apel, M.A., and von Poser, G.L. (2012). Chemical composition and amoebicidal activity of *Croton pallidulus, Croton ericoides,* and *Croton isabelli* (Euphorbiaceae) essential oils. Parasitol. Res. *111*, 961–966. https://doi.org/10.1007/s00436-012-2918-6

Index

www.ingramcontent.com/pod-product-compliance
Ingram Content Group UK Ltd.
Pitfield, Milton Keynes, MK11 3LW, UK
UKHW022034041026
15591UKWH00012B/673